# Lecture Notes in Computer Science

Edited by G. Goos, J. Hartmanis and J. van Leeuwen

# Springer

*Berlin*
*Heidelberg*
*New York*
*Barcelona*
*Hong Kong*
*London*
*Milan*
*Paris*
*Singapore*
*Tokyo*

Amihood Amir   Gad M. Landau (Eds.)

# Combinatorial Pattern Matching

12th Annual Symposium, CPM 2001
Jerusalem, Israel, July 1-4, 2001
Proceedings

 Springer

Series Editors

Gerhard Goos, Karlsruhe University, Germany
Juris Hartmanis, Cornell University, NY, USA
Jan van Leeuwen, Utrecht University, The Netherlands

Volume Editors

Amihood Amir
Bar-Ilan University, Department of Computer Science
52900 Ramat-Gan, Israel; E-mail: amir@cs.biu.ac.il
and
Georgia Tech, Atlanta, Georgia 30332-0280, USA

Gad M. Landau
University of Haifa, Department of Computer Science
31905 Haifa, Israel; E-mail: landau@cs.haifa.ac.il
and
Polytechnic University, Brooklyn, NY 11201, USA

Cataloging-in-Publication Data applied for

Die Deutsche Bibliothek - CIP-Einheitsaufnahme

Combinatorial pattern matching : 12th annual symposium ; proceedings / CPM
2001, Jerusalem, Israel, July 1 - 4, 2001. Amihood Amir ; Gad. M. Landau
(ed.). - Berlin ; Heidelberg ; New York ; Barcelona ; Hong Kong ; London ;
Milan ; Paris ; Singapore ; Tokyo : Springer, 2001
    (Lecture notes in computer science ; Vol. 2089)
    ISBN 3-540-42271-4

CR Subject Classification (1998): F.2.2, I.5.4, I.5.0, I.7.3, H.3.3, E.4, G.2.1

ISSN 0302-9743
ISBN 3-540-42271-4 Springer-Verlag Berlin Heidelberg New York

Springer-Verlag Berlin Heidelberg New York
a member of BertelsmannSpringer Science+Business Media GmbH

http://www.springer.de

© Springer-Verlag Berlin Heidelberg 2001
Printed in Germany

Typesetting: Camera-ready by author, data conversion by Christian Grosche, Hamburg
Printed on acid-free paper       SPIN: 10839362       06/3142       5 4 3 2 1 0

# Foreword

The papers contained in this volume were presented at the 12th Annual Symposium on Combinatorial Pattern Matching, held July 1–4, 2001 at the *Dan Panorama Hotel* in Jerusalem, Israel. They were selected from 35 abstracts submitted in response to the call for papers. In addition, there were invited lectures by Aviezri Fraenkel (*Weizmann Institute of Science*), Zvi Galil (*Columbia*), Rao Kosaraju (*Johns Hopkins University*), and Uzi Vishkin (*Technion and U. Maryland*). This year the call for papers invited short (poster) presentations. They also appear in the proceedings.

Combinatorial Pattern Matching (CPM) addresses issues of searching and matching strings and more complicated patterns such as trees, regular expressions, graphs, point sets, and arrays, in various formats. The goal is to derive non-trivial combinatorial properties of such structures and to exploit these properties in order to achieve superior performance for the corresponding computational problems. On the other hand, an important aim is to analyze and pinpoint the properties and conditions under which searches can not be performed efficiently.

Over the past decade a steady flow of high quality research on this subject has changed a sparse set of isolated results into a full-fledged area of algorithmics. This area is continuing to grow even further due to the increasing demand for speed and efficiency that stems from important applications such as the World Wide Web, computational biology, computer vision, and multimedia systems. These involve requirements for information retrieval in heterogeneous databases, data compression, and pattern recognition. The objective of the annual CPM gathering is to provide an international forum for the presentation of research results in combinatorial pattern matching and related applications.

The first 11 meetings were held in Paris, London, Tucson, Padova, Asilomar, Helsinki, Laguna Beach, Aarhus, Piscataway, Warwick, and Montreal, over the years 1990–2000. After the first meeting, a selection of papers appeared as a special issue of *Theoretical Computer Science* in volume 92. The proceedings of the 3rd to 11th meetings appeared as volumes 644, 684, 807, 937, 1075, 1264, 1448, 1645, and 1848 of the Springer LNCS series. Selected papers of the 12th meeting will appear in a special issue of *Discrete Applied Mathematics*.

The general organization and orientation of the CPM conferences is coordinated by a steering committee composed of Alberto Apostolico (*Padova and Purdue*), Maxime Crochemore (*Marne-la-Vallée*), Zvi Galil (*Columbia*) and Udi Manber (*Yahoo!*).

April 2001

Amihood Amir
Gad M. Landau

# Program Committee

Amihood Amir, co-chair,
    *Bar-Ilan & Georgia Tech*
Setsuo Arikawa, *Kyushu*
Gary Benson, *Mt. Sinai*
Andrei Broder, *Altavista*
Maxime Crochemore, *Marne-la-Vallée*
Leszek Gasieniec, *Liverpool*
Raffaele Giancarlo, *Palermo*
Costas S. Iliopoulos,
    *King's College, London*
Tomi Klein, *Bar-Ilan*
Gad Landau, co-chair,
    *Haifa & Polytechnic, New York*

Thierry Lecroq, *Rouen*
Moshe Lewenstein, *IBM Yorktown*
Yoelle Maarek, *IBM Haifa*
Kunsoo Park, *Seoul National*
Hershel Safer, *Compugen*
David Sankoff, *Montréal*
Jeanette Schmidt, *Incyte*
Dafna Sheinwald, *IBM Haifa*
Divesh Srivastava, *AT&T*
Naftali Tishbi, *Hebrew University*
Ian Witten, *Waikato*

# Local Organization

Local arrangements were made by the program committee co-chairs. The conference Web site was created and maintained by Revital Erez. Organizational help was provided by Gal Goldschmidt, Noa Gur, Libi Raz, and Daphna Stern.

# Sponsoring Institutions

- The Caesarea Edmond Benjamin de Rothschild Foundation, Institute for Interdisciplinary Applications of Computer Science, Haifa University.
- Bar Ilan University.
- Haifa University.

# List of Additional Reviewers

Jacques Desarmenien
Paolo Ferragina
Franya Franek
Dora Giammarresi
Jan Holub
Jesper Jansson
Dong Kyue Kim
Ralf Klasing
Andrzej Lingas

Gonzalo Navarro
Igor Potapov
Tomasz Radzik
Mathieu Raffinot
Rajeev Raman
Marie-France Sagot
Marinella Sciortino
Patrice Seebold
Ayumi Shinohara

Noam Slonim
W.F. Smyth
Dina Sokol
Shigeru Takano
Masayuki Takeda
Mutsunori Yagiura
Michal Ziv-Ukelson

# Table of Contents

# Regular Expression Searching over Ziv–Lempel Compressed Text

Gonzalo Navarro*

Dept. of Computer Science, University of Chile
Blanco Encalada 2120, Santiago, Chile
gnavarro@dcc.uchile.cl

**Abstract.** We present a solution to the problem of regular expression searching on compressed text. The format we choose is the Ziv–Lempel family, specifically the LZ78 and LZW variants. Given a text of length $u$ compressed into length $n$, and a pattern of length $m$, we report all the $R$ occurrences of the pattern in the text in $O(2^m + mn + Rm \log m)$ worst case time. On average this drops to $O(m^2 + (n + R) \log m)$ or $O(m^2 + n + Ru/n)$ for most regular expressions. This is the first nontrivial result for this problem. The experimental results show that our compressed search algorithm needs half the time necessary for decompression plus searching, which is currently the only alternative.

## 1  Introduction

The need to search for regular expressions arises in many text-based applications, such as text retrieval, text editing and computational biology, to name a few. A *regular expression* is a generalized pattern composed of (i) basic strings, (ii) union, concatenation and Kleene closure of other regular expressions [1]. The problem of regular expression searching is quite old and has received continuous attention since the sixties until our days (see Section 2.1).

A particularly interesting case of text searching arises when the text is compressed. Text compression [6] exploits the redundancies of the text to represent it using less space. There are many different compression schemes, among which the Ziv–Lempel family [32,33] is one of the best in practice because of its good compression ratios combined with efficient compression and decompression times. The compressed matching problem consists of searching a pattern on a compressed text without uncompressing it. Its main goal is to search the compressed text faster than the trivial approach of decompressing it and then searching. This problem is important in practice. Today's textual databases are an excellent example of applications where both problems are crucial: the texts should be kept compressed to save space and I/O time, and they should be efficiently searched. Surprisingly, these two combined requirements are not easy to achieve together, as the only solution before the 90's was to process queries by uncompressing the texts and then searching into them.

---

* Partially supported by Fondecyt grant 1-990627.

A. Amir and G.M. Landau (Eds.): CPM 2001, LNCS 2089, pp. 1–17, 2001.

Since then, a lot of research has been conducted on the problem. A wealth of solutions have been proposed (see Section 2.2) to deal with simple, multiple and, very recently, approximate compressed pattern matching. Regular expression searching on compressed text seems to be the last goal which still defies the existence of any nontrivial solution.

This is the problem we solve in this paper: we present the first solution for compressed regular expression searching. The format we choose is the Ziv–Lempel family, focusing in the LZ78 and LZW variants [33,29]. Given a text of length $u$ compressed into length $n$, we are able to find the $R$ occurrences of a regular expression of length $m$ in $O(2^m + mn + Rm \log m)$ worst case time, needing $O(2^m + mn)$ space. We also propose two modifications which achieve $O(m^2 + (n + R) \log m)$ or $O(m^2 + n + Ru/n)$ average case time and, respectively, $O(m + n \log m)$ or $O(m + n)$ space, for "admissible" regular expressions, i.e. those whose automaton runs out of active states after reading $O(1)$ text characters. These results are achieved using bit-parallelism and are valid for short enough patterns, otherwise the search times have to be multiplied by $\lceil m/w \rceil$, where $w$ is the number of bits in the computer word.

We have implemented our algorithm on LZW and compared it against the best existing algorithms on uncompressed text, showing that we can search the compressed text twice as fast as the naive approach of uncompressing and then searching.

## 2 Related Work

### 2.1 Regular Expression Searching

The traditional technique [26] to search a regular expression of length $m$ (which means $m$ letters, not counting the special operators such as "*", "|", etc.) in a text of length $u$ is to convert the expression into a nondeterministic finite automaton (NFA) with $O(m)$ nodes. Then, it is possible to search the text using the automaton at $O(mu)$ worst case time. The cost comes from the fact that more than one state of the NFA may be active at each step, and therefore all may need to be updated.

On top of the basic algorithm for converting a regular expression into an NFA, we have to add a self-loop at the initial state which guarantees that it keeps always active, so it is able to detect a match starting anywhere in the text. At each text position where a final state gets active we signal the end point of an occurrence.

A more efficient choice [1] is to convert the NFA into a deterministic finite automaton (DFA), which has only one active state at a time and therefore allows searching the text at $O(u)$ cost, which is worst-case optimal. The cost of this approach is that the DFA may have $O(2^m)$ states, which implies a preprocessing cost and extra space exponential in $m$.

An easy way to obtain a DFA from an NFA is via *bit-parallelism*, which is a technique to code many elements in the bits of a single computer word and manage to update all them in a single operation. In this case, the vector of active and inactive states is stored as the bits of a computer word. Instead of (ala Thompson [26]) examining the active states one by one, the whole computer word is used to index a table which, given the current text character, provides the new set of active states (another computer word). This can be considered either as a bit-parallel simulation of an NFA, or as an implementation of a DFA (where the identifier of each deterministic state is the bit mask as a whole). This idea was first proposed by Wu and Manber [31,30].

Later, Navarro and Raffinot [23] used a similar procedure, this time using Glushkov's [7] construction of the NFA. This construction has the advantage of producing an automaton of exactly $m + 1$ states, while Thompson's may reach $2m$ states. A drawback is that the structure is not so regular and therefore a table $D : 2^{m+1} \times (\sigma + 1) \rightarrow 2^{m+1}$ is required, where $\sigma$ is the size of the pattern alphabet $\Sigma$. Thompson's construction, on the other hand, is more regular and only needs a table $D : 2^{2m} \rightarrow 2^{2m}$ for the $\varepsilon$-transitions. It has been shown [23] that Glushkov's construction normally yields faster search time. In any case, if the table is too big it can be split horizontally in two or more tables [31]. For example, a table of size $2^m$ can be split into 2 subtables of size $2^{m/2}$. We need to access two tables for a transition but need only the square root of the space.

Some techniques have been proposed to obtain a tradeoff between NFAs and DFAs. In 1992, Myers [19] presented a four-russians approach which obtains $O(mu/\log u)$ worst-case time and extra space. The idea is to divide the syntax tree of the regular expression into "modules", which are subtrees of a reasonable size. These subtrees are implemented as DFAs and are thereafter considered as leaf nodes in the syntax tree. The process continues with this reduced tree until a single final module is obtained.

The ideas presented up to now aim at a good implementation of the automaton, but they must inspect all the text characters. Other proposals try to skip some text characters, as it is usual for simple pattern matching. For example, Watson [28, chapter 5] presented an algorithm that determines the minimum length of a string matching the regular expression and forms a tree with all the prefixes of that length of strings matching the regular expression. A multipattern search algorithm like Commentz-Walter [8] is run over those prefixes as a filter to detect text areas where a complete occurrence may start. Another technique of this kind is used in *Gnu Grep 2.0*, which extracts a set of strings which must appear in any match. This string is searched for and the neighborhoods of its occurrences are checked for complete matches using a lazy deterministic automaton.

The most recent development, also in this line, is from Navarro and Raffinot [23]. They invert the arrows of the DFA and make all states initial and the initial state final. The result is an automaton that recognizes all the reverse prefixes of strings matching the regular expression. The idea is in this sense similar to that of Watson, but takes less space. The search method is also different: instead of a Boyer-Moore like algorithm, it is based on BNDM [23].

## 2.2   Compressed Pattern Matching

The *compressed matching problem* was first defined in the work of Amir and Benson [2] as the task of performing string matching in a compressed text without decompressing it. Given a text $T$, a corresponding compressed string $Z = z_1 \ldots z_n$, and a pattern $P$, the compressed matching problem consists in finding all occurrences of $P$ in $T$, using only $P$ and $Z$. A naive algorithm, which first decompresses the string $Z$ and then performs standard string matching, takes time $O(m+u)$. An optimal algorithm takes worst-case time $O(m+n+R)$, where $R$ is the number of matches (note that it could be that $R = u > n$).

Two different approaches exist to search compressed text. The first one is rather practical. Efficient solutions based on Huffman coding [10] on words have been presented by Moura et al. [18], but they need that the text contains natural language and is large (say, 10 Mb or more). Moreover, they allow only searching for whole words and phrases. There are also other practical ad-hoc methods [15], but the compression they obtain is poor. Moreover, in these compression formats $n = \Theta(u)$, so the speedups can only be measured in practical terms.

The second line of research considers Ziv–Lempel compression, which is based on finding repetitions in the text and replacing them with references to similar strings previously appeared. LZ77 [32] is able to reference any substring of the text already processed, while LZ78 [33] and LZW [29] reference only a single previous reference plus a new letter that is added.

String matching in Ziv–Lempel compressed texts is much more complex, since the pattern can appear in different forms across the compressed text. The first algorithm for exact searching is from 1994, by Amir, Benson and Farach [3], who search in LZ78 needing time and space $O(m^2 + n)$.

The only search technique for LZ77 is by Farach and Thorup [9], a randomized algorithm to determine in time $O(m + n \log^2(u/n))$ whether a pattern is present or not in the text.

An extension of the first work [3] to multipattern searching was presented by Kida et al. [13], together with the first experimental results in this area. They achieve $O(m^2 + n)$ time and space, although this time $m$ is the total length of all the patterns.

New practical results were presented by Navarro and Raffinot [24], who proposed a general scheme to search on Ziv–Lempel compressed texts (simple and extended patterns) and specialized it for the particular cases of LZ77, LZ78 and a new variant proposed which was competitive and convenient for search purposes. A similar result, restricted to the LZW format, was independently found and presented by Kida et al. [14]. The same group generalized the existing algorithms and nicely unified the concepts in a general framework [12]. Recently, Navarro and Tarhio [25] presented a new, faster, algorithm based on Boyer-Moore.

Approximate string matching on compressed text aims at finding the pattern where a limited number of differences between the pattern and its occurrences are permitted. The problem, advocated in 1992 [2], had been solved for Huffman coding of words [18], but the solution is limited to search a whole word and retrieve whole words that are similar. The first true solutions appeared very recently, by Kärkkäinen et al. [11], Matsumoto et al. [16] and Navarro et al. [22].

## 3    The Ziv–Lempel Compression Formats LZ78 and LZW

The general idea of Ziv–Lempel compression is to replace substrings in the text by a pointer to a previous occurrence of them. If the pointer takes less space than the string it is replacing, compression is obtained. Different variants over this type of compression exist, see for example [6]. We are particularly interested in the LZ78/LZW format, which we describe in depth.

The Ziv–Lempel compression algorithm of 1978 (usually named LZ78 [33]) is based on a dictionary of blocks, in which we add every new block computed. At the beginning of the compression, the dictionary contains a single block $b_0$ of length 0. The current step of the compression is as follows: if we assume that a prefix $T_{1...j}$ of $T$ has been already compressed in a sequence of blocks $Z = b_1 ... b_r$, all them in the dictionary, then we look for the longest prefix of the rest of the text $T_{j+1...u}$ which is a block of the dictionary. Once we found this block, say $b_s$ of length $\ell_s$, we construct a new block $b_{r+1} = (s, T_{j+\ell_s+1})$, we write the pair at the end of the compressed file $Z$, i.e $Z = b_1 ... b_r b_{r+1}$, and we add the block to the dictionary. It is easy to see that this dictionary is prefix-closed (i.e. any prefix of an element is also an element of the dictionary) and a natural way to represent it is a tree.

We give as an example the compression of the word *ananas* in Figure 1. The first block is $(0, a)$, and next $(0, n)$. When we read the next $a$, $a$ is already the block 1 in the dictionary, but $an$ is not in the dictionary. So we create a third block $(1, n)$. We then read the next $a$, $a$ is already the block 1 in the dictionary, but $as$ do not appear. So we create a new block $(1, s)$.

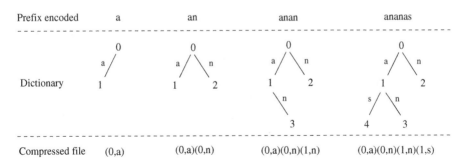

**Fig. 1.** Compression of the word *ananas* with the algorithm LZ78.

The compression algorithm is $O(u)$ time in the worst case and efficient in practice if the dictionary is stored as a tree, which allows rapid searching of the new text prefix (for each character of $T$ we move once in the tree). The decompression needs to build the same dictionary (the pair that defines the block $r$ is read at the $r$-th step of the algorithm), although this time it is not convenient to have a tree, and an array implementation is preferable. Compared to LZ77, the compression is rather fast but decompression is slow.

Many variations on LZ78 exist, which deal basically with the best way to code the pairs in the compressed file, or with the best way to cope with limited memory for compression. A particularly interesting variant is from Welch, called LZW [29]. In this case, the extra letter (second element of the pair) is not coded, but it is taken as the first letter of the next block (the dictionary is started with one block per letter). LZW is used by Unix's *Compress* program.

In this paper we do not consider LZW separately but just as a coding variant of LZ78. This is because the final letter of LZ78 can be readily obtained by keeping count of the first letter of each block (this is copied directly from the referenced block) and then looking at the first letter of the next block.

## 4    A Search Algorithm

We present now our approach for regular expression searching over a text $Z = b_1 \ldots b_n$, that is expressed as a sequence of $n$ *blocks*. Each block $b_r$ represents a substring $B_r$ of $T$, such that $B_1 \ldots B_n = T$. Moreover, each block $B_r$ is formed by a concatenation of a previously seen block and an explicit letter. This comprises the LZ78 and LZW formats. Our goal is to find the positions in $T$ where the pattern occurrences end, using $Z$.

Our approach is to modify the DFA algorithm based on bit-parallelism, which is designed to process $T$ character by character, so that it processes $T$ block by block using the fact that blocks are built from previous blocks and explicit letters. We assume that Glushkov's construction [7] is used, so the NFA has $m+1$ states. So we start by building the DFA in $O(2^m)$ time and space.

Our bit masks will denote sets of NFA states, so they will be of width $m + 1$. For clarity we will write the sets of states, keeping in mind that we can compute $A \cup B$, $A \cap B$, $A^c$, $A = B$, $A \leftarrow B$, $a \in A$ in constant time (or, for long patterns, in $O(\lceil m/w \rceil)$ time, where $w$ is the number of bits in the computer word). Another operation we will need to perform in constant time is to select any element of a set. This can be achieved with "bit magic", which means precomputing the table storing the position of, say, the highest bit for each possible bit mask of length $m + 1$, which is not much given that we already store $\sigma$ such tables.

About our automaton, we assume that the states are numbered $0 \ldots m$, being $0$ the initial state. We call $F$ the bit mask of final states and the transition function is $D : bitmasks \times \Sigma \to bitmasks$.

The general mechanism of the search is as follows: we read the blocks $b_r$ one by one. For each new block $b$ read, representing a string $B$, and where we have already processed $T_{1\ldots j}$, we update the state of the search so that after working on the block we have processed $T_{1\ldots j+|B|} = T_{1\ldots j}B$. To process each block, three steps are carried out: (1) its *description* is computed and stored, (2) the occurrences ending inside the block $B$ are reported, and (3) the state of the search is updated.

Say that block $b$ represents the text substring $B$. Then the *description* of $b$ is formed by

- a number $len(b) = |B|$, its length;
- a block number $ref(b)$, the referenced block;
- a vector $tr_{0...m}$ of bit masks, where $tr_i$ gives the states of the NFA that remain active after reading $B$ if only the $i$-th state of the NFA is active at the beginning;
- a bit mask $act = \cup \{i, tr_i \neq \emptyset\}$, which indicates which states of the NFA may yield any surviving state after processing $B$;
- a bit mask $fin$, which indicates which states, if active before processing $B$, produce an occurrence inside $B$ (after processing at least one character of $B$); and
- a vector $mat_{0...m}$ of block numbers, where $mat_i$ gives the most recent (i.e. longest) block $b'$ in the referencing chain $b, ref(b), ref(ref(b)), \ldots$ such that $i \in fin(b')$, or a null value if there is no such block.

The state of the search consists of two elements

- the last text position considered, $j$ (initially 0);
- a bit mask $S$ of $m + 1$ bits, which indicates which states are active after processing $T_{1...j}$. Initially, $S$ has active only its initial state, $S = \{0\}$.

As we show next, the total cost to search for all the occurrences with this scheme is $O(2^m + mn + Rm \log m)$ in the worst case. The first term corresponds to building the DFA from the NFA, the second to computing block descriptions and updating the search state, and the last to report the occurrences. The existence problem is solved in time $O(2^m + mn)$. The space requirement is $O(2^m + mn)$. We recall that patterns longer than the computer word $w$ get their search cost multiplied by $\lceil m/w \rceil$.

## 4.1  Computing Block Descriptions

We show how to compute the description of a new block $b'$ that represents $B' = Ba$, where $B$ is the string represented by a previous block $b$ and $a$ is an explicit letter. An initial block $b_0$ represents the string $\varepsilon$, and its description is: $len(b_0) = 0$; $tr_i(b_0) = \{i\}$; $act(b_0) = \{0 \ldots m\}$; $fin(b_0) = \emptyset$; $mat_i(b_0) = $ a null value. We give now the update formulas for $B' = Ba$.

- $len(b') \leftarrow len(b) + 1$.
- $ref(b') \leftarrow b$.
- $tr_i(b') \leftarrow D(tr_i(b), a)$ (we only need to do this for $i \in act(b)$).
- $act(b') \leftarrow \{i \in act(b), tr_i(b') \neq \emptyset\}$.
- $fin(b') \leftarrow fin(b) \cup \{i \in act(b'), tr_i(b') \cap F \neq \emptyset\}$.
- $mat_i(b') \leftarrow mat_i(b)$ if $tr_i(b') \cap F = \emptyset$, and $b'$ otherwise.

In the worst case we have to update all the cells of $tr$ and $mat$, so we pay $O(mn)$ time (recall that bit parallelism permits performing set operations in constant time). The space required for the block descriptions is $O(mn)$ as well.

## 4.2   Reporting Matches and Updating the Search State

The $fin(b')$ mask tells us whether there are any occurrences to report depending on the active states at the beginning of the block. Therefore, our first action is to compute $S \cap fin(b')$, which tells us which of the currently active states will produce occurrences inside $B'$. If this mask turns out to be null, we can skip the process of reporting matches.

If there are states in the intersection then we will have matches to report inside $B'$. Now, each state $i$ in the intersection produces a list of positions which can be retrieved in decreasing order using $mat_i(b')$, $mat_i(ref(mat_i(b')))$, …. If $B'$ starts at text position $j$, then we have to report the text positions $j + len(mat_i(b')) - 1$, $j + len(mat_i(ref(mat_i(b')))) - 1$, …. These positions appear in decreasing order, but we have to merge the decreasing lists of all the states in $S \cap fin(b')$. A priority queue can be used to obtain each position in $O(\log m)$ time. If there are $R$ occurrences overall, then in the worst case each occurrence can be reported $m$ times (reached from each state), which gives a total cost of $O(Rm \log m)$.

Finally, we update $S$ in $O(m)$ time per block with $S \leftarrow \cup_{i \in S \cap act(b')} tr_i(b')$.

# 5   A Faster Algorithm on Average

An average case analysis of our algorithm reveals that, except for $mat$, all the other operations can be carried out in linear time. This leads to a variation of the algorithm that is linear time on average.

The main point is that, on average, $|act(b)| = |tr_i(b)| = O(1)$, that is, the number of states of the automaton which can survive after processing a block is constant. We prove in the Appendix that this holds under very general assumptions and for "admissible" regular expressions (i.e. those whose automata run out of active states after processing $O(1)$ text characters). Note that, thanks to the self loop in the initial state 0, this state is always in $act(b)$ and in $tr_0(b)$.

*Constant Time Operations.* Except for $mat$, all the computation of the block description is proportional to the size of $act$ and hence it takes $O(n)$ time (see Section 4.1): $tr_i(b')$ needs to be computed only for those $i \in act(b)$; and $act(b')$ and $fin(b')$ can also be computed in time proportional to $|act(b)|$ or $|act(b')|$. The update to $S$ (see Section 4.2) needs only to consider the states in $act(b')$. Each active bit in $act$ is obtained in constant time by bit magic.

*Updating the mat Vector.* What we need is a mechanism to update $mat$ fast. Note that, despite that $mat_i(b')$ is null if $i \notin fin(b')$, it may not be true that $|fin(b')| = O(1)$ on average, because as soon as a state belongs to $fin(b)$, it belongs to all its descendants in the LZ78 tree.

However, it is still true that just $O(1)$ values of $mat(b)$ change in $mat(b')$, where $ref(b') = b$, since $mat$ changes only on those $\{i, \ tr_i(b') \cap F \neq \emptyset\} \subseteq act(b')$, and $|act(b')| = O(1)$.

Hence, we do not represent a new *mat* vector for each block, but only its differences with respect to the referenced block. This must be done such that (*i*) the *mat* vector of the referenced block is not altered, as it may have to be used for other descendants; and (*ii*) we are able to quickly find $mat_i$ for any $i$.

A solution is to represent *mat* as a complete tree (i.e. perfectly balanced), which will always have $m + 1$ nodes and associates the keys $\{0 \ldots m\}$ to their value $mat_i$. This permits obtaining in $O(\log m)$ time the value $mat_i$. We start with a complete tree, and later need only to modify the values associated to tree keys, but never add or remove keys (otherwise an AVL would have been a good choice). When a new value has to be associated to a key in the tree of the referenced block in order to obtain the tree of the referencing block, we find the key in the old tree and create of copy of the path from the root to the key. Then we change the value associated to the new node holding the key. Except when the new nodes are involved, the created path points to the same nodes where the old paths points, hence sharing part of the tree. The new root corresponds to the modified tree of the new block. The cost of each such modification is $O(\log m)$. We have to perform this operation $O(1)$ times on average per block, yielding $O(n \log m)$ time.

Figure 2 illustrates the idea. This kind of technique is usual when implementing the logical structure of WORM (write once read many) devices, in order to reflect the modifications of the user on a medium that does not permit alterations.

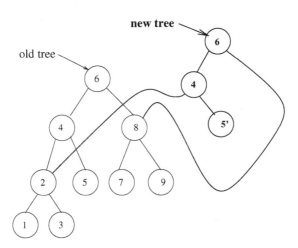

**Fig. 2.** Changing node 5 to 5' in a read-only tree.

*Reporting Matches.* We have to add now the cost to report the $R$ matches. Since $|tr_i(b)| = O(1)$ on average, there are only $O(1)$ states able to trigger an occurrence at the end of a block, and hence each occurrence is triggered by $O(1)$

states on average. The priority queue gives us those positions in $O(\log m)$ time per position, so the total cost to trigger occurrences is on average $O(R \log m)$.

*Lowering Space and Preprocessing Costs.* The fact that $|tr_i(b)| = O(1)$ on average shows another possible improvement. We have chosen a DFA representation of our automaton which needs $O(2^m)$ space and preprocessing time. Instead, an NFA representation would require $O(m^2)$. The problem with the NFA is that, in order to build $tr_i(b')$ for $b' = (b, a)$, we need to make the union of the NFA states reachable via the letter $a$ from each state in $tr(b)$. This has a worst case of $O(m)$, yielding $O(m^2)$ worst case search time to update a block. However, on average this drops to $O(1)$ since only $O(1)$ states $i$ have $tr_i(b) \neq \emptyset$ (because $|act(b)| = O(1)$) and each such $tr_i(b)$ has constant size.

Therefore, we have obtained average complexity $O(m^2 + (n + R) \log m)$. The space requirements are lowered as well. The NFA requires only $O(m)$ space. The block descriptions take $O(n)$ space because there are only $O(1)$ nonempty $tr_i$ masks. With respect to the *mat* trees, we have that there are on average $O(1)$ modifications per block and each creates $O(\log m)$ new nodes, so the space required for *mat* is on average $O(n \log m)$. Hence the total space is $O(m + n \log m)$.

If $R$ is really small we may prefer an alternative implementation. Instead of representing *mat*, we store for each block a bit mask *ffin*, which tells whether there is a match exactly at the end of the block. While *fin* is active we go backward in the referencing chain of the block reporting all those blocks whose *ffin* mask is active in a state of $S$. This yields $O(m^2 + n + Ru/n)$ time on average instead of $O(m^2 + (n + R) \log m)$. The space becomes $O(m + n)$.

# 6    Experimental Results

We have implemented our algorithm in order to determine its practical value. We chose to use the LZW format by modifying the code of Unix's *uncompress*, so our code is able to search files compressed with *compress* (.Z). This implies some small changes in the design, but the algorithm is essentially the same. We have used bit parallelism, with a single table (no horizontal partitioning) and map (at search time) the character set to an integer range representing the different pattern characters, to reduce space. Finally, we have chosen to use the *ffin* masks instead of representing *mat*.

We ran our experiments on an Intel Pentium III machine of 550 MHz and 64 Mb of RAM. We have compressed 10 Mb of Wall Street Journal articles, which gets compressed to 42% of its original size with *compress*. We measure user time, as system times are negligible. Each data point has been obtained by repeating the experiment 10 times.

In the absence of other algorithms for compressed regular expression searching, we have compared our algorithm against the naive approach of decompressing and searching. The WSJ file needed 3.58 seconds to be decompressed with *uncompress*. After decompression, we run two different search algorithms. A first one, *DFA*, uses a bit-parallel DFA to process the text. This is interesting because it is the algorithm we are modifying to work on compressed text. A second one, the software *nrgrep* [20], uses a character skipping technique for searching [23], which is much faster. In any case, the time to uncompress is an order of magnitude higher than that to search the uncompressed text, so the search algorithm used does not significantly affect the results.

A major problem when presenting experiments on regular expressions is that there is not a concept of "random" regular expression, so it is not possible to search, say, 1,000 random patterns. Lacking such good choice, we fixed a set of 7 patterns which were selected to illustrate different interesting cases. The patterns are given in Table 1, together with some parameters and the obtained search times. We use the normal operators to denote regular expressions plus some extensions, such as "[a-z]" = $(a|b|c|...|z)$ and "." = all the characters. Note that the 7th pattern is not "admissible" and the search time gets affected.

**Table 1.** The patterns used on Wall Street Journal articles and the search times in seconds.

| No. | Pattern | $m$ | $R$ | Ours | Uncompress + Nrgrep | Uncompress + DFA |
|-----|---------|-----|-----|------|---------------------|------------------|
| 1 | `American|Canadian` | 17 | 1801 | 1.81 | 3.75 | 3.85 |
| 2 | `Amer[a-z]*can` | 9 | 1500 | 1.79 | 3.67 | 3.74 |
| 3 | `Amer[a-z]*can|Can[a-z]*ian` | 16 | 1801 | 2.23 | 3.73 | 3.87 |
| 4 | `Ame(i|(r|i)*)can` | 10 | 1500 | 1.62 | 3.70 | 3.72 |
| 5 | `Am[a-z]*ri[a-z]*an` | 9 | 1504 | 1.88 | 3.68 | 3.72 |
| 6 | `(Am|Ca)(er|na)(ic|di)an` | 15 | 1801 | 1.70 | 3.70 | 3.75 |
| 7 | `Am.*er.*ic.*an` | 12 | 92945 | 2.74 | 3.68 | 3.74 |

As the table shows, we can actually improve over the decompression of the text followed by the application of any search algorithm (indeed, just the decompression takes much more time). In practical terms, we can search the original file at about 4–5 Mb/sec. This is about half the time necessary for decompression plus searching with the best algorithm.

We have used *compress* because it is the format we are dealing with. In some scenarios, LZW is the preferred format because it maximizes compression (e.g. it compressed DNA better than LZ77). However, we may prefer a decompress plus search approach under the LZ77 format, which decompresses faster. For example, Gnu *gzip* needs 2.07 seconds for decompression in our machine. If we compare our search algorithm on LZW against decompressing on LZ77 plus searching, we are still 20% faster.

# 7  Conclusions

We have presented the first solution to the open problem of regular expression searching over Ziv–Lempel compressed text. Our algorithm can find the $R$ occurrences of a regular expression of length $m$ over a text of size $u$ compressed by LZ78 or LZW into size $n$ in $O(2^m + mn + Rm \log m)$ worst-case time and, for most regular expressions, $O(m^2 + (n + R) \log m)$ or $O(m^2 + n + Ru/n)$ average case time. We have shown that this is also of practical interest, as we are able to search on compressed text twice as fast as decompressing plus searching.

An interesting question is whether we can improve the search time using character skipping techniques [28,23]. The first would have to be combined with multipattern search techniques on LZ78/LZW [13]. For the second type of search (BNDM [23]), there is no existing algorithm on compressed text yet. We are also pursuing on extending these ideas to other compression formats, e.g. a Ziv–Lempel variant where the new block is the concatenation of the previous and the current one [17]. The existence problem seems to require $O(m^2 n)$ time for this format.

# References

1. A. Aho, R. Sethi, and J. Ullman. *Compilers: Principles, Techniques and Tools.* Addison-Wesley, 1985.
2. A. Amir and G. Benson. Efficient two-dimensional compressed matching. In *Proc. DCC'92*, pages 279–288, 1992.
3. A. Amir, G. Benson, and M. Farach. Let sleeping files lie: Pattern matching in Z-compressed files. *J. of Comp. and Sys. Sciences*, 52(2):299–307, 1996. Earlier version in *Proc. SODA'94*.
4. R. Baeza-Yates. *Efficient Text Searching*. PhD thesis, Dept. of Computer Science, Univ. of Waterloo, May 1989. Also as Research Report CS-89-17.
5. R. Baeza-Yates and G. Gonnet. Fast text searching for regular expressions or automaton searching on a trie. *J. of the ACM*, 43(6):915–936, 1996.
6. T. Bell, J. Cleary, and I. Witten. *Text Compression*. Prentice Hall, 1990.
7. G. Berry and R. Sethi. From regular expression to deterministic automata. *Theoretical Computer Science*, 48(1):117–126, 1986.
8. B. Commentz-Walter. A string matching algorithm fast on the average. In *Proc. ICALP'79*, LNCS v. 6, pages 118–132, 1979.
9. M. Farach and M. Thorup. String matching in Lempel-Ziv compressed strings. *Algorithmica*, 20:388–404, 1998.
10. D. Huffman. A method for the construction of minimum-redundancy codes. *Proc. of the I.R.E.*, 40(9):1090–1101, 1952.
11. J. Kärkkäinen, G. Navarro, and E. Ukkonen. Approximate string matching over Ziv-Lempel compressed text. In *Proc. CPM'2000*, LNCS 1848, pages 195–209, 2000.
12. T. Kida, Y. Shibata, M. Takeda, A. Shinohara, and S. Arikawa. A unifying framework for compressed pattern matching. In *Proc. SPIRE'99*, pages 89–96. IEEE CS Press, 1999.
13. T. Kida, M. Takeda, A. Shinohara, M. Miyazaki, and S. Arikawa. Multiple pattern matching in LZW compressed text. In *Proc. DCC'98*, pages 103–112, 1998.

14. T. Kida, M. Takeda, A. Shinohara, M. Miyazaki, and S. Arikawa. Shift-And approach to pattern matching in LZW compressed text. In *Proc. CPM'99*, LNCS 1645, pages 1–13, 1999.

15. U. Manber. A text compression scheme that allows fast searching directly in the compressed file. *ACM Trans. on Information Systems*, 15(2):124–136, 1997.

16. T. Matsumoto, T. Kida, M. Takeda, A. Shinohara, and S. Arikawa. Bit-parallel approach to approximate string matching in compressed texts. In *Proc. SPIRE'2000*, pages 221–228. IEEE CS Press, 2000.

17. V. Miller and M. Wegman. Variations on a theme by Ziv and Lempel. In *Combinatorial Algorithms on Words*, volume 12 of *NATO ASI Series F*, pages 131–140. Springer-Verlag, 1985.

18. E. Moura, G. Navarro, N. Ziviani, and R. Baeza-Yates. Fast and flexible word searching on compressed text. *ACM Trans. on Information Systems*, 18(2):113–139, 2000.

19. G. Myers. A four-russian algorithm for regular expression pattern matching. *J. of the ACM*, 39(2):430–448, 1992.

20. G. Navarro. Nr-grep: A fast and flexible pattern matching tool. Technical Report TR/DCC-2000-3, Dept. of Computer Science, Univ. of Chile, August 2000.

21. G. Navarro. A guided tour to approximate string matching. *ACM Computing Surveys*, 2001. To appear.

22. G. Navarro, T. Kida, M. Takeda, A. Shinohara, and S. Arikawa. Faster approximate string matching over compressed text. In *Proc. 11th IEEE Data Compression Conference (DCC'01)*, 2001. To appear.

23. G. Navarro and M. Raffinot. Fast regular expression search. In *Proceedings of the 3rd Workshop on Algorithm Engineering (WAE'99)*, LNCS 1668, pages 198–212, 1999.

24. G. Navarro and M. Raffinot. A general practical approach to pattern matching over Ziv-Lempel compressed text. In *Proc. CPM'99*, LNCS 1645, pages 14–36, 1999.

25. G. Navarro and J. Tarhio. Boyer-Moore string matching over Ziv-Lempel compressed text. In *Proc. CPM'2000*, LNCS 1848, pages 166–180, 2000.

26. K. Thompson. Regular expression search algorithm. *Comm. of the ACM*, 11(6):419–422, 1968.

27. J. Vitter and P. Flajolet. Average-case analysis of algorithms and data structures. In *Handbook of Theoretical Computer Science*, chapter 9. Elsevier Science, 1990.

28. B. Watson. *Taxonomies and Toolkits of Regular Language Algorithms*. Phd. dissertation, Eindhoven University of Technology, The Netherlands, 1995.

29. T. Welch. A technique for high performance data compression. *IEEE Computer Magazine*, 17(6):8–19, June 1984.

30. S. Wu and U. Manber. Agrep – a fast approximate pattern-matching tool. In *Proc. of USENIX Technical Conference*, pages 153–162, 1992.

31. S. Wu and U. Manber. Fast text searching allowing errors. *Comm. of the ACM*, 35(10):83–91, 1992.

32. J. Ziv and A. Lempel. A universal algorithm for sequential data compression. *IEEE Trans. Inf. Theory*, 23:337–343, 1977.

33. J. Ziv and A. Lempel. Compression of individual sequences via variable length coding. *IEEE Trans. Inf. Theory*, 24:530–536, 1978.

# Appendix: Average Number of Active Bits

The goal of this Appendix is to show that, on average, $|act(b)| = |tr_i(b)| = O(1)$. In this section $\sigma$ denotes the size of the text alphabet.

Let us consider the process of generating the LZ78/LZW tree. A string from the text is read and the current tree is followed, until the new string read "falls out" of the tree. At that point we add a new node to the tree and restart reading the text. It is clear that, at least for Bernoulli sources, the resulting tree is the same as the result of inserting $n$ random strings of infinite length.

Let us now consider initializing our NFA with just state $i$ active. Now, we backtrack on the LZ78 tree, entering into all possible branches and feeding the automaton with the corresponding letter. We stop when the automaton runs out of active states.

The total amount of tree nodes touched in this process is exactly the amount of text blocks $b$ whose $i$-th bit in $act(b)$ is active, i.e. the blocks such that if we start with state $i$ active, we finish the block with some active state. Hence the total amount of states in $act$ over all the blocks of the text corresponds to the sum of tree nodes touched when starting the NFA initialized with each possible state $i$.

As shown by Baeza-Yates and Gonnet [4,5], the cost of backtracking on a tree of $n$ nodes with a regular expression is $O(\text{polylog}(n)n^\lambda)$, where $0 \leq \lambda < 1$ depends on the structure of the regular expression. This result applies only to random tries over a uniformly distributed alphabet and for an arbitrary regular expression which has no outgoing edges from final states. We remark that the letter probabilities on the LZ78 tree are more uniform than on the text, so even on biased text the uniform model is not so bad approximation. In any case the result can probably be extended to biased cases.

Despite being suggestive, the previous result cannot be immediately applied to our case. First, it is not meaningful to consider such a random text in a compression scenario, since in this case compression would be impossible. Even a scenario where the text follows a biased Bernoulli or Markov model can be restrictive. Second, our DFAs can perfectly have outgoing transitions from the final states (the previous result is relevant because as soon as a final state is reached they report the whole subtrie). On the other hand, we cannot afford an arbitrary text and pattern simultaneously because it will always be possible to design a text tailored to the pattern that yields a low efficiency. Hence, we consider the most general scenario which is reasonable to face:

**Definition 1.** *Our* arbitrariness assumption *states that text and pattern are arbitrary but independent, in the sense that there is zero correlation between text substrings and strings generated by the regular expression.*

The arbitrariness assumption permits us extending our analysis to any text and pattern, under the condition that the text cannot be especially designed for the pattern. Our second step is to set a reasonable condition over the pattern.

The number of strings of length $\ell$ that are accepted by an automaton is [27]

$$N(\ell) \;\; = \;\; \sum_j \pi_j \omega_j^\ell \;\; = \;\; O(c^\ell)$$

where the sum is finitary and $\pi_j$ and $\omega_j$ are constants. The result is simple to obtain with generating functions: for each state $i$ the function $f_i(z)$ counts the number of strings of each length that can be generated from state $i$ of the DFA, so if edges labeled $a_1 \ldots a_k$ reach states $i_1 \ldots i_k$ from $i$ we have $f_i(z) = z(f_{i_1}(z) + \ldots + f_{i_k}(z) + 1 \cdot [i \text{ final}])$, which leads to a system of equations formed by polynomials and possibly fractions of the form $1/(1-z)$. The solution to the system is a rational function, i.e. a quotient between polynomials $P(z)/Q(z)$, which corresponds to a sequence of the form $\sum_j \pi_j \omega_j^\ell$. We are ready now to establish our condition over the admissible regular expressions.

**Definition 2.** *A regular expression is* admissible *if the number of strings of length $\ell$ that it generates is at most $c^\ell$, where $c < \sigma$, for any $\ell = \omega(1)$.*

Unadmissible regular expressions are those which basically match all the strings of every length, e.g. $a(a|b)^*a$ over the alphabet $\{a, b\}$, which matches $2^\ell/4 = \Theta(2^\ell)$ strings of length $\ell$. However, there are other cases. For example, pattern matching allowing $k$ errors can be modeled as a regular expression which matches every string for $\ell = O(k)$ [21]. As we see shortly, we can handle some unadmissible regular expressions anyway.

If a regular expression is admissible and the arbitrariness assumption holds, then if we feed it with characters from a random text position the automaton runs out of active states after $O(1)$ iterations. The reason is that the automaton recognizes $c^\ell$ strings of length $\ell$, out of the $\sigma^\ell$ possibilities. Since text and pattern are uncorrelated, the probability that the automaton recognizes the selected text substring after $\ell$ iterations is $O((c/\sigma)^\ell) = O(\alpha^\ell)$, where we have defined $\alpha = c/\sigma < 1$. Hence the expected amount of steps until the automaton runs out of active states is $\sum_{\ell>=0} \alpha^\ell = 1/(1-\alpha) = O(1)$.

Let us consider a perfectly balanced tree of $n$ nodes obtained from the text, of height $h = \log_\sigma n$. If we start an automaton at the root of the trie, it will touch $O(c^\ell)$ nodes at the tree level $\ell$. This means that the total number of nodes traversed is

$$O\left(c^h\right) \;\; = \;\; O\left(c^{\log_\sigma n}\right) \;\; = \;\; O\left(n^{\log_\sigma c}\right) \;\; = \;\; O\left(n^\lambda\right)$$

for $\lambda < 1$. So in this particular case we repeat the result that exists for random tries, which is not surprising. Let us now consdier the LZ78 tree of an *arbitrary* text, which has $f(\ell)$ nodes at depth $\ell$, where

$$\sum_{\ell=0}^{h} f(\ell) = n \quad \text{and} \quad f(0) = 1, \; f(\ell-1) \le f(\ell) \le \sigma^\ell$$

By the arbitrariness assumption, those $f(\ell)$ strings cannot have correlation with the pattern, so the traversal of the tree touches $\alpha^\ell f(\ell)$ of those nodes at level $\ell$.

Therefore the total number of nodes traversed is

$$C \;=\; \sum_{\ell=0}^{h} \alpha^{\ell} f(\ell)$$

Let us now start with an arbitrary tree and try to modify it in order to increase the number of traversed nodes while keeping the same total number of nodes $n$. Let us move a node from level $i$ to level $j$. The new cost is $C' = C - \alpha^{i} + \alpha^{j}$. Clearly we increase the cost by moving nodes upward. This means that the worst possible tree is the perfectly balanced one, where all nodes are as close to the root as possible. On the other hand, LZ78 tries obtained from texts tend to be quite balanced, so the worst and average case are quite close anyway. As an example of the other extreme, consider a LZ78 tree with maximum unbalancing (e.g. for the text $a^{u}$). In this case the total number of nodes traversed is $O(1)$.

So we have that, under the arbitrariness assumption, the total number of tree nodes traversed by an admissible regular expression is $O(n^{\lambda})$ for some $\lambda < 1$. We use now this result for our analysis.

It is clear that if we take our NFA and make state $i$ the initial state, the result corresponds to a regular expression because any NFA can be converted into a regular expression. So the total amount of states in $act$ is

$$O\left(n^{\lambda_0} \;+\; n^{\lambda_1} \;+\; \ldots \;+\; n^{\lambda_m}\right)$$

where $\lambda_i$ corresponds to taking $i$ as the initial state. We say that a state is admissible if, when that state is considered as the initial state, the regular expression becomes admissible.

Note that, given the self-loop we added at state 0, we have $\alpha_0 = 1$, i.e. state 0 is unadmissible. However, all the other states must be admissible because otherwise the original regular expression would not be admissible. That is, there is a fixed probability $p$ of reaching the unadmissible state and from there the automaton recognizes all the $\sigma^{\ell}$ strings, which gives at least $p\sigma^{\ell} = \Theta(\sigma^{\ell})$ strings recognized.

Hence, calling

$$\lambda \;=\; \max(\alpha_1, \ldots, \alpha_m) \;<\; 1$$

we have that the total number of active states in all the $act$ bit masks is

$$O\left(n \;+\; mn^{\lambda}\right) \;=\; O(n)$$

where we made the last simplification considering that $m = O(\text{polylog}(n))$, which is weaker than usual assumptions and true in practice. Therefore, we have proved that, under mild restrictions (much more general than the usual randomness assumption), the amortized number of active states in the $act$ masks is $O(1)$.

Note that we can afford even that the unadmissible states are reachable only from $O(1)$ other states, and the result still holds. For example, if our regular expression is $a(a|b)^*a^m$ we have only $O(1)$ initial states that yield unadmissible expressions, and our result holds. On the other hand, if we have $a^m(a|b)^*a$ then the unadmissible state can be reached from $\Theta(m)$ other states and our result does not hold.

We focus now on the size of the $tr_i(b)$ sets for admissible regular expressions. Let us consider the text substring $B$ corresponding to a block $b$.

We first consider the initial state, which is always active. How many states can get activated from the initial state? At each step, the initial state may activate $O(\sigma)$ admissible states, but given the arbitrariness assumption, the probability of each such state being active $\ell$ steps later is $O(\alpha^\ell)$. While processing $B_{1..k}$, the initial state is always active, so at the end of the processing we have $\sum_{\ell=0}^{k} \sigma\alpha^\ell = O(1)$ active states (the term $\alpha^\ell$ corresponds to the point where we were processing $B_{k-\ell}$).

We consider now the other $m$ admissible states, whose activation vanishes after examining $O(1)$ text positions. In their case the probability of yielding an active state after processing $B$ is $O(\alpha^k)$. Hence they totalize $O(m\alpha^k)$ active states. As before, the worst tree is the most balanced one, in which case there are $\sigma^k$ blocks of lengths 0 to $h = \log_\sigma n$. The total number of active states totalizes

$$\sum_{\ell=0}^{h} \sigma^\ell m\alpha^\ell \;=\; O(mc^h) \;=\; O\left(mn^\lambda\right)$$

Hence, we have in total $O(n + mn^\lambda) = O(n)$ active bits in the $tr_i$ sets, where the $n$ comes from the $O(1)$ states activated from the initial state and the $mn^\lambda$ from the other states.

# Parallel Lempel Ziv Coding

## (Extended Abstract)

Shmuel Tomi Klein[1] and Yair Wiseman[2]

[1] Dept. of Math. & CS, Bar Ilan University
Ramat-Gan 52900, Israel
`tomi@cs.biu.ac.il`
[2] Dept. of Math. & CS, Bar Ilan University
Jerusalem College of Technology
`wiseman@cs.biu.ac.il`

**Abstract.** We explore the possibility of using multiple processors to improve the encoding and decoding tasks of Lempel Ziv schemes. A new layout of the processors is suggested and it is shown how LZSS and LZW can be adapted to take advantage of such parallel architectures. Experimental results show an improvement in compression and time over standard methods.

## 1 Introduction

Compression methods are often partitioned into static and dynamic methods. The *static* methods assume that the file to be compressed has been generated according to a certain model which is fixed in advance and known to both compressor and decompressor. The model could be based on the probability distribution of the different characters or more generally of certain variable length substrings that appear in the file, combined with a procedure to parse the file into a well determined sequence of such elements. The encoded file can then be obtained by applying some statistical encoding function, such as Huffman or arithmetic coding. Information about the model is either assumed to be known (such as the distribution of characters in English text), or may be gathered in a first pass over the file, so that the compression process may only be performed in a second pass.

Many popular compression methods, however, are *adaptive* in nature. The underlying model is not assumed to be known, but discovered during the sequential processing of the file. The encoding and decoding of the $i$-th element is based on the distribution of the $i - 1$ preceding ones, so that compressor and decompressor can work in synchronization without requiring the transmittal of the model itself. Examples of adaptive methods are the Lempel-Ziv (LZ) methods and their variants, but there are also adaptive versions of Huffman and arithmetic coding.

We wish to explore the possibility of using multiple processors to improve the encoding and decoding tasks. In [7] this has been done for static Huffman coding,

A. Amir and G.M. Landau (Eds.): CPM 2001, LNCS 2089, pp. 18–30, 2001.

focusing in particular on the decoding process. The current work investigates how parallel processing could be made profitable for Lempel Ziv coding.

Previous work on parallelizing compression includes [1,2,3], which deal with LZ compression, [5], relating to Huffman and arithmetic coding, and [4]. A parallel method for the construction of Huffman trees can be found in [8]. Our work concentrates on LZ methods, in particular a variant of LZ77 [13] known as LZSS, and a variant of LZ78 [14] known as LZW. In LZSS [9], the encoded file consists of a sequence of items each of which is either a single character, or a pointer of the form (*off, len*) which replaces a string of length *len* that appeared *off* characters earlier in the file. Decoding of such a file is thus a very simple procedure, but for the encoding there is a need to locate longest reoccurring strings, for which sophisticated data structures like hash tables or binary trees have been suggested. In LZW [10], the encoded file consists of a sequence of pointers to a *dictionary*, each pointer replacing a string of the input file that appeared earlier and has been put into the dictionary. Encoder and decoder must therefore construct identical copies of the dictionary.

The basic idea of parallel coding is partitioning the input file of size $N$ into $n$ blocks of size $N/n$ and assigning each block to one of the $n$ available processors. For static methods the encoding is then straightforward, but for the decoding, it is the compressed file that is partitioned into equi-sized blocks, so there might be a problem of synchronization at the block boundaries. This problem may be overcome by inserting dummy bits to align the block boundaries with codeword boundaries, which causes a negligible overhead if the block size is large enough. Alternatively, in the case of static Huffman codes, one may exploit their tendency to resynchronize quickly after an error, to devise a parallel decoding procedure in which each processor decodes one block, but is allowed to overflow into one or more following blocks until synchronization is reached [7].

For dynamic methods one is faced with the additional problem that the encoding and decoding of elements in the $i$-th block may depend on elements of some previous blocks. Even if one assumes a CREW architecture, in which all the processors share some common memory space which can be accessed in parallel, this would still be essentially equivalent to a sequential model. This is so because elements dealt with by processor $i$ at the beginning of block $i$ may rely upon elements at the end of block $i - 1$ which have not been processed yet by processor $i - 1$; thus processor $i$ can in fact start its work only after processor $i - 1$ has terminated its own.

The easiest way to implement parallelization in spite of the above problem is to let each processor work independently of the others. The file is thus partitioned into $n$ blocks which are encoded and decoded without any transfer of data between the processors. If the block size is large enough, this solution may even be recommendable: most LZ methods put a bound on the size of the history taken into account for the current item, and empirical tests show that the additional compression, obtained by increasing this history beyond some reasonable size, rapidly tends to zero. The cost of parallelization would therefore be a small deterioration in compression performance at the block boundaries, since each processor has to "learn" the main features of the file on its own, but this loss will often be tolerated as it may allow to cut the processing time by a factor of $n$. In [6] the authors suggest letting each processor keep the last characters of the previous block and thereby improve the encoding speed, but each block must then be larger than the size of the history window. On the other hand, putting a lower bound on the size $N/n$ of each block effectively puts an upper bound on the number of processors $n$ which can be used for a given file of size $N$, so we might not fully take advantage of all the available computing power.

We therefore turn to the question how to use $n$ processors, even when the size of each block is not very large. In the next section we propose a new parallel coding algorithm, based on a time versus compression efficiency tradeoff which is related to the degree of parallelization. On the one extreme, for full parallelization, each of the $n$ processors works independently, which may sharply reduce the compression gain if the size of the blocks is small. On the other extreme, all the processors may communicate, forcing delays that make this variant as time consuming as a sequential algorithm. The suggested tradeoff is based on a hierarchical structure of the connections between the processors, each of which depending at most on $\log n$ others. The task can be performed in parallel by $n$ processors in $\log n$ sequential stages. There will be a deterioration in the compression ratio, but the loss will be inferior to that incurred when all $n$ processors are independent.

In contrast to Huffman coding, for which parallel decoding could be applied regardless of whether the possibility having multiple processors at decoding time was known at the time of encoding, there is a closer connection between encoding and decoding for LZ schemes. We therefore need to deal also with the parallel encoding scheme, and we assume that the same number of processors is available for both tasks.

Note, however, that one cannot assume simultaneously equi-sized blocks for both encoding and decoding. If encoding is done with blocks of fixed size, the resulting compressed blocks are of variable lengths. So one either has to store a vector of indices to the starting point of each processor in the compressed file, which adds an unnecessary storage overhead, or one performs *a priori* the compression on blocks of varying size, such that the resulting compressed blocks are all of roughly the same size. To get blocks of exactly the same size and to achieve byte alignment, one then needs to pad each block with a small number of bits, but in this case the loss of compression due to this padding is generally negligible. Moreover, the second alternative is also the preferred choice for many specific applications. For instance, in an Information Retrieval system built on a large static database, compression is done only once, so the speedup of parallelization may not have any impact, whereas decompression of selected parts is required for each query to be processed, raising the importance of parallel decoding.

## 2 A Tree-Structured Hierarchy of Processors

The suggested form of the hierarchy is that of a full binary tree, similarly to a binary heap. This basic form has already been mentioned in [6], but the way to use it as presented here is new. The input file is partitioned into $n$ blocks $B_1, \ldots, B_n$, each of which is assigned to one of the available processors. Denote the $n$ processors by $P_1, \ldots, P_n$, and assume, for the ease of description, that $n + 1$ is a power of 2, that is $n = 2^k - 1$ for some $k$. Processor $P_1$ is at the root of the tree and deals with the first block. As there is no need to "point into the future", communication lines between the processors may be unidirectional, permitting a processor with higher index to access processors with lower index, but not vice versa. Restricting this to a tree layout yields a structure in which $P_{2i}$ and $P_{2i+1}$ can access the memory of $P_i$, for $1 \leq i \leq (n - 1)/2$. Figure 1 shows this layout for $n = 15$, the arrows indicating the dependencies between the processors. The numbers indicate both the indices of the blocks and of the corresponding processors.

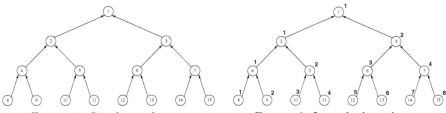

FIGURE 1: *Simple tree layout.*          FIGURE 2: *Layer-by-layer layout.*

The compression procedure for LZSS works as follows: $P_1$ starts at the beginning of block $B_1$, which is stored in its memory. Once this is done, $P_2$ and $P_3$ start simultaneously their work on $B_2$ and $B_3$ respectively, both searching for reoccurring strings first within the block they have been assigned to, and then extending the search back into block $B_1$. In general, after $P_i$ has finished the processing of block $B_i$, processors $P_{2i}$ and $P_{2i+1}$ start scanning simultaneously their corresponding blocks. The compression of the file is thus not necessarily done layer by layer, e.g., $P_{12}$ and $P_{13}$ may start compressing blocks $B_{12}$ and $B_{13}$, even if $P_5$ is not yet done with $B_5$.

Note that while the blocks $B_2$ and $B_1$ are contiguous, this is not the case for $B_3$ and $B_1$, so that the $(off, len)$ pairs do not necessarily point to *close* previous occurrences of a given string. This might affect compression efficiency, as one of the reasons for the good performance of LZ methods is the tendency of many files to repeat certain strings within the close vicinity of their initial occurrences. For processors and blocks with higher indices, the problem is even aggravated. The experimental section below brings empirical estimates of the resulting loss.

The layout suggested in Figure 1 is obviously wasteful, as processors of the higher layers stay idle after having compressed their assigned block. The number of necessary processors can be reduced by half, or, which is equivalent, the block size for a given number of processors may be doubled, if one allows a processor to deal with multiple blocks. The easiest way to achieve this is displayed in Figure 2, where the numbers in the nodes are the indices of the blocks, and the boldface numbers near the nodes refer to the processors. Processors $1, \ldots, 2^m$ are assigned sequentially, from left to right, to the blocks of layer $m$, $m = 0, 1, \ldots, k-1$. This simple way of enumerating the blocks has, however, two major drawbacks: refer, e.g., to block $B_9$ which should be compressed by processor $P_2$. First, it might be that $P_1$ finishes the compression of blocks $B_2$ and $B_4$, before $P_2$ is done with $B_3$. This causes an unnecessary delay, $B_9$ having to wait until $P_2$ processes both $B_3$ and $B_5$, which could be avoided if another processor would have been assigned to $B_9$, for example one of those that has not been used in the upper layers. Moreover, the problem is not only one of wasted time: $P_2$ stores in its memory information about the blocks it has processed, namely $B_3$ and $B_5$. But the compression of $B_9$ does not depend on these blocks, but only on $B_4$, $B_2$ and $B_1$. The problem thus is that the hierarchical structure of the tree is not inherited by the dependencies between the processors.

To correct this deficiency of the assignment scheme, each processor will continue working on one of the offsprings of its current block. For example, one could consistently assign a processor to the left child block of the current block, whereas the right child block is assigned to the next available newly used processor. More formally, let $S_j^i$ be the index of the processor assigned to block $j$ of layer $i$, where $i = 0, \ldots, k-1$ and $j = 1, \ldots, 2^i$, then $S_1^0 = 1$ and for $i > 0$,

$$S_{2j-1}^i = S_j^{i-1} \qquad \text{and} \qquad S_{2j}^i = 2^{i-1} + j.$$

The first layers are thus processed, from left to right, by processors with indices: (1), (1,2), (1, 3, 2, 4), (1, 5, 3, 6, 2, 7, 4, 8), etc. Figure 3(a) depicts the new layout of the blocks, the rectangles indicating the sets of blocks processed by the same processor. This structure induces a corresponding tree of processors, depicted in Figure 3(b).

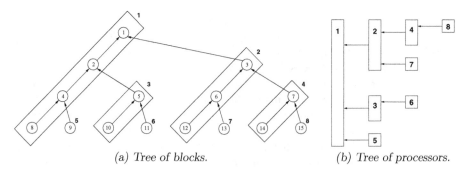

(a) Tree of blocks.          (b) Tree of processors.

FIGURE 3: *New hierarchical structure.*

As a results of this method, processor $P_i$ will start its work with block $B_{2i-1}$, and then continue with $B_{4i-2}$, $B_{8i-4}$, etc. In each layer, the evenly indexed blocks inherit their processors from their parent block, and each of the oddly indexed blocks starts a new sequence of blocks with processors that have not been used before.

The memory requirements of the processors have also increased by this new scheme, and space for the data of up to $\log_2 n$ blocks have to be stored. However, most of the processors deal only with a few blocks, and the average number of blocks to be memorized, when amortized over the $n$ processors is

$$\frac{1}{n} \sum_{i=1}^{\log_2 n} i \frac{n}{2^i} = 2 - \frac{\log_2 n + 2}{n} < 2.$$

For the encoding and decoding procedures, we need a fast way to convert the index of a block into the index of the corresponding processor, i.e., a function $f$, such that $f(i) = j$ if block $B_i$ is coded by processor $P_j$. Define $r(i)$ as the largest power of 2 that divides the integer $i$, that is, $r(i)$ is the length of the longest suffix consisting only of zeros of the binary representation of $i$.

CLAIM:
$$f(i) = \frac{1}{2} \left( \frac{i}{2^{r(i)}} + 1 \right).$$

PROOF: By induction on $i$. For $i = 1$, we get $f(1) = 1$, which is correct. Assume the claim is true up to $i - 1$. If $i$ is odd, $r(i) = 0$ and the formula gives $f(i) = (i+1)/2$. As has been mentioned above, any oddly indexed block is the starting point of a new processor and indeed processor $P_{(i+1)/2}$ starts at block $B_i$. If $i$ is even, block $B_i$ is coded by the same processor as its parent block $B_{i/2}$, for which the inductive assumption applies, and we get

$$f(i) = f(i/2) = \frac{1}{2} \left( \frac{i/2}{2^{r(i/2)}} + 1 \right) = \frac{1}{2} \left( \frac{i}{2\,2^{r(i)-1}} + 1 \right) = \frac{1}{2} \left( \frac{i}{2^{r(i)}} + 1 \right),$$

so that the formula holds also for $i$. ∎

## 2.1   Parallel Coding for LZSS

We now turn to the implementation details of the encoding and decoding procedures for LZSS. Since the coding is done by stages, the parallel co-routines will invoke themselves the depending offsprings. For the encoding, the procedure PLZSS-encode($i, j$) given in Figure 4 will process block $B_i$ with processor $P_j$, where $j = f(i)$. The whole process is initialized by a call to PLZSS-encode($1,1$) from the main program.

Each routine starts by copying the text of the current block into the memory of the processor, possibly adding to texts of previous blocks that have been stored there. As in the original LZSS, the longest substring in the history is sought that matches the suffix of the block starting at the current position. The search for this substring can be accelerated by several techniques, and one of the fastest is by use of a hash table [12]. The longest substring is then replaced by

$PLZSS\text{-}encode(i, j)$

```
{
        append text of B_i to memory of P_j
        cur ⟵ 1
        while cur < |B_i|
        {
                S ⟵ suffix of B_i starting at cur
                ind ⟵ i
                while ind > 0
                {
                        access memory of P_{f(ind)} and
                        record occurrences in B_{ind} matching a prefix of S
                        ind ⟵ ⌊ind/2⌋
                }

                if longest occurrence not long enough
                {       encode single character        cur ⟵ cur + 1   }
                else
                {       encode as (off, len)        cur ⟵ cur + len   }
        }

        perform in    ⎧ if 2i ≤ n       PLZSS-encode (2i, j)
        parallel      ⎩ if 2i + 1 ≤ n   PLZSS-encode (2i + 1, i + 1)
}
```

FIGURE 4:   *Parallel LZSS encoding for block $B_i$ by processor $P_j$.*

a pair (*offset, length*), unless *length* is too small (2 or 3 in implementations of [12], such as the patent [11], which is the basis of Microsoft's DoubleSpace), in which case a single character is sent to output and the window is shifted by one.

In our case, the search is not limited to the current block, but extends backwards to the parent blocks in the hierarchy, up to the root. For example, referring to Figure 3, the encoding of block $B_{13}$ will search also through $B_6$, $B_3$ and $B_1$, and thus access the memory of the processors $P_7$, $P_2$, $P_2$ and $P_1$, respectively. Note that the size of the history window is usually limited by some constant $W$. We do not impose any such limit, but in fact, the encoding of any element is based on a history of size at most $\log_2 n \times$ the block size.

For the decoding, recall that we assume that the encoded blocks are of equal size *Blocksize*. The decoding routine can thus address earlier locations as if the blocks, that are ancestors of the current block in the tree layout, were stored contiguously. Any element of the form (*off, len*) in block $B_i$ can point back into a block $B_j$, with $j = \lfloor i/2^b \rfloor$ for $b = 0, 1, \ldots, \lfloor \log_2 i \rfloor$, and the index of this block can be calculated by

$$b \longleftarrow \lceil (\text{off} - \text{cur} + 1)/\text{Blocksize} \rceil,$$

where *cur* is the index of the current position in block $B_i$. The formal decoding procedure is given in Figure 5.

*PLZSS-decode*$(i, j)$

{
       $cur \longleftarrow 1$
      while there are more items to decode
        {
              if next item is a character
              {        store the character      $cur \longleftarrow cur + 1$  }
              else     // the item is (*off*, *len*)
              {
                    if $off < cur$     // pointer within block $B_i$
                            copy *len* characters, starting at position $cur-off$
                    else     // pointer to earlier block
                    {
                          $b \longleftarrow \lceil (off - cur + 1)/Blocksize \rceil$
                          $t \longleftarrow (off - cur) \bmod Blocksize$
                          copy *len* characters, starting at position $t$
                                in block $B_{\lfloor i/2^b \rfloor}$ which is stored in $P_{f(\lfloor i/2^b \rfloor)}$
                  }
                  $cur \longleftarrow cur + len$
              }
        }
      perform in   $\begin{cases} \text{if } 2i \leq n & \textit{PLZSS-decode}(2i, j) \\ \text{if } 2i+1 \leq n & \textit{PLZSS-decode}(2i+1, i+1) \end{cases}$
        parallel
}

FIGURE 5:   *Parallel LZSS decoding for block $B_i$ on processor $P_j$.*

The input of the decoding routine is supposed to be a file consisting of a sequence of items, each being either a single character or a pointer of the form (*off*, *len*); *cur* is the current index in the reconstructed text file.

## 2.2   Parallel Coding for LZW

Encoding and decoding for LZW is similar to that of LZSS, with a few differences. While for LZSS, the "dictionary" of previously encountered strings is in fact the text itself, LZW builds a continuously growing table *Table*, which need not be transmitted, as it is synchronously reconstructed by the decoder. The table is initialized to include the set of single characters composing the text, which is often assumed to be ASCII. If, as above, we denote by $S$ the suffix of the text in block $B_i$ starting at the current position, then the next encoded element will be the index of the longest prefix $R$ of $S$ for which $R \in Table$, and the next element to be adjoined to *Table* will be the shortest prefix $R'$ of $S$ for which $R' \notin Table$; $R$ is a prefix of $R'$ and $R'$ extends $R$ by one additional character.

During the encoding process of $B_i$, one therefore needs to access the tables in $B_i$ itself and in the blocks which are ancestors of $B_i$ in the tree layout, but the order of access has to be top down rather than bottom up as for LZSS. For each $i$, we therefore need a list $list_i$ of the indices of the blocks accessed on the way from the root to block $B_i$, that is, $list_i[ind]$ is the number whose binary representation is given by the *ind* leftmost bits of the binary representation of $i$. For example, $list_{13} = [1, 3, 6, 13]$.

To encode a new element $P$, it is first searched for in *Table* of $B_1$, and if not found there, then in *Table* of $B_{list_i[2]}$, which is stored in the memory of processor $P_{f(list_i[2])}$, etc. However, storing only the elements in the tables may lead to errors. To illustrate this, consider the following example, referring again to Figure 3.

Suppose that the longest prefix of the string abcde appearing in the *Table* of $B_1$ is abc. Suppose we later encounter abcd in the text of block $B_2$. The string abcd will thus be adjoined to the same *Table*, since both $B_1$ and $B_2$ are processed by the same processor $P_1$. Assume now that the texts of both blocks $B_5$ and $B_3$ start with abcde. While for $B_5$ it is correct to store abcde as the first element in its *Table*, the first element to be stored in the *Table* of $B_3$ should be abcd, since the abcd in the memory of $P_1$ was generated by block $B_2$, whereas $B_3$ only depends on $B_1$.

$PLZW\text{-}encode(i, j)$
{
      $\omega \longleftarrow B_i[1]$
      $cur \longleftarrow 2$
      while $cur \leq |B_i|$
        {
            $ind \longleftarrow 1$
            while $list_i[ind] \leq i$
              {
                  while    $cur < |B_i|$ and
                             $(\omega B_i[cur], ind) \in Table$ stored in $P_{f(list_i[ind])}$
                    {
                        $\omega \longleftarrow \omega B_i[cur]$
                        $cur \longleftarrow cur + 1$
                        $last \longleftarrow ind$
                  }
                $ind \longleftarrow ind + 1$
              }
            $indx \longleftarrow index(\omega)$ in $Table$ of $P_{f(list_i[last])}$
            store $(indx, last)$ in memory of $P_j$
            store $(\omega B_i[cur], ind)$ in $Table$ in memory of $P_j$
            $\omega \longleftarrow B_i[cur]$
            $cur \longleftarrow cur + 1$
        }
        perform in   $\Big\{$ if $2i \leq n$      $PLZW\text{-}encode(2i, j)$
          parallel       if $2i + 1 \leq n$   $PLZW\text{-}encode(2i + 1, i + 1)$
}

FIGURE 6:   *Parallel LZW encoding for block $B_i$ on processor $P_j$.*

To avoid such errors, we need a kind of a "time stamp", indicating at what stage an element has been added to a *Table*. If the elements are stored sequentially in these tables, one only needs to record the indices of the last element for each block. But implementations of LZW generally use hashing to maintain the tables, so one cannot rely on deducing information from its physical location, and each element has to be marked individually. The easiest way is to store with each string $P$ also the index $i$ of the block which caused the addition of $P$. This would require $\log_2 n$ bits for each entry. One can however take advantage of the fact that the elements stored by different blocks $B_i$ in the memory of a given processor correspond to different indices $ind$ in the corresponding lists $list_i$. It thus suffices to store with each element the index in $list_i$ rather than $i$ itself, so that only $\log_2\log_2 n$ bits are needed for each entry. The formal encoding and decoding procedures are given in Figures 6 and 7, respectively.

The parallel LZW encoding refers to the characters in the input block as belonging to a vector $B_i[cur]$, with $cur$ giving the current index. If $x$ and $y$ are strings, then $xy$ denotes their concatenation. As explained above, since the *Table* corresponding to block $B_i$ is stored in the memory of a processor which is also accessed by other blocks, each element stored in the *Table* needs an identifier indicating the block from which is has been generated. The elements in the *Table* are therefore of the form (*string, identifier*).

The output of LZW encoding is a sequence of pointers, which are the indices of the encoded elements in the *Table*. In our case, these pointers are of the form (*index, identifier*). There is, however, no deterioration in the compression efficiency, as the additional bits needed for the identifier are saved in the representation of the index, which addresses a smaller range.

For simplicity, we do not go into details of handling the incremental encoding of the indices, and overflow conditions when the *Table* gets full. It can be done as for the serial LZW.

The parallel LZW decode routine assumes that its input is a sequence of elements of the form (*index, identifier*). The empty string is denoted by $\Lambda$.

The algorithm in Figure 7 is a simplified version of the decoding, which does not work in case the current element to be decoded was the last one to be added to the *Table*. This is also a problem in the original LZW decoding and can be solved here in the same way. The details have been omitted to keep the emphasis on the parallelization.

## 3   Experimental Results

We now report on some experiments on files in different languages: the Bible (King James Version) in English, the Bible in Hebrew and the *Dictionnaire philosophique* of Voltaire in French. Table 1 first brings the sizes of the files in MB and to what size they can be reduced by LZSS and LZW, expressed in percent of the sizes of the original files. We consider three algorithms: the serial one, using a single processor and yielding the compressed sizes in Table 1, but being slow; a parallel algorithm we refer to as *standard*, where each block is treated

PLZW-decode(i, j)

```
{
        cur  ⟵  1
        old  ⟵  Λ
        while cur ≤ number of items in block Bᵢ
        {
                (indx, ind)  ⟵  Bᵢ[cur]
                access Table in P_{f(listᵢ[ind])} at index indx
                        and send string str found there to output
                if old ≠ Λ
                        store (old first[str], ⌈log₂(i + 1)⌉) in Table of Pⱼ
                old  ⟵  str
                cur  ⟵  cur + 1
        }

        perform in    ⎰ if 2i ≤ n        PLZW-decode(2i, j)
          parallel    ⎱ if 2i + 1 ≤ n    PLZW-decode(2i + 1, i + 1)
}
```

FIGURE 7:   *Parallel LZW decoding for block $B_i$ on processor $P_j$.*

independently of the others; and the *new* parallel algorithm presented herein, which exploits the hierarchical layout. The columns headed Time in Table 1 compare the new algorithm with the serial one. The time measurements were taken on a Sun 450 with four UltraSPARC–II 248 MHz processors, which allowed a layout with 7 blocks. The values are in seconds and correspond to LZW, which turned out to give better compression performance than LZSS in our case. The improvement is obviously not expected to be 4-fold, due to the overhead of the parallelization, but on the examples the time is generally cut to less than half.

TABLE 1:   *Size and time measurements on test files.*

|  | Size | | | Time | | | |
| --- | --- | --- | --- | --- | --- | --- | --- |
|  | Full | compressed by | | compression | | decompression | |
|  |  | LZSS | LZW | Serial | New | Serial | New |
| English Bible | 3.860 | 41.6 | 36.6 | 5.508 | 2.296 | 3.653 | 1.504 |
| Hebrew Bible | 1.471 | 51.7 | 44.7 | 2.134 | 0.853 | 1.488 | 0.566 |
| Voltaire | 0.529 | 49.0 | 40.6 | 0.770 | 0.380 | 0.456 | 0.310 |

For the compression performance, we compare the two parallel versions. Both are equivalent to the serial algorithm if the block size is chosen large enough, as in [6]. The graphs in Figure 8 show the sizes of the compressed files in MB as functions of the block size (in bytes), for both LZSS and LZW. We see that for large enough blocks (about 64K for LZSS and 128K for LZW) the loss relative to a serial algorithm with a single processor is negligible (about 1%) for both the standard and the new methods. However, when the blocks become shorter, the compression gain in the independent model almost vanishes, whereas with the new processor layout the decrease in compression performance is much slower. For blocks as small as 128 bytes, running a standard parallel compression achieves only about 1–4% compression for LZSS and about 12–15% for LZW, while with the new layout this might be reduced by some additional 30–40%.

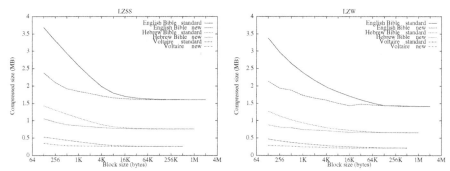

FIGURE 8:    *Size of compressed file as function of block size.*

We conclude that the simple hierarchical layout might allow us to considerably reduce the size of the blocks that are processed in parallel without paying too high a price in compression performance. As a consequence, if a large number of processors is available, it enables a better utilization of their full combined computing power.

# References

1. DE AGOSTINO S., STORER J.A., Near Optimal Compression with Respect to a Static Dictionary on a Practical Massively Parallel Architecture, *IEEE Computer Society Press* (1995) 172–181.

2. DE AGOSTINO S., STORER J.A., Parallel Algorithms for Optimal Compression using Dictionaries with the Prefix Property, *IEEE Computer Society Press* (1992) 52–61.

3. GONZALEZ SMITH M.E., STORER J.A., Parallel Algorithms for Data Compression, *Journal of the ACM* **32**(2) (1985) 344–373.

4. HIRSCHBERG D.S., STAUFFER L.M., Parsing Algorithms for Dictionary Compression on the PRAM, *IEEE Computer Society Press* (1994) 136–145.

5. HOWARD P.G., VITTER J.S., Parallel lossless image compression using Huffman and arithmetic coding, *Proc. Data Compression Conference DCC–92*, Snowbird, Utah (1992) 299–308.

6. IWATA K., MORII M., UYEMATSU T., OKAMOTO E., A simple parallel algorithm for the Ziv-Lempel encoding, *IEICE Trans. Fundamentals* **E81–A** (1998) 709–712.

7. KLEIN S.T., WISEMAN Y., Parallel Huffman decoding, *Proc. Data Compression Conference DCC–2000*, Snowbird, Utah (2000) 383–392.

8. LAWRENCECE L.L., PRZYTYCKA T.M., Constructing Huffman Trees in Parallel, *SIAM Journal of Computing* **24**(6) (1995) 1163–1169.

9. STORER J.A., SZYMANSKI, T.G., Data compression via textual substitution, *J. ACM* **29** (1982) 928–951.

10. WELCH T.A., A technique for high-performance data compression, *IEEE Computer* **17** (June 1984) 8–19.

11. WHITING D.L., GEORGE G.A., IVEY G.E., Data Compression Apparatus and Method, U.S. Patent 5,126,739 (1992).

12. WILLIAMS R.N., An extremely fast Ziv-Lempel data compression algorithm, *Proc. Data Compression Conference DCC–91*, Snowbird, Utah (1991) 362–371.

13. ZIV J., LEMPEL A., A universal algorithm for sequential data compression, *IEEE Trans. on Inf. Th.* **IT–23** (1977) 337–343.

14. ZIV J., LEMPEL A., Compression of individual sequences via variable-rate coding, *IEEE Trans. on Inf. Th.* **IT–24** (1978) 530–536.

# Approximate Matching of Run-Length Compressed Strings

Veli Mäkinen[1][*], Gonzalo Navarro[2][**], and Esko Ukkonen[1][*]

[1] Department of Computer Science, P.O Box 26 (Teollisuuskatu 23)
FIN-00014 University of Helsinki, Finland
{vmakinen,ukkonen}@cs.helsinki.fi
[2] Department of Computer Science, University of Chile
Blanco Encalada 2120, Santiago, Chile
gnavarro@dcc.uchile.cl

**Abstract.** We focus on the problem of approximate matching of strings that have been compressed using run-length encoding. Previous studies have concentrated on the problem of computing the longest common subsequence (LCS) between two strings of length $m$ and $n$, compressed to $m'$ and $n'$ runs. We extend an existing algorithm for the LCS to the Levenshtein distance achieving $O(m'n+n'm)$ complexity. This approach gives also an algorithm for approximate searching of a pattern of $m$ letters ($m'$ runs) in a text of $n$ letters ($n'$ runs) in $O(mm'n')$ time, both for LCS and Levenshtein models. Then we propose improvements for a greedy algorithm for the LCS, and conjecture that the improved algorithm has $O(m'n')$ expected case complexity. Experimental results are provided to support the conjecture.

## 1 Introduction

The problem of *compressed pattern matching* is, given a compressed text $T$ and a (possibly compressed) pattern $P$, find all occurrences of $P$ in $T$ without decompressing $T$ (and $P$). The goal is to search faster than by using the basic scheme: decompression followed by a search.

In the basic approach, we are interested in reporting only the *exact* occurrences, i.e. the locations of the substrings of $T$ that match exactly pattern $P$. We can loosen the requirement of exact occurrences to *approximate occurrences* by introducing a distance function to measure the similarity between $P$ and a substring of $T$. Now, we want to find all the approximate occurrences of $P$ in $T$, where the distance between $P$ and a substring of $T$ is at most a given error threshold $k$. Often a suitable distance measure between two strings is the *edit distance*, where the minimum amount of character insertions, deletions, and replacements, that are needed to make the two strings equal, is calculated. For this distance we are interested in $k < |P|$ errors.

---

[*] Supported by the Academy of Finland under grant 22584.
[**] Supported in part by Fondecyt grants 1-990627 and 1-000929.

A. Amir and G.M. Landau (Eds.): CPM 2001, LNCS 2089, pp. 31–49, 2001.
© Springer-Verlag Berlin Heidelberg 2001

Many studies have been made around the subject of compressed pattern matching over different compression formats, starting with the work of Amir and Benson [1], e.g. [2,8,10,9]. The only works addressing the approximate variant of the problem have been [11,13,15], on Ziv-Lempel [20].

Our focus is approximate matching over *run-length encoded* strings. In run-length encoding a string that consists of repetitions of letters is compressed by encoding each repetition as a pair ("letter","length of the repetition"). For example, string *aaabbbbccaab* is encoded as a sequence $(a, 3)(b, 4)(c, 2)(a, 2)(b, 1)$. This technique is widely used especially in image compression, where repetitions of pixel values are common. This is particularly interesting for fax transmissions and bilevel images. Approximate matching on images can be a useful tool to detect distortions. Even a one-dimensional compressed approximate matching algorithm would be useful to speed up existing two-dimensional approximate matching algorithms, e.g. [5].

Exact pattern matching over run-length encoded text can be done optimally in $O(m' + n')$ time, where $m'$ and $n'$ are the compressed sizes of the pattern and the text [1]. Approximate pattern matching over run-length encoded text has not been considered before this study, but there has been work on the distance calculation, namely, given two strings of length $m$ and $n$ that are run-length compressed to lengths $m'$ and $n'$, calculate their distance using the compressed representations of the strings. This problem was first posed by Bunke and Csirik [6]. They considered the version of edit distance without the replacement operation, that is related to the problem of calculating the longest common subsequence (LCS) of two strings. They gave an $O(m'n')$ time algorithm for a special case of the problem, where all run-lengths are of equal size. Later, they gave an $O(m'n + n'm)$ time algorithm for the general case [7]. A major improvement over the previous results was due to Apostolico, Landau, and Skiena [3]; they first gave a basic $O(m'n'(m' + n'))$ algorithm, and further improved it to $O(m'n' \log(m'n'))$. Mitchell [14] gave an algorithm with the same time comlexity in the worst case, but faster with

some inputs; its time complexity is $O((p + m' + n') \log(p + m' + n'))$, where $p$ is the amount of pairs of compressed characters that match ($p$ equals to the amount of equal letter boxes, see the definition in Sect. 2.2).

All these algorithms were limited to the LCS distance, although, Mitchell's method [14] could be applied when different costs are assigned to the insertion and deletion operations. It still remain an open question (as posed by Bunke and Csirik) whether similar improvements could be found for a more general set of edit operations and their costs.

We give an algorithm for matching run-length encoded strings under *Levenshtein* distance [12]. In the Levenshtein distance a unit cost is assigned to each of the three edit operations. The algorithm is an extension of the $O(m'n + n'm)$ algorithm of Bunke and Csirik [7]; we keep the same cost but generalize the algorithm to handle a more complex distance model. Independently from our work, Arbell, Landau, and Mitchell have found a similar algorithm [4].

We modify our algorithm to work in a context of approximate pattern matching, and achieve $O(mm'n')$ time for searching a pattern of length $m$ that is run-length compressed to length $m'$, in a run-length compressed text of length $n'$. This algorithm works for both Levenshtein and LCS distance models.

We also study the LCS calculation. First, we give a greedy algorithm for the LCS that works in $O(m'n'(m' + n'))$ time. Adapting the well known diagonal method [17], we are able to improve the greedy method to work in $O(d^2 min(n', m'))$ time, where $d$ is the edit distance between the two strings (under insertions and deletions with the unit cost model).

Then we present improvements for the greedy method for the LCS, that do not however affect the worst case, but do have effect on the average case. We end up conjecturing that our improved algorithm is $O(m'n')$ time on average. As we are unable to prove it, we provide instead experimental evidence to support the conjecture.

## 2    Edit Distance on Run-Length Compressed Strings

### 2.1    Edit Distance

Let $\Sigma$ be a finite set of symbols, called an *alphabet*. A *string* $A$ of length $|A| = m$ is a sequence of symbols in $\Sigma$, denoted by $A = A_{1...m} = a_1 a_2 \ldots a_m$, where $a_i \in \Sigma$ for every $i$. If $|A| = 0$, then $A = \lambda$ is an empty string. A *subsequence* of $A$ is any sequence $a_{i_1} a_{i_2} \ldots a_{i_k}$, where $1 \leq i_1 < i_2 \cdots < i_k \leq m$.

The *edit distance* can be used to measure the similarity between two strings $A = a_1 a_2 \ldots a_m$ and $B = b_1 b_2 \ldots b_n$ by calculating the minimum cost of edit operations that are needed to convert $A$ into $B$ [12,19,16]. The usual edit operations are *substitution* (convert $a_i$ into $b_j$, denoted by $a_i \rightarrow b_j$), *insertion* ($\lambda \rightarrow b_j$), and *deletion* ($a_i \rightarrow \lambda$). Different costs for edit operations can be given. For *Levenshtein distance* (denoted by $D_L(A, B)$) [12], we assign costs $w(a \rightarrow a) = 0$, $w(a \rightarrow b) = 1$, $w(a \rightarrow \lambda) = 1$, and $w(\lambda \rightarrow a) = 1$, for all $a, b \in \Sigma$, $a \neq b$. If substitutions are forbidden, i.e. $w(a \rightarrow b) = \infty$, we get the distance $D_{ID}(A, B)$.

Distance $D_L(A, B)$ can be calculated by using dynamic programming [16]; evaluate an $(m + 1) \times (n + 1)$ matrix $(d_{ij})$, $0 \leq i \leq m$, $0 \leq j \leq n$, using the recurrence

$$
\begin{aligned}
d_{i,0} &= i, \quad 0 \leq i \leq m, \\
d_{0,j} &= j, \quad 0 \leq j \leq n, \\
d_{i,j} &= min(\textbf{if } a_i = b_j \textbf{ then } d_{i-1,j-1} \textbf{ else } d_{i-1,j-1} + 1, \\
& \qquad d_{i-1,j} + 1, d_{i,j-1} + 1), \quad \text{otherwise.}
\end{aligned}
\tag{1}
$$

The matrix $(d_{ij})$ can be evaluated row-by-row or column-by-column in $O(mn)$ time, and the value $d_{mn}$ equals $D_L(A, B)$.

A similar method can be used to calculate the distance $D_{ID}(A, B)$. Now, the recurrence is

$$
\begin{aligned}
d_{i,0} &= i, & 0 \leq i \leq m, \\
d_{0,j} &= j, & 0 \leq j \leq n, \\
d_{i,j} &= min(\text{if } a_i = b_j \text{ then } d_{i-1,j-1} \text{ else } \infty, \\
& \qquad d_{i-1,j} + 1, d_{i,j-1} + 1), & \text{otherwise.}
\end{aligned} \tag{2}
$$

The problem of calculating the *longest common subsequence* of strings $A$ and $B$ (denoted by $LCS(A, B)$), is related to the distance $D_{ID}(A, B)$. It is easy to see that $2 * |LCS(A, B)| = m + n - D_{ID}(A, B)$.

## 2.2   Dividing the Edit Distance Matrix into Boxes

A *run-length* encoding of the string $A = a_1 a_2 \ldots a_m$ is $A' = (a_1, p_1)(a_{p_1+1}, p_2)$ $(a_{p_1+p_2+1}, p_3) \ldots (a_{m-p_{m'}+1}, p_{m'}) = (a_{i_1}, p_1)(a_{i_2}, p_2) \ldots (a_{i_{m'}}, p_{m'})$, where $(a_{i_k}, p_k)$ denotes a sequence $\alpha_k = a_{i_k} a_{i_k} \ldots a_{i_k} = a_{i_k}^{p_k}$ of length $|\alpha_k| = p_k$. We also call $(a_{i_k}, p_k)$ a *run* of $a_{i_k}$. String $A$ is *optimally run-length encoded* if $a_{i_k} \neq a_{i_{k+1}}$ for all $1 \leq k < m'$.

In the next sections, we will show how to speed up the evaluation of values $d_{mn}$ for both distances $D_L(A, B)$ and $D_{ID}(A, B)$ when both the strings $A$ and $B$ are run-length encoded. In both methods, we use the following notation to divide the matrix $(d_{ij})$ into submatrices (see Fig. 1).

DP matrix

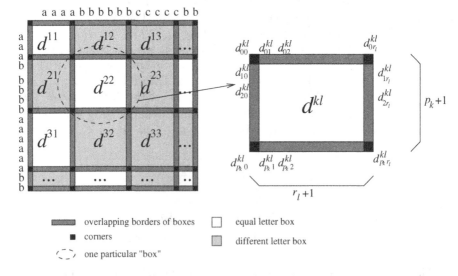

**Fig. 1.** A dynamic programmig matrix split into run-length blocks.

Let $A' = (a_{i_1}, p_1)(a_{i_2}, p_2) \ldots (a_{i_{m'}}, p_{m'})$ and $B' = (b_{j_1}, r_1), (b_{j_2}, r_2) \ldots$ $(b_{j_{n'}}, p_{n'})$ be the run-length encoded representations of strings $A$ and $B$. The rows and columns that correspond to the ends of runs in $A$ and $B$ separate the edit distance matrix $(d_{ij})$ into submatrices. To ease the notation later on, we define the submatrices so that they overlap on the borders. Formally, each pair of runs $(a_{i_k}, p_k), (b_{j_\ell}, r_\ell)$ defines a $(p_k + 1) \times (r_\ell + 1)$ submatrix $(d_{s,t}^{k,\ell})$ such that

$$d_{s,t}^{k,\ell} = d_{i_k + s - 1, j_\ell + t - 1}, \qquad 0 \le s \le p_k, 0 \le t \le r_\ell. \tag{3}$$

We will call submatrices $(d_{s,t}^{k,\ell})$ *boxes*. If a pair of runs corresponding to a box contain equal letters (i.e. $a_{i_k} = b_{j_\ell}$), then $(d_{s,t}^{k,\ell})$ is called an *equal letter box*. Otherwise we call $(d_{s,t}^{k,\ell})$ a *different letter box*. Adjacent boxes can form *runs of different letter boxes* along rows and columns. We assume that both strings are optimally run-length encoded, and hence runs of equal letter boxes can not occur.

## 3  An $O(mn' + m'n)$ Algorithm for the Levenshtein Distance

Bunke and Csirik [7] gave an $O(mn' + m'n)$ time algorithm for computing the LCS between two strings of lengths $n$ and $m$ run-length compressed to $n'$ and $m'$. They pose as an open problem extending their algorithm to the Levenshtein distance. This is what we do in this section, without increasing the complexity to compute the new distance $D_L$. Arbell, Landau, and Mitchell [4] have independently found a similar algorithm. Their solution is also based on the same idea of extending the $O(mn' + m'n)$ LCS algorithm to the Levenshtein distance.

Compared to the LCS-related distance $D_{ID}$, the Levenshtein distance $D_L$ permits an additional character substitution operation, at cost 1. We compute $D_L(A, B)$ by filling all the borders of all the boxes $(d_{s,t}^{k,\ell})$ (see Fig. 1). We manage to fill each cell in constant time, which adds up the promised $O(mn' + m'n)$ complexity. The space complexity can be made $O(n + m)$ by processing the matrix row-wise or column-wise.

### 3.1  The Basic Algorithm

We start with two lemmas that characterize the relationships between the border values in the boxes $(d_{s,t}^{k,\ell})$. First, we consider the equal letter boxes:

**Lemma 1 (Bunke and Csirik [7])** *The recurrences (1) and (2) can be replaced by*

$$d_{s,t}^{k,\ell} = \textbf{if } s \le t \textbf{ then } d_{0,t-s}^{k,\ell} \textbf{ else } d_{s-t,0}^{k,\ell}, \tag{4}$$

*where $1 \le s \le p_k$ and $1 \le t \le r_\ell$, for values $d_{s,t}^{k,\ell}$ in an equal letter box.*   □

Note that Lemma 1 holds for both Levenshtein and LCS distance models, because formulas (1) and (2) are equal when $a_i = b_j$. Since we are computing all the cells in the borders of the boxes, Lemma 1 permits computing new box borders in constant time using those of previous boxes.

The difficult part lies in the different letter boxes.

**Lemma 2** *The recurrence (1) can be replaced by*

$$d_{s,t}^{k,\ell} = 1 + min \ ( \ t-1 + min_{max(0,s-t) \leq q \leq s} \ d_{q,0}^{k,\ell} \ ,$$
$$s-1 + min_{max(0,t-s) \leq q \leq t} \ d_{0,q}^{k,\ell}), \tag{5}$$

*where $1 \leq s \leq p_k$ and $1 \leq t \leq r_\ell$, for values $d_{s,t}^{k,\ell}$ in a different letter box.*

*Proof.* We use induction on $s + t$. If $s + t = 2$ the formula (5) becomes $d_{1,1}^{k,\ell} = 1 + min(d_{0,0}^{k,\ell}, d_{1,0}^{k,\ell}, d_{0,1}^{k,\ell})$, which matches recurrence (1). In the inductive case we have

$$d_{s,t}^{k,\ell} = 1 + min(d_{s-1,t-1}^{k,\ell}, d_{s-1,t}^{k,\ell}, d_{s,t-1}^{k,\ell})$$

by recurrence (1), and using the induction hypothesis we get

$$\begin{aligned}
d_{s,t}^{k,\ell} = 2 + min(min( \ &t-2 + min_{max(0,s-t) \leq q \leq s-1} \ d_{q,0}^{k,\ell}, \\
&s-2 + min_{max(0,t-s) \leq q \leq t-1} \ d_{0,q}^{k,\ell}), \\
min( \ &t-1 + min_{max(0,s-1-t) \leq q \leq s-1} \ d_{q,0}^{k,\ell}, \\
&s-2 + min_{max(0,t-s+1) \leq q \leq t} \ d_{0,q}^{k,\ell}), \\
min( \ &t-2 + min_{max(0,s-t+1) \leq q \leq s} \ d_{q,0}^{k,\ell}, \\
&s-1 + min_{max(0,t-1-s) \leq q \leq t-1} \ d_{0,q}^{k,\ell})) \\
= 1 + min( &t-1+ min_{max(0,s-t) \leq q \leq s} \ d_{q,0}^{k,\ell}, \\
&s-1+ min_{max(0,t-s) \leq q \leq t} \ d_{0,q}^{k,\ell}),
\end{aligned}$$

where we have used the property that consecutive cells in the $(d_{ij})$ matrix differ at most by 1 [18]. Note that we have assumed $s > 1$ and $t > 1$. The particular cases $s = 1$ or $t = 1$ are easily derived as well, for example for $s = 1$ and $t > 1$ we have

$$\begin{aligned}
d_{1,t}^{k,\ell} &= 1 + min(d_{0,t-1}^{k,\ell}, d_{0,t}^{k,\ell}, d_{1,t-1}^{k,\ell}) \\
&= 1 + min(d_{0,t-1}^{k,\ell}, d_{0,t}^{k,\ell}, \\
&\quad 1 + min(t-2 + min_{max(0,2-t) \leq q \leq 1} \ d_{q,0}^{k,\ell}, min_{max(0,t-2) \leq q \leq t-1} \ d_{0,q}^{k,\ell})) \\
&= 1 + min \left( d_{0,t-1}^{k,\ell}, d_{0,t}^{k,\ell}, t-1 + min(d_{0,0}^{k,\ell}, d_{1,0}^{k,\ell}), 1 + min(d_{0,t-2}^{k,\ell}, d_{0,t-1}^{k,\ell}) \right) \\
&= 1 + min \left( t-1 + min(d_{0,0}^{k,\ell}, d_{1,0}^{k,\ell}), min(d_{0,t-1}^{k,\ell}, d_{0,t}^{k,\ell}) \right),
\end{aligned}$$

which is the particularization of formula (5) for $s = 1$. □

Formula (5) relates the values at the right and bottom borders of a box to its left and top borders. Yet it is not enough to compute the cells in constant time. Although we cannot compute one cell in $O(1)$ time, we can compute all the $p_k$ (or $r_\ell$) cells in overall $O(p_k)$ (or $O(r_\ell)$) time.

Fig. 2 shows the algorithm. We use a data structure (which in the pseudocode is represented just as a set $M_*$) able to handle a multiset of elements starting with a single element, adding and deleting elements, and delivering its minimum value at any time. It will be used to maintain and update the minima $min_{max(0,s-t)\leq q\leq s}\ d^{k,\ell}_{q,0}$ and $min_{max(0,t-s)\leq q\leq t}\ d^{k,\ell}_{0,q}$, used in the formula (5). We see later that in our particular application all those operations can be performed in constant time.

In the code we use $dr^{k,\ell}_s = d^{k,\ell}_{s,r_\ell}$ for the rightmost column and $db^{k,\ell}_t = d^{k,\ell}_{p_k,t}$ for the bottom row. Their update formulas are derived from the formula (5):

$$dr^{k,\ell}_s = 1 + min(\ r_\ell - 1 + min_{max(0,s-r_\ell)\leq q\leq s}\ dr^{k,\ell-1}_q\ ,$$
$$s - 1 + min_{max(0,r_\ell-s)\leq q\leq r_\ell}\ db^{k-1,\ell}_q),$$
$$db^{k,\ell}_t = 1 + min(\ t - 1 + min_{max(0,p_k-t)\leq q\leq p_k}\ dr^{k,\ell-1}_q\ ,$$
$$p_k - 1 + min_{max(0,t-p_k)\leq q\leq t}\ db^{k-1,\ell}_q).$$

The whole algorithm can be made $O(n+m)$ space by noting that in a column-wise traversal we need, when computing cell $(kl)$, to store only $dr^{k-1,\ell}$ and $db^{k,\ell-1}$, so the space is that for storing one complete column $(m)$ and a row whose width is one box (at most $n$). Our multiset data structure does not increase this space complexity. Hence we have

**Theorem 3** *Given strings $A$ and $B$ of lengths $m$ and $n$ that are run-length encoded to lengths $m'$ and $n'$, there is an algorithm to calculate $D_L(A,B)$ in $O(m'n + n'm)$ time and $O(m+n)$ space in the worst case.*

□

## 3.2    The Multiset Data Structure

What is left is to describe our data structure to handle a multiset of natural numbers. We exploit the fact that consecutive cells in $(d_{ij})$ differ by at most 1 [18]. Our data structure represents the multiset $S$ as a triple $(min(S), max(S), V_{min(S)...max(S)} \rightarrow \mathbb{N})$. That is, we store the minimum and maximum value of the multiset and a vector of counters $V$, which stores at $V_i$ the number of elements equal to $i$ in $S$. Given the property that consecutive cells differ by at most 1, we have that no value $V_i$ is equal to zero. This is proved in the following lemma.

**Lemma 4** *No value $V_i$ for $min(S) \leq i \leq max(S)$ is equal to zero when $S$ is a set of consecutive values in $(d_{ij})$ (i.e., $S$ contains a contiguous part of a row or a column of the matrix $(d_{ij})$).*

*Proof.* The lemma is trivially true for the extremes $i = min(S)$ and $i = max(S)$. Let us now suppose that $V_i = 0$ for an intermediate value. Let us assume that

**Levenshtein** $(A' = (a_{i_1}, p_1)(a_{i_2}, p_2) \ldots (a_{i_{m'}}, p_{m'}), B' = (b_{j_1}, r_1)(b_{j_2}, r_2) \ldots (b_{j_{n'}}, r_{n'}))$

1.          /* We fill the topmost row and leftmost column first */
2.      $dr_0^{0,0} \leftarrow 0, \quad db_0^{0,0} \leftarrow 0$
3.      **For** $k \in 1 \ldots m'$ **Do**
4.          **For** $s \in 0 \ldots p_k$ **Do** $dr_s^{k,0} \leftarrow dr_{p_{k-1}}^{k-1,0} + s$
5.          $db_0^{k,0} \leftarrow dr_{p_k}^{k,0}$
6.      **For** $\ell \in 0 \ldots n'$ **Do**
7.          **For** $t \in 0 \ldots r_\ell$ **Do** $db_t^{0,\ell} \leftarrow db_{r_{\ell-1}}^{0,\ell-1} + t$
8.          $dr_0^{0,\ell} \leftarrow db_{r_\ell}^{0,\ell}$
9.          /* Now we fill the rest of the matrix */
10.     **For** $\ell \in 1 \ldots m'$ **Do** /* column-wise traversal */
11.         **For** $k \in 1 \ldots n'$ **Do**
12.             **If** $a_k = b_\ell$ **Then** /* equal letter box */
13.                 **For** $s \in 1 \ldots p_k$ **Do**
14.                     **If** $s \le r_\ell$ **Then** $dr_s^{k,\ell} \leftarrow db_{r_\ell-s}^{k-1,\ell}$ **Else** $dr_s^{k,\ell} \leftarrow dr_{s-r_\ell}^{k,\ell-1}$
15.                 **For** $t \in 1 \ldots r_\ell$ **Do**
16.                     **If** $p_k \le t$ **Then** $db_t^{k,\ell} \leftarrow db_{t-p_k}^{k-1,\ell}$ **Else** $db_t^{k,\ell} \leftarrow dr_{p_k-t}^{k,\ell-1}$
17.             **Else** /* different letter box */
18.                 $M_r \leftarrow \{dr_0^{k,\ell-1}\}, \quad M_b \leftarrow \{db_{r_\ell}^{k-1,\ell}\}$
19.                 $dr_0^{k,\ell} \leftarrow dr_{p_{k-1}}^{k-1,\ell}$
20.                 **For** $s \in 1 \ldots p_k$ **Do**
21.                     $M_r \leftarrow M_r \cup \{dr_s^{k,\ell-1}\}$
22.                     **If** $s > r_\ell$ **Then** $M_r \leftarrow M_r - \{dr_{s-r_\ell-1}^{k,\ell-1}\}$
23.                     **If** $r_\ell \ge s$ **Then** $M_b \leftarrow M_b \cup \{db_{r_\ell-s}^{k-1,\ell}\}$
24.                     $dr_s^{k,\ell} \leftarrow 1 + min(r_\ell - 1 + min(M_r), s - 1 + min(M_b))$
25.                 $M_r \leftarrow \{dr_{p_k}^{k,\ell-1}\}, \quad M_b \leftarrow \{db_0^{k-1,\ell}\}$
26.                 $db_0^{k,\ell} \leftarrow db_{r_\ell-1}^{k,\ell-1}$
27.                 **For** $t \in 1 \ldots r_\ell$ **Do**
28.                     **If** $p_k \ge t$ **Then** $M_r \leftarrow M_r \cup \{dr_{p_k-t}^{k,\ell-1}\}$
29.                     $M_b \leftarrow M_b \cup \{db_t^{k-1,\ell}\}$
30.                     **If** $t > p_k$ **Then** $M_b \leftarrow M_b - \{db_{t-p_k-1}^{k-1,\ell}\}$
31.                     $db_t^{k,\ell} \leftarrow 1 + min(t - 1 + min(M_r), p_k - 1 + min(M_b))$
32.     **Return** $dr_{p_{m'}}^{m'n'}$ /* or $db_{r_{n'}}^{m'n'}$ */

**Fig. 2.** The $O(m'n + n'm)$ time algorithm to compute the Levenshtein distance between $A$ and $B$, coded as a run-length sequence of pairs (*letter*, *run_length*).

the value $min(S)$ is achieved at cell $d_{i,j}$ and that the value $max(S)$ is achieved at cell $d_{i',j'}$. Since all the intermediate cell values are also in $S$ by hypothesis, and consecutive cells differ by at most 1, it follows that any value between $min(S)$ and $max(S)$ exists in a path that goes from $d_{i,j}$ to $d_{i',j'}$.                □

Fig. 3 shows the detailed algorithms. When we initialize the data structure with the single element $S = \{x\}$ we represent the situation as $(x, x, V_x = 1)$. When we have to add an element $y$ to $S$, we check whether $y$ is outside the range $min(S) \ldots max(S)$, and in that case we extend the range. In any case we increment $V_y$. Note that the domain extension is never by more than one cell, as there cannot appear empty cells in between by Lemma 4. When we have to remove an element $z$ from $S$ we simply decrement $V_z$. If $V_z$ becomes zero, Lemma 4 implies that this is because $z$ is either the minimum or the maximum of the set. So we reduce the domain of $V$ by one. Finally, the operation $min(S)$ is trivial as we have it already precomputed.

---

**Create** $(x)$
1.      **Return** $(x, x, V_x = 1)$

**Add** $((minS, maxS, V), y)$
2.      **If** $y < minS$ **Then**
3.          $minS \leftarrow y$
4.          add new first cell $V_y = 0$
5.      **Else If** $y > maxS$ **Then**
6.          $maxS \leftarrow y$
7.          add new last cell $V_y = 0$
8.      $V_y \leftarrow V_y + 1$
9.      **Return** $(minS, maxS, V)$

**Remove** $((minS, maxS, V), z)$
10.     $V_z \leftarrow V_z - 1$
11.     **If** $V_z = 0$ **Then**
12.         **If** $z = minS$ **Then**
13.             remove first cell from $V$
14.             $minS \leftarrow minS + 1$
15.         **Else** /* $z = maxS$ */
16.             remove last cell from $V$
17.             $maxS \leftarrow maxS - 1$
18.     **Return** $(minS, maxS, V)$

**Min** $((minS, maxS, V))$
19.     **Return** $minS$

---

**Fig. 3.** The multiset data structure implementation.

It is easily seen that all the operations take constant time. As a practical matter, we note that it is a good idea to keep $V$ in a circular array so that it can grow and shrink by any extreme. Its maximum size corresponds to $p_k$ (for $M_r$) or $r_\ell$ (for $M_b$), which are known at the time of **Create**.

## 4    Approximate Searching

Let us now consider a problem related to computing the LCS or the Levenshtein distance. Assume that string $A$ is a short pattern and string $B$ is a long text (so $m$ is much smaller than $n$), and that we are given a threshold parameter $k$. We are interested in reporting all the "approximate occurrences" of $A$ in $B$, that is, all the positions of text substrings which are at distance $k$ or less from the pattern $A$. In order to ensure a linear size output, we content ourselves with reporting the ending positions of the occurrences (which we call "matches").

The classical algorithm to find all the matches [16] computes a matrix exactly like those of recurrences (2) and (1), with the only difference that $d_{0,j} = 0$. This permits the occurrences to start at any text position. The last row of the matrix $d_{mj}$ is examined and every text position $j$ such that $d_{m,j} \leq k$ is reported as a match.

Our goal now is to devise a more efficient algorithm when pattern and text are run-length compressed. A trivial $O(m^2 n' + R)$ algorithm (where $R$ is the size of the output) is obtained as follows. We start filling the matrix only at beginnings of text runs, and complete the first $2m$ columns only (at $O(m^2)$ cost). The rest of the columns of the run are equal to the $2m$-th because no optimal path can be longer than $2m - 1$ under the LCS or Levenshtein models. We later examine the last row of the matrix and report every text position with value $\leq k$. If the run is longer than $2m$, then we have not produced the whole last row but only the first $2m$ cells of it. In this case we report the positions $2m + 1 \ldots r_\ell$ of the $\ell$-th run if and only if the position $2m$ was reported.

We improve now the trivial algorithm. A first attempt is to apply our algorithms directly using the new base value $d_{0,j} = 0$. This change does not present complications.

Let us first concentrate on the Levenshtein distance. Our algorithm obtains $O(m'n + n'm)$ time, which may or may not be better than the trivial approach. The problem is that $O(m'n)$ may be too much in comparison to $O(m^2 n')$, especially if $n$ is much larger than $m$. We seek for an algorithm proportional to the compressed text size. We divide the text runs in *short* (of length at most $2m$) and *long* (longer than $2m$) runs. We apply our Levenshtein algorithm on the text runs, filling the matrix column-wise. If we have a short run $(a_{i_\ell}, r_\ell)$, $r_\ell \leq 2m$, we compute all the $m' + 1$ horizontal borders plus its final vertical border (which becomes the initial border of the next column). The time to achieve this is $O(m'r_\ell + m)$. For an additional $O(r_\ell)$ cost we examine all the cells of the last row and report all the text positions $i_\ell + t$ such that $d_{p_{m'},t}^{m',\ell} \leq k$.

If we have a long run $(a_{i_\ell}, r_\ell)$, $r_\ell > 2m$, we limit its length to $2m$ and apply the same algorithm, at $O(m'm + m + m)$ cost. The columns $2m + 1 \ldots r_\ell$ of that run are equal to the $2m$-th, so we just need to examine the last row of the $2m$-th column, and report all the text positions up to the end of the run, $i_\ell + 2m + 1 \ldots i_\ell + p_k$, if $d_{p_{m'},2m}^{m',\ell} \leq k$.

This algorithm is $O(n'm'm + R)$ time in the worst case, where $R$ is the number of occurrences reported. For the LCS model we have the same upper bound, so we achieve the same complexity. Our $O(m'n'(m'+n'))$ algorithm does not yield a good complexity here. The space is that to compute one text run limited to length $2m$, i.e. $O(m'm)$.

Note that if we are allowed to represent the occurrences as a sequence of *runs* of consecutive text positions (all of which match), then the $R$ extra term of the search cost disappears.

**Theorem 5** *Given a pattern $A$ and a text $B$ of lengths $m$ and $n$ that are runlength encoded to lengths $m'$ and $n'$, there is an algorithm to find all the ending points of the approximate occurrences of $A$ in $B$, either under the LCS or Levenshtein model, in $O(m'mn')$ time and $O(m'm)$ space in the worst case.*

$\square$

# 5    Improving a Greedy Algorithm for the LCS

The idea in our algorithm for the Levenshtein distance $D_L$ in Sect. 3 was to fill all the borders of all the boxes $(d_{s,t}^{k,\ell})$. The natural way to reduce the complexity would be to fill only the corners of the boxes (see Fig. 1). For the $D_L$ distance this seems difficult to obtain, but for the $D_{ID}$ distance there is an obvious greedy algorithm that achieves this goal; in different letter boxes, we can calculate the corner values in constant time, and in equal letter boxes we can trace an optimal path to a corner in $O(m'+n')$ time. Thus, we can calculate all the corner values in $O(m'n'(m'+n'))$ time[1].

It turns out that we can improve the greedy algorithm significantly by fairly simple means. We notice that the diagonal method of [17] can be applied, and achieve an $O(d^2 min(n'm'))$ algorithm. We give also other improvements that do not affect the worst case, but are significant in the average case and in practice. We end the section conjecturing that our improved algorithm runs in $O(m'n')$ time in the average. As we are unable to prove this conjecture, we provide experimental evidence to support it.

## 5.1    Greedy Algorithm for the LCS

Calculating the corner value $d_{p_k,r_\ell}^{k,\ell}$ in a different letter box is easy, because it can be retrieved from the values $d_{0,r_\ell}^{k,\ell} = d_{p_{k-1},r_\ell}^{k-1,\ell}$ and $d_{p_k,0}^{k,\ell} = d_{p_k,r_{\ell-1}}^{k,\ell-1}$, which

---

[1] Apostolico et. al. [3] also gave a basic $O(m'n'(m'+n'))$ algorithm for the LCS, which they then improved to $O(m'n'\log(m'n'))$. Their basic algorithm differs from our greedy algorithm in that they were using the recurrence for calculating the LCS directly, and we are calculating the distance $D_{ID}$. Also, they traced a specific optimal path (which was the property that they could use to achieve the $O(m'n'\log(m'n'))$ algorithm).

are calculated earlier during the dynamic programming. This follows from the lemma:

**Lemma 6 (Bunke and Csirik [7])** *The recurrence (2) can be replaced by the recurrence*

$$d_{s,t}^{k,\ell} = min(d_{s,0}^{k,\ell} + t, d_{0,t}^{k,\ell} + s),  \tag{6}$$

*where $1 \leq s \leq p_k$ and $1 \leq t \leq r_\ell$, for values $d_{s,t}^{k,\ell}$ in a different letter box.*  □

In contrast to the $D_L$ distance, the difficult part in $D_{ID}$ distance lies in equal letter boxes. As noted earlier, Lemma 1 applies also for the $D_{ID}$ distance. From Lemma 1 we can see that the corner values are retrieved along the diagonal, and those values may not have been calculated earlier. However, if $p_k = r_\ell$ in all equal letter boxes, then each corner $d_{p_k,r_\ell}^{k,\ell}$ can be calculated in constant time. This gives an $O(m'n')$ algorithm for a special case, as previously noted in [6].

What follows is an algorithm to retrieve the value $d_{p_k,r_\ell}^{k,\ell}$ in an equal letter box in $O(m' + n')$ time. The idea is to trace an optimal path to the cell $d_{p_k,r_\ell}^{k,\ell}$. This can be done by using lemmas 1 and 6 recursively. Assume that $d_{p_k,r_\ell}^{k,\ell} = d_{0,r_\ell-p_k}^{k,\ell}$ by Lemma 1 (case $d_{p_k,r_\ell}^{k,\ell} = d_{p_k-r_\ell,0}^{k,\ell}$ is symmetric). If $k = 1$, then the value $d_{0,r_\ell-p_k}^{1,\ell}$ corresponds to a value in the first row (0) of the matrix $(d_{ij})$ which is known. Otherwise, the box $(d_{s,t}^{k-1,\ell})$ is a different letter box, and using the definition of overlapping boxes and Lemma 6 it holds

$$d_{0,r_\ell-p_k}^{k,\ell} = d_{p_k-1,r_\ell-p_k}^{k-1,\ell} = min(d_{p_k-1,0}^{k-1,\ell} + r_\ell - p_k, d_{0,r_\ell-p_k}^{k-1,\ell} + p_{k-1}).$$

Now, the value $d_{p_k-1,0}^{k-1,\ell}$ is calculated during the dynamic programming, so we can continue on tracing value $d_{0,r_\ell-p_k}^{k-1,\ell}$ using lemmas 1 and 6 recursively until we meet a value that has already been calculated during dynamic programming (including the first row and the first column of the matrix $(d_{ij})$. The recursion never branches, because Lemma 1 defines explicitly the next value to trace, and one of the two values (from which the minimum is taken over in Lemma 6) is always known (that is because we enter the different letter boxes at the borders, and therefore the other value is from a corner that is calculated during the dynamic programming). We call the path described by the recursion a *tracing path*.

Tracing the value $d_{p_k,r_\ell}^{k,\ell}$ in an equal letter box may take $O(m' + n')$ time, because we are skipping one box at a time, and there are at most $m' + n'$ boxes in the tracing path. Therefore, we get an $O(m'n'(m'+n'))$ algorithm to calculate $D_{ID}(A, B)$. A worst case example that actually achieves the bound is $A = a^n$ and $B = (ab)^{n/2}$.

The space requirement of the algorithm is $O(m'n')$, because we need to store only the corner value in each box, and the $O(m' + n')$ space for the stack is not needed, because the recursion does not branch.

We also achieve the $O(m'n + n'm)$ bound, because the corner values $d_{p_k,r_\ell}^{k,\ell}$ of equal letter boxes define distinct tracing paths, and therefore each cell in the borders of the boxes can be visited only once. To see this observe that each border cell reached by a tracing path uniquely determines the border cell it comes from along the tracing path, and therefore no two different paths can meet in a border cell. The only exception is a corner cell, but in this case all the tracing paths end there immediately.

**Theorem 7** *Given strings $A$ and $B$ of lengths $m$ and $n$ that are run-length encoded to lengths $m'$ and $n'$, there is an algorithm to calculate $D_{ID}(A, B)$ in $O(min(m'n'(m' + n'), m'n + n'm))$ time and $O(m'n')$ space.*                                                     □

## 5.2  Diagonal Algorithm

The diagonal method [17] provides an $O(d\, min(m, n))$ algorithm for calculating the distance $d = D_{ID}(A, B)$ (or $D_L$ as well) between strings $A$ and $B$ of length $m$ and $n$, respectivily. The idea is the following: The value $d_{mn} = D_{ID}(A, B)$ in the $(d_{ij})$ matrix of (2) defines a diagonal band, where the optimal path must lie. Thus, if we want to check whether $D_{ID} < k$, we can limit the calculation to the diagonal band defined by value $k$ (consisting of $O(k)$ diagonals). Starting with $k = |n - m| + 1$, we can double the value $k$ and run in each step the recurrence (2) on the increasing diagonal band. As soon as $d_{mn} < k$, we have found $D_{ID}(A, B) = d_{mn}$, and we can stop the doubling. The total number of diagonals evaluated is at most $2\, D_{ID}(A, B)$, and there are at most $min(m, n)$ cells in each diagonal. Therefore, the total cost of the algorithm is $O(d\, min(m, n))$, where $d = D_{ID}(A, B)$.

We can use the diagonal method with our greedy algorithm as follows: We calculate only the corner values that are inside the diagonal band defined by value $k$ in the above doubling algorithm. The corner values in equal letter boxes inside the diagonal band can be retrieved in $O(k)$ time. That is because we can limit the length of the tracing paths with the value $2k + 1$ (between two equal letter boxes there is a different letter box that contributes at least 1 to the value that we are tracing, and we are not interested in corner values that are greater than $k$). Therefore, we get the total cost $O(d^2 min(m', n'))$, where $d = D_{ID}(A, B)$.

## 5.3  Faster on Average

There are some practical refinements for the greedy algorithm that do not improve its worst case behavior, but do have an impact on its average case.

First of all, the runs of different letter boxes can be skipped in the tracing paths.

Consider two consecutive different letter boxes $(d_{s,t}^{k,\ell})$ and $(d_{s,t}^{k+1,\ell})$. By Lemma 6 it holds for the values $1 \leq t \leq r_\ell$,

$$
\begin{aligned}
d_{p_{k+1},t}^{k+1,\ell} &= min \left( d_{0t}^{k+1,\ell} + p_{k+1}, d_{p_{k+1},0}^{k+1,\ell} + t \right) \\
&= min \left( d_{p_k,t}^{k,\ell} + p_{k+1}, d_{p_{k+1},0}^{k+1,\ell} + t \right) \\
&= min \left( d_{0t}^{k,\ell} + p_k + p_{k+1}, d_{p_k,0}^{k,\ell} + p_{k+1} + t, d_{p_{k+1},0}^{k+1,\ell} + t \right) \\
&= min \left( d_{0t}^{k,\ell} + p_k + p_{k+1}, d_{p_{k+1},0}^{k+1,\ell} + t \right).
\end{aligned}
$$

The above result can be extended to the following lemma by using induction:

**Lemma 8** *Let* $((d_{s,t}^{k',\ell}), (d_{s,t}^{k'+1,\ell}), \ldots, (d_{s,t}^{k,\ell}))$ *and* $((d_{s,t}^{k,\ell'}), (d_{s,t}^{k,\ell'+1}), \ldots, (d_{s,t}^{k,\ell}))$ *be vertical and horizontal runs of different letter boxes. When* $1 \leq t \leq r_\ell$ *and* $1 \leq s \leq p_k$, *the recurrence (4) can be replaced by the recurrences*

$$
d_{p_k,t}^{k,\ell} = min \left( d_{p_k,0}^{k,\ell} + t, d_{0,t}^{k',\ell} + \sum_{s=k'}^{k} p_s \right) \qquad 1 \leq t \leq r_\ell,
$$

$$
d_{s,r_\ell}^{k,\ell} = min \left( d_{0,r_\ell}^{k,\ell} + s, d_{s,0}^{k,\ell'} + \sum_{t=\ell'}^{\ell} r_t \right) \qquad 1 \leq s \leq p_k.
$$

□

Now it is obvious how to speed up the retrieval of values $d_{p_k,r_\ell}^{k,\ell}$ in the equal letter boxes. During dynamic programming, we can maintain pointers in each different letter box to the last equal letter box encountered in the direction of the row and the column. When we enter a different letter box while tracing the value of $d_{p_k,r_\ell}^{k,\ell}$ in an equal letter box, we can use Lemma 8 to calculate the minimum over the run of different letter boxes at once, and continue on tracing from the equal letter box preceding the run of different letter boxes. (Note that in order to use the summations of Lemma 8 we should better store the cumulative $i_k$ and $j_\ell$ values instead of $p_k$ and $r_\ell$.) Therefore we get the following result:

**Theorem 9** *Given strings $A$ and $B$ of lengths $m$ and $n$ that are run-length encoded to lengths $m'$ and $n'$, such that all the runs of different letters over an alphabet of size $|\Sigma|$ are equally likely and in random order, there is an algorithm to calculate $D_{ID}(A,B)$ in $O(m'n'(1 + (m' + n')/|\Sigma|^2))$ time in the average.*

*Proof.* (Sketch) The first part of the cost, $O(m'n')$ comes from the constant time computation of all the different letter boxes. On the other hand, there are on the average $O(m'n'/|\Sigma|)$ equal letter boxes. Between two runs of a letter $\sigma \in \Sigma$, there are on the average $|\Sigma| - 1$ runs of other letters. This holds both for strings $A$ and $B$. In other words, the expected length of a run of different letter boxes is $|\Sigma| - 1$. Therefore the retrieval of the value $d_{p_k,r_\ell}^{k,\ell}$ in an equal letter box takes time at most $O((m' + n')/|\Sigma|)$ in the average.                □

The second improvement to the greedy algorithm is to limit the length of the tracing paths. In the greedy algorithm the tracing is continued until a value is reached that has been calculated during the dynamic programming. However, there are more known values than those that have been explicitly calculated. Consider value $d^{k,\ell}_{p_k,t}$, $1 \le t \le r_\ell$ (or symmetrically $d^{k,\ell}_{s,r_\ell}$, $1 \le s \le p_k$) in the border of a different letter box. If $d^{k,\ell}_{p_k,r_\ell} = d^{k,\ell}_{p_k,0} + r_\ell$ then it must hold $d^{k,\ell}_{p_k,t} = d^{k,\ell}_{p_k,0} + t$, otherwise we get a contradiction: $d^{k,\ell}_{p_k,r_\ell} < d^{k,\ell}_{p_k,0} + r_\ell$.

We call the above situation a horizontal (vertical) *bridge*. Note that from Lemma 6 it follows that there is either a vertical or a horizontal bridge in each different letter box. When we enter a different letter box in the recursion, we can check whether the bridge property holds at the border we entered, using the corner values that are calculated during the dynamic programming. Thus, we can stop the recursion at the first bridge encountered. To combine this improvement with the algorithm that skips runs of different letter boxes, we need Lemma 10 below that states that the bridges propagate along runs of different letter boxes. Therefore we only need to check whether the last different letter box has a bridge to decide whether we have to skip to the next equal letter box. The resulting algorithm is given in pseudo-code in Fig. 4.

**Lemma 10** *Let* $((d^{k',\ell}_{s,t}),(d^{k'+1,\ell}_{s,t}),\dots,(d^{k,\ell}_{s,t}))$ *be a vertical run of different letter boxes. If there is a horizontal bridge* $d^{k',\ell}_{p_{k'},r_\ell} = d^{k',\ell}_{p_{k'},0}+r_\ell$ *then there is a horizontal bridge* $d^{k'',\ell}_{p_{k''},r_\ell} = d^{k'',\ell}_{p_{k''},0} + r_\ell$ *for all* $k' < k'' \le k$. *The symmetric result holds for horizontal runs of different letter boxes.*

*Proof.* We use the counter-argument that $d^{k'',\ell}_{p_{k''},r_\ell} = d^{k'',\ell}_{p_{k''},0} + r_\ell$ does not hold for some $k' < k'' \le k$. Then by Lemma 8 and by the bridge assumption it holds

$$d^{k'',\ell}_{p_{k''},r_\ell} = d^{k'+1,\ell}_{0,r_\ell} + \sum_{s=k'+1}^{k''} p_s = d^{k'+1,l}_{0,0} + r_\ell + \sum_{s=k'+1}^{k''} p_s.$$

On the other hand, using the counter-argument and the fact that consecutive cells in the $(d_{ij})$ matrix differ at most by 1 [18], we get

$$d^{k'',\ell}_{p_{k''},r_\ell} < d^{k'',\ell}_{p_{k''},0} + r_\ell \le d^{k'+1,\ell}_{0,0} + \left( \sum_{s=k'+1}^{k''} p_s \right) + r_\ell,$$

which is a contradiction and so the the original proposition holds.    □

Lemma 10 has a corollary: if the last different letter box in a run does not have a horizontal (vertical) bridge, then none of the boxes in the same run have a horizontal (vertical) bridge and, on the other hand, all the boxes in the same run must have a vertical (horizontal) bridge.

Now, if two tracing paths cross inside a box (or run thereof), then one of them necessarily meets a bridge. In the average case, there are a lot of crossings

**LCS** $(A' = (a_{i_1}, p_1)(a_{i_2}, p_2) \ldots (a_{i_{m'}}, p_{m'}), B' = (b_{j_1}, r_1)(b_{j_2}, r_2) \ldots (b_{j_{n'}}, r_{n'}))$

1.　　　　/* We use structure $d^{k,\ell}$ to denote a box $(d^{k,\ell}_{s,t})$ as follows: */

2.　　$d^{k,\ell}.corner := d^{k,\ell}_{p_k, r_\ell}$

3.　　$d^{k,\ell}.jumptop :=$ "location of the next equal letter box above"

4.　　$d^{k,\ell}.jumpleft :=$ "location of the next equal letter box in the left"

5.　　$d^{k,\ell}.sumtop :=$ **If** $a_{i_k} \neq b_{j_{ell}}$ **Then** $\sum_{t=d^{k,\ell}.jumptop+1}^{k} p_t$

6.　　$d^{k,\ell}.sumleft :=$ **If** $a_{i_k} \neq b_{j_\ell}$ **Then** $\sum_{t=d^{k,\ell}.jumpleft+1}^{\ell} r_t$

7.　　　　/* Initialize first row and column (let $a_{i_0} = b_{j_0} = \epsilon, p_0 = r_0 = 1$) */

8.　　$d^{00}.corner \leftarrow 0$

9.　　**For** $k \in 1 \ldots n'$ **Do** $d^{k,0}.corner \leftarrow d^{k-1,0}.corner + r_{k-1}$

10.　　**For** $\ell \in 1 \ldots m'$ **Do** $d^{0,\ell}.corner \leftarrow d^{0,\ell-1}.corner + p_{\ell-1}$

11.　　**Calculate** values $d^{k,\ell}.(jumptop, jumpleft, sumtop, sumleft)$

12.　　　　/* Now we fill the rest of the corner values */

13.　　**For** $k \in 1 \ldots m'$ **Do**

14.　　　　**For** $\ell \in 1 \ldots n'$ **Do**

15.　　　　　　$(bridge, k', \ell', p, r, sum, d^{k,\ell}.corner) \leftarrow (false, k, \ell, p_k, r_\ell, 0, \infty)$

16.　　　　　　**If** $a_{i_k} \neq b_{j_\ell}$ **Then** /* Different letter box */

17.　　　　　　　　$d^{k,\ell}.corner \leftarrow min(d^{k-1,\ell}.corner + a_{i_k}, d^{k,\ell-1}.corner + b_{j_\ell})$

18.　　　　　　**Else While** $bridge = false$ **Do**

19.　　　　　　　　/* Equal letter box, trace $d^{k,\ell}.corner$ */

20.　　　　　　　　**If** $p = r$ **Then** /* Straight from the diagonal */

21.　　　　　　　　　　$d^{k,\ell}.corner \leftarrow min(d^{k,\ell}.corner, sum + d^{k'-1,\ell'-1}.corner)$

22.　　　　　　　　　　$bridge \leftarrow true$

23.　　　　　　　　**Else If** $p < r$ **Then** /* Diagonal up */

24.　　　　　　　　　　$(r, k') \leftarrow (r - p, k' - 1)$

25.　　　　　　　　　　$d^{k,\ell}.corner \leftarrow min(d^{k,\ell}.corner, sum + d^{k',\ell'-1}.corner + r)$

26.　　　　　　　　　　**If** $d^{k',\ell'}.corner = d^{k',\ell'-1}.corner + r_{\ell'}$ **Then** $bridge \leftarrow true$

27.　　　　　　　　　　**Else** /* Jump to the next equal letter box */

28.　　　　　　　　　　　　$(sum, k') \leftarrow (sum + d^{k',\ell'}.sumtop, d^{k',\ell'}.jumptop)$

29.　　　　　　　　　　　　$p \leftarrow p_{k'}$

30.　　　　　　　　　　　　**If** $k' = 0$ **Then** /* First row */

31.　　　　　　　　　　　　　　$d^{k,\ell}.corner \leftarrow min(d^{k,\ell}.corner,$
　　　　　　　　　　　　　　　　　　　　$sum + d^{k',\ell'-1}.corner + r)$

32.　　　　　　　　　　　　　　$bridge \leftarrow true$

33.　　　　　　　　**Else** /* Diagonal left similarly*/

34.　　**Return** $(m + n - d^{m',n'}.corner)/2$ /* return the length of the LCS */

**Fig. 4.** The improved greedy algorithm to compute the LCS between $A$ and $B$, coded as a run-length sequence of pairs *(letter, run_length)*.

of the tracing paths and the total cost for tracing the values in equal letter boxes decreases.

Another way to consider the average length of a tracing path is to think that every time a tracing path enters a different letter box, it has some probability to hit a bridge. If the bridges were placed randomly in the different letter boxes, then the probability to hit a bridge would be $\frac{1}{2}$. This would give immediately a constant expected length for a tracing path. However, the placing of the bridges depends on the computation of recurrence (2), and this makes the reasoning with probabilities much more complex. We are still confident that the following conjecture holds, although we are not (yet) able to prove it.

**Conjecture 11** *Let $A$ and $B$ be strings that are run-length encoded to lengths $m'$ and $n'$, such that the runs are equally distributed with the same mean in both strings. Under these assumptions the expected running time of the algorithm in Fig. 4 for calculating $D_{ID}(A, B)$ is $O(m'n')$.*

## 5.4   Experimental Results

To test the Conjecture 11, we ran the algorithm in Fig. 4 with the following settings:

1. $m' = n' = 2000, |\Sigma| = 2$, runs in $[1, x]$
   $x \in \{1, 10, 100, 1000, 10000, 100000, 1000000\}$.
2. $m' = 2000, n' \in \{1, 50, 100, 500, 1000, 1500, 2000\}, |\Sigma| = 2$, runs in $[1, 1000]$.
3. $m' = n' = 2000, |\Sigma| \in \{2, 4, 8, 16, 32, 64, 128, 256\}$, runs in $[1, 1000]$.
4. String $A$ was as in item 1 with runs in $[1, 1000]$. String $B$ was generated by applying $k$ random insertions/deletions on $A$, where $k \in \{0, 1, 10, 100, 1000, 10000, 100000\}$.
5. Real data: three different black/white images (printed lines from a book draft ($187 \times 591$), technical drawing ($160 \times 555$), and a signature ($141 \times 362$)). We ran the LCS algorithm on all pairs of lines in each image.

Table 1 shows the results. Different parameter choices are listed in the order they appear in the above listing (e.g. setting 1 in test 1 corresponds to $x = 1$, setting 2 corresponds to $x = 10$, etc.).

The average length $L$ of a tracing path (i.e. the amount of equal letter boxes visited by a tracing path) was smaller than 2 in tests 1-4 (slightly greater in test 5). That is, the running time was in practice $O(m'n')$ with a very small constant factor. Test 1 showed that when the mean length of the runs increases, then also $L$ increases, but not exceeding 2 ($L \in [1, 1.99]$). In test 2, the worst situation was with $n' = m'$ ($L = 1.98$). We tested the effect of the alphabet in test 3, and the worst was $|\Sigma| = 2$ ($L = 1.99$) and the best was $|\Sigma| = 256$ ($L = 1.13$). Test 4 was used to simulate a typical situation, in which the distance between the strings is small. The amount of errors did not have much influence ($L \in [1.71, 1.72]$). In real data (test 5), there were also pairs that were close to the worst case (close to $A = a^n, B = (ab)^{n/2}$), and therefore the results were slightly worse than with randomly generated data: $L \in \{2.00, 2.34, 2.31\}$ with the three images.

**Table 1.** The average length and the maximum length of a tracing path was measured in different test settings. The values of tests 1-4 are averages over 10-10000 trials (e.g. on small values of $n'$ in test 2, more trials were needed because of high variance, whereas otherwise the variance was small). Test 5 was deterministic (i.e. the values are from one trial).

| | Average length of a tracing path (maximum length) |
|---|---|
| test X | setting 1, setting 2, ... |
| test 1 | 1 (1), 1.71 (18), 1.96 (28), 1.98 (27), 1.98 (32), 1.99 (29), 1.98 (25) |
| test 2 | 1.73 (5), 1.77 (10), 1.74 (13), 1.80 (21), 1.90 (30), 1.97 (35), 1.98 (38) |
| test 3 | 1.99 (30), 1.77 (20), 1.60 (14), 1.45 (14), 1.33 (9), 1.24 (7), 1.17 (6), 1.13 (6) |
| test 4 | 1.71 (9), 1.71 (8), 1.71 (7), 1.71 (10), 1.72 (9), 1.72 (10), 1.72 (12) |
| test 5 | 2.00 (35), 2.34 (146), 2.32 (31) |

# 6  Conclusions

We have presented new algorithms to compute approximate matches between run-length compressed strings. The previous algorithms [7,3] permit computing their LCS. We have extended an LCS algorithm [7] to the Levenshtein distance without increasing the cost, and presented an algorithm with nontrivial complexity for approximate searching a run-length compressed pattern on a run-length compressed text under either model.

Future work involves adapting our algorithm to more complex versions of the Levenshtein distance, including at least different costs for the edit operations. This would be interesting for applications related to image compression, where the change from a pixel value to the next is smooth.

With respect to the original models, an interesting question is whether an algorithm can be obtained whose cost is just the product of the compressed lengths. Indeed, this seems possible in the average case, as demonstrated by the experiments with our improved algorithm for the LCS.

Finally, a combination of two-dimensional approximate pattern matching algorithm with two-dimensional run-length compression [5,1] seems extremely interesting.

# References

1. A. Amir and G. Benson. Efficient two-dimensional compressed matching. In *Proc. DCC'92*, pages 279–288, 1992.
2. A. Amir, G. Benson, and M. Farach. Let sleeping files lie: Pattern matching in Z-compressed files. *J. of Comp. and Sys. Sciences*, 52(2):299–307, 1996.
3. A. Apostolico, G. Landau, and S. Skiena. Matching for run-length encoded strings. *J. of Complexity*, 15:4–16, 1999.
   (Also at Sequences'97, Positano Italy, June 11–13, 1997).
4. O. Arbell, G. Landau, and J. Mitchell. Edit distance of run-length encoded strings. Submitted for publication, August 2000.

5. R. Baeza-Yates and G. Navarro. Fast two-dimensional approximate pattern matching. In *Proc. LATIN'98*, LNCS 1380, pages 341–351, 1998.
6. H. Bunke and J. Csirik. An algorithm for matching run-length coded strings. *Computing*, 50:297–314, 1993.
7. H. Bunke and J. Csirik. An improved algorithm for computing the edit distance of run-length coded strings. *Information Processing Letters*, 54(2):93–96, 1995.
8. M. Farach and M. Thorup. String matching in Lempel-Ziv compressed texts. *Algorithmica*, 20:388–404, 1998.
9. T. Kida, Y. Shibata, M. Takeda, A. Shinohara, and S. Arikawa. A unifying framework for compressed pattern matching. In *Proc. SPIRE'99*, pages 89–96. IEEE CS Press, 1999.
10. T. Kida, M. Takeda, A. Shinohara, M. Miyazaki, and S. Arikawa. Multiple pattern matching in LZW compressed text. In *Proc. DCC'98*, pages 103–112, 1998.
11. J. Kärkkäinen, G. Navarro, and E. Ukkonen. Approximate string matching over Ziv-Lempel compressed text. In *Proc. CPM'2000*, LNCS 1848, pages 195–209, 2000.
12. V. Levenshtein. Binary codes capable of correcting deletions, insertions and reversals. *Soviet Physics Doklady* 6:707–710, 1966.
13. T. Matsumoto, T. Kida, M. Takeda, A. Shinohara, and S. Arikawa. Bit-parallel approach to approximate string matching. In *Proc. SPIRE'2000*, IEEE CS Press, pages 221–228, 2000.
14. J. Mitchell. A geometric shortest path problem, with application to computing a longest common subsequence in run-length encoded strings. In *Technical Report*, Dept. of Applied Mathematics, SUNY Stony Brook, 1997.
15. G. Navarro, T. Kida, M. Takeda, A. Shinohara, and S. Arikawa. Faster Approximate String Matching over Compressed Text. In *Proc. 11th IEEE Data Compression Conference (DCC'01)*, 2001, To appear.
16. P. Sellers. The theory and computation of evolutionary distances: Pattern recognition. *J. of Algorithms*, 1(4):359–373, 1980.
17. E. Ukkonen. Algorithms for approximate string matching. *Information and Control* 64(1–3):100–118, 1985.
18. E. Ukkonen. Finding approximate patterns in strings. *J. of Algorithms* 6(1–3):132–137, 1985.
19. R. Wagner and M. Fisher. The string-to-string correction problem. *J. of the ACM* 21(1):168–173, 1974.
20. J. Ziv and A. Lempel. A universal algorithm for sequential data compression. *IEEE Trans. Inf. Theory*, 23:337–343, 1977.

# What to Do with All this Hardware?
## (Invited Lecture)

Uzi Vishkin

University of Maryland, USA and Technion - Israel Institute of Technology, Israel
vishkin@umiacs.umd.edu
WWW home page for the XMT project: http://www.umiacs.umd.edu/~vishkin/XMT

**Abstract.** The upcoming so-called "on-chip Billion transistor" era raises the question: What to do with all the on-chip hardware once the returns on adding more on-chip memory start to diminish?

Parallel computing has been a strategic area of growth for computer science since the 1940s. So far, parallel computing affected main stream computer science only in a limited way. The key problem with parallel computers has been their programmability.

The parallel algorithms research community has developed a theory of parallel algorithms for a very simple parallel computation model, the so-called PRAM (for parallel random-access machine, or model). That theory appears to be second in magnitude only to serial algorithmics.

However, the evolution of parallel computers never reached a situation where the PRAM algorithmic computation model offered effective abstraction for them. So, this elegant algorithmic theory remained in the ivory towers of theorists. Not only that it has not been matched with a real computer system, there has hardly been an experimental study of what works better, more refined performance measurements, and a broad study of applications. For example, the general question "how good parallel algorithms can really be" has remained generally open.

Explicit Multi-Threading (XMT) is a new fine-grained computation framework which tries to address the hardware opportunity using the PRAM parallel algorithmic knowledge base. XMT aims at faster single-task completion time by way of executing in parallel many instruction all within a single chip. Building on some key ideas of parallel computing, XMT covers the spectrum from algorithms through architecture to implementation; the main implementation related innovation in XMT was through the incorporation of low-overhead hardware mechanisms (for more effective fine-grained parallelism).

The two key research questions facing our *"PRAM-on-chip vision"* are: (i) "how to build?" an XMT computer, and (ii) "who cares?"; that is, what will be the key applications?

A. Amir and G.M. Landau (Eds.): CPM 2001, LNCS 2089, pp. 50–50, 2001.
© Springer-Verlag Berlin Heidelberg 2001

# Efficient Experimental String Matching
# by Weak Factor Recognition[*]

Cyril Allauzen[1], Maxime Crochemore[1], and Mathieu Raffinot[2]

[1] Institut Gaspard-Monge, Université de Marne-la-Vallée
Cité Descartes, Champs-sur-Marne
77454 Marne-la-Vallée Cedex 2, France
{allauzen,mac}@monge.univ-mlv.fr
[2] Equipe génome, cellule et informatique, Université de Versailles
45 avenue des Etats-Unis, 78035 Versailles Cedex, France
raffinot@genome.uvsq.fr

**Abstract.** We introduce a new notion of *weak factor recognition* that is the foundation of new data structures and *on-line* string matching algorithms. We define a new automaton built on a string $p = p_1 p_2 \ldots p_m$ that acts like an oracle on the set of factors $p_i \ldots p_j$. If a string is recognized by this automaton, it may be a factor of $p$. But, if it is rejected, it is surely not a factor. We call it *factor oracle*. More precisely, this automaton is acyclic, recognizes at least the factors of $p$, has $m + 1$ states and a linear number of transitions. We give a very simple sequential construction algorithm to build it. Using this automaton, we design an efficient experimental *on-line* string matching algorithm (we conjecture its optimality in regard to the experimental results) that is really simple to implement. We also extend the *factor oracle* to predict that a string could be a suffix (*i.e.* in the set $p_i \ldots p_m$) of $p$. We obtain the *suffix oracle*, that enables in some cases a tricky improvement of the previous string matching algorithm.
**Keywords**: Finite automaton, string matching, algorithm design, information retrieval.

## 1   Introduction

A *string* $p$ is a sequence $p = p_1 p_2 \ldots p_m$ of letters taken in a finite alphabet $\Sigma$. We keep the notation $p$ along this paper to denote the string we are working on. A factor of $p$ is a string $p_i \ldots p_j$, $1 \leq i \leq j \leq m$.

The basic string matching problem is to find all occurrences of a pattern string $p$ in a large text $T$. Efficient *on-line* string matching algorithms are based on indexes built on $p$.

Many indexing techniques exist for this purpose. The simplest methods use precomputed tables of $q$-grams while more advanced methods use more elaborated data structures. These classical structures are: suffix arrays, suffix trees, suffix automata or DAWGs, and factor automata (see [8] for a survey).

---

[*] Work partially supported by Wellcome Trust Foundation and by NATO Grant PST.CLG.977017.

A. Amir and G.M. Landau (Eds.): CPM 2001, LNCS 2089, pp. 51–72, 2001.

The notion on which these structures and these *on-line* string matching algorithms are based is *exact factor recognition*. It means that we need to know if a given string $u$ is or is not a factor of the pattern $p$. This notion leads to very time-efficient string matching algorithms, but presents two major drawbacks: (a) the structures require a fairly large amount of memory space (which implies a large number of memory page breaks when searching for the string in the text); (b) the algorithms are rather involved to implement.

It is considered, for example, that the implementation of suffix arrays can be achieved using five bytes per string character and that other structures need about twelve bytes per string character.

We propose in this paper a new approach based on the notion of *weak factor recognition*. The idea is to recognize more than the exact factors of $p$ to win in simplicity and memory requirements.

For this purpose, we build a new structure, called *factor oracle*, that can replace many of these indexes in *on-line* string matching algorithm. More precisely, this structure is an automaton (a) that is acyclic (b) that recognizes at least the factors of $p$ (c) that has the fewest states possible (*i.e.* $m + 1$) and (d) that has a linear number of transitions. The suffix and factor automata [5,7] satisfy (a)-(b)-(d) but not (c) whereas the sub-sequence automaton [4] satisfies (a)-(b)-(c) but not (d).

We give two different construction algorithms for the factor oracle, the first is only conceptual (and is used as definition), the second is a really simple practical sequential algorithm.

The relations between our factor oracle and the suffix automaton allow us to define another new structure: the *suffix oracle*.

From a theoretical point of view, these two structures are of interest. They represent the first attempt to formalize the notion of *weak factor recognition*. Although their constructions are very simple, their properties are rather difficult to establish and many points remain open and require further studies.

We use these two new structures to design new experimental *on-line* string matching algorithms. These algorithms have a very good average behavior that we conjecture optimal. The main advantages of these new algorithms are (1) that they are easy to implement for an optimal behavior and (2) that they are in pratice as fast as the fastest ones. A preliminary abstract version of this paper appears in [2].

The factor oracle can be extended to a set of strings, and be used in multi string matching algorithms, that leads to very promising experimental results [3].

We now define the notions and the notations we need along this paper. We denote Fact($p$) the set of all the factors of string $p$. A factor $x$ of $p$ is a *prefix* (resp. a *suffix*) of $p$ if $p = xu$ (resp. $p = ux$) with $u \in \Sigma^*$. The set of all the prefixes of $p$ is denoted by Pref($p$) and the one of all the suffixes Suff($p$). We say that $x$ is a *proper factor* (resp. *proper prefix*, *proper suffix*) of $p$ if $x$ is a factor (resp. prefix, suffix) of $p$ distinct from $p$ and from the empty string $\epsilon$. We denote $\text{pref}_p(i)$ the prefix of length $i$ of $p$ for $0 \leq i \leq |p|$. We denote for $u \in \text{Fact}(p)$, $\text{poccur}(u, p) = \min\{|z| \, , \, z = wu \text{ and } p = wuv\}$, the ending position of the first occurrence of $u$ in $p$. Finally, we define for $u \in \text{Fact}(p)$ the set $\text{endpos}_p(u) = \{i \mid p = wup_{i+1} \ldots p_m\}$.

## 2    Factor Oracle

### 2.1    Construction Algorithm

---

**Build_Oracle**$(p = p_1 p_2 \ldots p_m)$
1.    For $i$ from $0$ to $m$
2.        Create a new state $i$
3.    For $i$ from $0$ to $m - 1$
4.        Build a new transition from $i$ to $i + 1$ by $p_{i+1}$
5.    For $i$ from $0$ to $m - 1$
6.        Let $u$ be a minimal length word in state $i$
7.        For all $\sigma \in \Sigma, \sigma \neq p_{i+1}$
8.        If $u\sigma \in \text{Fact}(p_{i-|u|+1} \ldots p_m)$
9.        Build a new transition from $i$ to $i + \text{poccur}(u\sigma, p_{i-|u|+1} \ldots p_m)$ by $\sigma$

---

**Fig. 1.** High-level construction algorithm of Oracle$(p)$.

**Definition 1.** *The* factor oracle *of a string* $p = p_1 p_2 \ldots p_m$ *is the automaton build by the algorithm* Build_Oracle *(Figure 1) on the string* $p$, *where all the states are terminal. It is denoted by* Oracle$(p)$.

A string $w$ is recognized in state $i$ by the factor oracle if it labels a path from state $0$ to state $i$. The factor oracle of the string $p = abbbaab$ is given as an example in Figure 2. On this example, it can be noticed that the string $aba$ is recognized whereas it is not a factor of $p$.

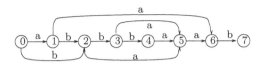

**Fig. 2.** Factor oracle of *abbbaab*. The word *aba* is recognized whereas it is not a factor.

Note: all the transitions that reach state $i$ of Oracle$(p)$ are labeled by $p_i$.

**Lemma 1.** *Let* $u \in \Sigma^*$ *be a minimal length string among the strings recognized in state* $i$ *of* Oracle$(p)$. *Then,* $u \in \text{Fact}(p)$ *and* $i = \text{poccur}(u, p)$.

*Proof.* By induction on the number of the state $i$. It is true for states 0 and 1. Assume that it is true for all states $0 \leq j \leq i - 1$. Let $u$ be a minimal length string among the strings recognized in state $i$. We consider the last transition on the path labeled by $u$ leading from 0 to $i$.

(i) This transition was built at line 2. Then $u = zp_i$ with $z \in \Sigma^*$. As $u$ is a minimal length string of state $i$, $z$ is a minimal length string for state $i - 1$. By the induction hypothesis, $z \in \text{Fact}(p)$ and $i - 1 = \text{poccur}(z, p)$. And therefore, $u = zp_i \in \text{Fact}(p)$ and $i = \text{poccur}(u, p)$.

(ii) This transition was built at line 7. Then it leads from $j$ to $i$ labeled by $p_i$ with $0 < j < i - 1$. So $u = zp_i$ and $z$ is a minimal length string for state $j$. By the induction hypothesis, $z \in \text{Fact}(p)$ and $j = \text{poccur}(z, p)$. And on account of the construction of the transition line 7, $u = zp_i \in \text{Fact}(p)$ and $i = \text{poccur}(u, p)$.

□

**Corollary 1.** *Let $u \in \Sigma^*$ be a minimal length string among the strings recognized in state $i$ of Oracle($p$), $u$ is unique.*

We denote $\min(i)$ the minimal length string recognized in state $i$.

**Corollary 2.** *Let $i$ and $j$ be two states of Oracle($p$) such as $j < i$. Let $u = \min(i)$ and $v = \min(j)$, $u$ can not be a suffix of $v$.*

*Proof.* Assume that $u$ is a suffix of $v$. In this case, $\text{poccur}(u, p) \leq \text{poccur}(v, p)$ which is a contradiction of (according to lemma 1) $j < i$. □

**Lemma 2.** *Let $i$ be a state of Oracle($p$) and $u = \min(i)$. $u$ is a suffix of any string $c \in \Sigma^*$ recognized in state $i$.*

*Proof.* By induction on the number of the state $i$. It is true for states 0 and 1. Assume that it is true for all states $0 \leq j \leq i - 1$. We consider state $i$ and let $u = \min(i)$.

Let $c$ be a path leading to state $i$, $c = c_1 a$ where $c_1$ leads to state $j < i$. Let $v = \min(j)$. Then

- $|va| \geq |u|$ because $u$ is the minimum of $i$ and $va$ leads to $i$.
- according to the construction (Figure 1 line 7), $i \in \text{endpos}_p(va)$ because $v = \min(j)$ and there is a transition from $j$ to $i$ by $a$.
- lemma 1 sets $i \in \text{endpos}_p(u)$.

From these three facts, it comes that $u$ is a suffix of $va$. As $j < i$, by the induction hypothesis, $v$ is a suffix of $c_1$ and therefore $u$ is a suffix of $c$. □

**Lemma 3.** *Let $w \in \text{Fact}(p)$. $w$ is recognized by Oracle($p$) in a state $j \leq \text{poccur}(w, p)$.*

*Proof.* By induction on the length of $w = w_0 w_1 \ldots w_f$. We denote $i = \text{poccur}(w, p)$.

- There is a transition from 0 by $w_0$ leading to a state $k_0 \leq i - f$.
- Assume that there is a path labeled by $w_0 \ldots w_j$ leading to a state $k_j \leq i - f + j$. Let $u = \min(k_j)$. According to lemma 2, $u = w_{j-|u|+1} \ldots w_j$. As $uw_{j+1}$ is a factor of $p$ and as $i - f + (j + 1) \in \text{endpos}_p(uw_{j+1})$, there is a transition from $k_j$ labeled $w_{j+1}$ leading to a state $k_{j+1} \leq i - f + (j + 1)$.

□

**Corollary 3.** *Let $w \in \text{Fact}(p)$. Every string $v \in \text{Suff}(w)$ is recognized by Oracle(p) in a state $j \leq \text{poccur}(w)$.*

**Lemma 4.** *Let $i$ be a state of Oracle(p) and $u = \min(i)$. Any path ending by $u$ leads to a state $j \geq i$.*

*Proof.* By induction on the number of the state $i$. It is true for states 0 and 1. Assume that it is true for all states $0 \leq j \leq i - 1$. We consider state $i$. Let $u$ be a minimal length string among the strings recognized in state $i$.

Consider the minimal length path labeled by $u$ leading to $i$. We denote $j$ the state preceding $i$ along this path, and $v = \min(j)$. We have $u = va$ by construction. Consider a path $c = c_1 u$ leading to a state $k$. Assume that $k < i$.

By the induction hypothesis, the path $c_1 v$ leads to a state $l \geq j$. If $l = j$, then $k = i$ which is a contradiction with $k < i$. We have now $j < l < k < i$. Let $w = \min(l)$.

- If $|w| > |v|$, then $v$ is a suffix of $w$ (lemma 2). In that case, $u = va$ is a suffix of $wa$, so that $k \in \text{endpos}_p(u)$ which is a contradiction with $k < i = \text{poccur}(u, p)$ (lemma 1).
- If $|w| \leq |v|$, then $w$ is a suffix of $v$ (lemma 2). This is a a contradiction with $j < l$ (corollary 2).

In both cases, we reach a contradiction, so that $k \geq i$ and the induction hypothesis is verified for $i$. □

**Lemma 5.** *Let $w \in \Sigma^*$ be a string recognized by Oracle(p) in $i$, then any suffix of $w$ is recognized in a state $j \leq i$.*

*Proof.* By induction on $|w|$. It is true if $|w| = 0$ or $|w| = 1$. Assume that it is true for all the strings $\zeta$ such that $|\zeta| < |w|$. We show that it is also true for $w$, recognized in $i$.

Let $w = \zeta a$, $\zeta$ is recognized in $k < i$. Consider a proper suffix of $w$. It can be written $va$ where $v$ is a proper suffix of $\zeta$.

According to the induction hypothesis, $v$ is recognized in $l \leq k$. Let $\bar{\zeta} = \min(k)$ and $\bar{v} = \min(l)$. The lemma 2 implies that $\bar{\zeta}$ is a suffix of $\zeta$ and that $\bar{v}$ is a proper suffix of $v$. The corollary 2 sets that $\bar{v}$ is a suffix of $\bar{\zeta}$. As $i \in \text{endpos}_p(\bar{\zeta}a)$ (by construction of the transition by $a$), $i \in \text{endpos}_p(\bar{v}a)$. So here is a transition from $l$ by $a$ leading to a state $j \leq i$. Thus, the proper suffix $va$ is recognized in state $j$. □

The number of states of Oracle(p) with $p = p_1 p_2 \ldots p_m$ is $m + 1$. We now consider the number of transitions.

**Lemma 6.** *The number $T_{Or}(p)$ of transitions in $Oracle(p = p_1p_2 \ldots p_m)$ satisfies $m \le T_{Or}(p) \le 2m - 1$.*

*Proof.* There is always $m$ transitions $i \to i + 1$ labeled by $p_{i+1}$. As the strings $a^m$ only have these transitions, $m$ is the minimal number of transitions.

Consider now the transitions $i \to j$ where $j > i+1$. We build an injective function that maps each of these transitions to a proper suffix of $p$. Each transition $i \to j$ with $j > i + 1$ labeled by $\sigma$ is mapped to the string $\min(i)\sigma p_{j+1} \ldots p_m$. By construction of $Oracle(p)$, this string is a proper suffix of $p$. We show *a contrario* that this function is injective. Assume that there are two distinct transitions $i_1 \to j_1$ and $i_2 \to j_2$ respectively labeled by $\sigma_1$ and $\sigma_2$ such that $\min(i_1)\sigma_1 p_{j_1+1} \ldots p_m = \min(i_2)\sigma_2 p_{j_2+1} \ldots p_m$.

Assume that $i_1 \ge i_2$. We consider three cases. If $j_1 = j_2$, then we have $\sigma_1 = \sigma_2$ and the equality implies $\min(i_1) = \min(i_2)$ and $i_1 = i_2$. If $j_1 > j_2$, $\min(i_2)\sigma_2$ is a proper prefix of $\min(i_1)$. Let $\delta = |\min(i_1)| - |\min(i_2)\sigma_2|$, then we have $j_2 = j_1 - \delta - 1$. As an occurrence of $\min(i_1)$ ends in $i_1$ (according to lemma 1), an occurrence of $\min(i_2)\sigma_2$ ends in $i_1 - \delta < j_1 - \delta - 1 = j_2$. This is a contradiction with the construction of $Oracle(p)$. If $j_1 < j_2$, $\min(i_1)$ is a prefix of $\min(i_2)$, so there is an occurrence of $\min(i_1)$ ending before $i_2 \le i_1$. This is a contradiction with lemma 1.

The three other cases, for $i_1 \le i_2$, are resolved in a symmetric way. The function is indeed injective, and as the set of proper suffixes of $p = p_1p_2 \ldots p_m$ is of size $m - 1$, $T_{Or}(p) \le m + m - 1 = 2m - 1$. This maximum is reached for the strings $a^{m-1}b$. $\square$

The factor automaton can be coded in a memory efficient simple way. We do not have to code the states since they are positions in the string. We just have to code the external transitions. Their labels do not have to be coded since they are fixed by their arrival states.

## 2.2   Sequential Algorithm

This section presents a sequential construction of the automaton $Oracle(p)$, that means a way of building the automaton by reading the letters of $p$ one by one from left to right, upgrading the automaton at each step.

We denote $repet_p(i)$ the longest suffix of $pref_p(i)$ that appears at least twice in $pref_p(i)$. We define a function $S_p$ on the states of the automaton, called supply function, that maps each state $i > 0$ of $Oracle(p)$ to state $j$ in which the reading of $repet_p(i)$ ends. We arbitrarily set $S_p(0) = -1$. Notice that $S_p(i)$ is well defined for every state $i$ of $Oracle(p)$ (Corollary 3), and that for any state $i$ of $Oracle(p)$, $i > S_p(i)$ (lemma 3). We denote $k_0 = m$, $k_i = S_p(k_{i-1})$ for $i \ge 1$. The sequence of the $k_i$ is finite, strictly decreasing and ends in state 0. We denote $CS_p = \{k_0 = m, k_1, \ldots, k_t = 0\}$ the suffix path of $p$ in $Oracle(p)$.

**Lemma 7.** *Let $k > 0$ be a state of $Oracle(p)$ such that $s = S_p(k)$ is strictly positive. We denote $w_k = repet_p(k)$ and $w_s = repet_p(s)$. Then $w_s$ is a suffix of $w_k$.*

*Proof.* Let $j$ be the state of Oracle($p$) reached after the reading of $w_s$. Let $v = \min(s)$.

- $j < s$. Lemma 3 implies that $j \leq \mathrm{poccur}(w_s)$ which is by definition strictly less than $s$.
- $|w_s| < |v|$. Assume the contrary. Then there is a path ending by $v = \min(s)$ leading to $j < s$. This is a contradiction with lemma 4.

Lemma 1 implies $s = \mathrm{poccur}(v, p)$. As $w_s$ is a suffix of $\mathrm{pref}_p(s)$ of lenght less than $|v|$, $w_s$ is a proper suffix of $v$, and, as $v$ is himself a suffix of $w_k$ (lemma 2), $w_s$ also is. $\square$

**Corollary 4.** *Let $CS_p = \{k_0, k_1 \dots, k_t = 0\}$ be the suffix path of $p$ in Oracle($p$) and let $w_i = \mathrm{repet}_p(k_{i-1})$ for $1 \leq i \leq t$ and $w_0 = p$. Then, for $0 < l \leq t$, $w_l$ is a suffix of all the $w_i$, $0 \leq i < l \leq t$.*

We now consider for a string $p = p_1 p_2 \dots p_m$ and a letter $\sigma \in \Sigma$ the construction of Oracle($p\sigma$) from Oracle($p$). We denote Oracle($p$)$+\sigma$ the automaton Oracle($p$) on which a transition by $\sigma$ from state $m$ to state $m + 1$ is added. We already notice that a transition that exists in Oracle($p$)$+\sigma$ also exists in Oracle($p\sigma$), so that the difference between the two automata only rely on transitions by $\sigma$ to state $m + 1$ that have to be added to Oracle($p$)$+\sigma$ in order to get Oracle($p\sigma$). We are investigating states from which there may exist transitions by $\sigma$ to state $m + 1$.

**Lemma 8.** *Let $k$ be a state of Oracle($p$)$+\sigma$ such that there is a transition from $k$ by $\sigma$ to $m + 1$ in Oracle($p\sigma$). Then $k$ has to be one of the states of the suffix path $CS_p$ in Oracle($p$)$+\sigma$.*

*Proof.* On the contrary, assume that there exists a state $k_{i+1} < k < k_i$ such that there is a transition to $m + 1$ by $\sigma$ in Oracle($p\sigma$). We denote $w_j = \mathrm{repet}_p(k_{j-1})$ for $1 \leq j \leq t$, and $w_0 = p$. We have $m \in \mathrm{endpos}_p(w_j)$. Let $v = \min(k)$. As there is a transition by $\sigma$ from $k$ to $m + 1$, $m \in \mathrm{endpos}_p(v)$, and $v$ is comparable to the factors $w_j$ (corollary 4). The factor $v$ must satisfy:

(i) $|v| < |w_i|$. Assume, on the contrary, that $|v| \geq |w_i|$. $|v| > |w_i|$, or else there will be two path labeled by $v$ leading to two different states. Consider the greatest $0 \leq d < i$ such that $|w_{d+1}| < |v| < |w_d|$. The factor $v$ is a suffix of $w_d$. As $v = \min(k)$, according to lemma 1, it occurred in $k < k_d$, and as it is also a suffix of $w_d$, it occurred at least twice in $\mathrm{pref}_p(k_d)$. In that case, by definition of the $k_j$, $k_{d+1}$ can not be $S_p(d)$, and there is a contradiction. So $|v| < |w_i|$ and $v$ is a proper suffix of $w_i$.
(ii) $|v| > |w_{i+1}|$. Assume on the contrary that $|v| \leq |w_{i+1}|$. $|v|$ is then a proper suffix of $w_{i+1}$ and the path labeled $w_{i+1}$ leading to $k_{i+1} < k$ ends by $v = \min(k)$. This is a contradiction with lemma 4.
(iii) $|v| < |\min(k_i)|$. Assume on the contrary that $|v| \geq |\min(k_i)|$. The factor $\min(k_i)$ is a suffix of $w_i$ (lemma 2) of which $v$ is also a suffix (by (i)). So $\min(k_i)$ is a suffix of $v = \min(k)$ and we get a contradiction with corollary 2.

As $v$ is a suffix of $\min(k_i)$ (by (iii)), $k_i \in \text{endpos}_p(v)$ (lemma 1). But $k_i > k \in \text{endpos}_p(v)$ and $v$ occurs at least twice in $\text{pref}_p(k_i)$. As (by (ii))$|v| > |w_{i+1}|$, there is a contradiction with $w_{i+1} = \text{repet}_p(k_i)$. $\square$

Among the states on the suffix path of $p$, every state that has no transition by $\sigma$ in $\text{Oracle}(p) + \sigma$ must have one in $\text{Oracle}(p\sigma)$. More formally, the following lemma sets this fact.

**Lemma 9.** *Let $k_l < m$ be a state on the suffix path $CS_p$ of state $m$ in $\text{Oracle}(p = p_1p_2\ldots p_m) + \sigma$. If $k_l$ does not have a transition by $\sigma$ in $\text{Oracle}(p)$, then there is a transition by $\sigma$ from $k_l$ to $m + 1$ in $\text{Oracle}(p\sigma)$.*

*Proof.* Let $v = \min(k_l)$, then $v$ is a suffix of $w_l = \text{repet}_p(k_{l-1})$ (lemma 2). As $w_l$ is a suffix of $p$, $w_l\sigma$ is a suffix of $p\sigma$ and $m + 1 = \text{poccur}(w_l\sigma)$. According to the construction of $\text{Oracle}(p\sigma)$, there is a transition by $\sigma$ from $k_l$ to $m + 1$. $\square$

**Lemma 10.** *Let $k_l < m$ be a state on the suffix path $CS_p = \{k_0 = m, k_1 \ldots, k_t = 0\}$ of $m$ in $\text{Oracle}(p = p_1p_2\ldots p_m)+\sigma$. If $k_l$ has a transition by $\sigma$ in $\text{Oracle}(p)+\sigma$, then all the states $k_i$, $l \leq i \leq t$ also have a transition by $\sigma$ in $\text{Oracle}(p) + \sigma$.*

*Proof.* Let $w_l = \text{repet}_p(k_{l-1})$. All the $w_i = \text{repet}_p(k_{i-1})$, $0 \leq i \leq l$ are suffixes of $w_l$. As $w_l\sigma$ is recognized by $\text{Oracle}(p) + \sigma$, by lemma 5, all its suffixes also are. $\square$

The idea of the sequential construction algorithm is the following. According to the three lemmas 8, 9, 10, to transform $\text{Oracle}(p) + \sigma$ in $\text{Oracle}(p\sigma)$ we only have to go down the suffix path $CS_p = \{k_0 = m, k_1, \ldots, k_t = 0\}$ of state $m$ and while the current state $k_l$ does not have an exiting transition by $\sigma$, a transition by $\sigma$ to $m + 1$ should be added (lemma 9). If $k_l$ already has one, the process ends because, according to lemma 10, all the states $k_j$ after $k_l$ on the suffix path already have a transition by $\sigma$.

To add a single letter, the preceding algorithm is enough. But, as we build the automaton by adding the letters of $p$ the one after the other, we must update the supply function $S_{p\sigma}$ of the new automaton $\text{Oracle}(p\sigma)$. As (according to the definition of $S_p$), the supply function of states $0 \leq i \leq m$ does not change from $\text{Oracle}(p)$ to $\text{Oracle}(p\sigma)$, the only thing to do is to compute $S_{p\sigma}(m + 1)$. This is done with the following lemma.

**Lemma 11.** *If there is a state $k_d$ which is the greatest element of $CS_p = \{k_0 = m, k_1, \ldots, k_t = 0\}$ in $\text{Oracle}(p)$ such that there is a transition by $\sigma$ from $k_d$ to a state $s$ in $\text{Oracle}(p)$, then $S_{p\sigma}(m + 1) = s$ in $\text{Oracle}(p\sigma)$. Else $S_{p\sigma} = 0$.*

*Proof.* Let $w = \text{repet}_{p\sigma}(m + 1)$. First assume that there is no such state. As $0 \in CS_P$ in $\text{Oracle}(p)$, there is no transition by $\Sigma$ leaving 0 in $\text{Oracle}(p)$, so $\sigma$ does not occur in $p$ and $w = \epsilon$ and $S_{p\sigma} = 0$.

We now assume that there is a such state $k_d$. Then $w$ is not the empty string and so we can write $w = \alpha\sigma$. Furthermore $k_d < m$ because $m$ is the last state of $\text{Oracle}(p)$. Let $w_j = \text{repet}_p(k_{i-1})$ for $0 < j \leq t$ and $w_0 = p$. We first prove the two following points:

(1) $|\alpha| < |w_{d-1}|$. Conversely assume that $|\alpha| \geq |w_{d-1}|$, the $w_{d-1}$ is a suffix of $\alpha$. As $\alpha\sigma$ is a factor of $p$ (it occurs twice in $p\sigma$), $\alpha\sigma$ is recognized by Oracle($p$) (lemma 5) and $w_{d-1}\sigma$ also is. This contradicts the fact that $d$ is the greatest such that $k_d$ has a transition by $\sigma$ in Oracle($p$).

(2) Let $i$ be the state recognizing $\alpha$ in Oracle($p$); $i$ is strictly inferior to $k_{d-1}$ according to lemma 5 and to the fact that $i$ has a transition by $\sigma$ whereas $k_d$ has not.

We now compare $\alpha$ and $w_d$. The one is a suffix of the other. We get the two following cases:

(1) Assume that $|\alpha| \geq |w_d|$. $i \geq k_d$ because $w_d$ is a suffix of $\alpha$. We conversely prove that $k_d = i$. Assume that $k_d < i$. Then $|w_d| < |\min(i)|$ because otherwise the path labeled by $w_d$ will end by $\min(i)$ and will leads to state strictly before $i$ which will contradicts lemma 4. But $\min(i)$ also occurs in $k_{d-1}$ because: $\min(i)$ and $\min(k_{d-1})$ are comparable (by suffix relation) and on account of corollary 2 $\min(k_{d-1})$ can not be a suffix of $\min(i)$ so $\min(i)$ is a suffix of $\min(k_{d-1})$. So that $\min(i)$ is a suffix of $w_{d-1}$ which occurs twice in $\text{pref}_p(k_{d-1})$ and which is strictly longer than $w_d = \text{repet}_p(k_{d-1})$. This contradicts the definition of $w_d$. So $k_d = i$ with the result that $\alpha\sigma$ leads to the same state as $w_d\sigma$: $s$.

(2) Now assume that $|\alpha| < |w_d|$ then $i \leq k_d$ because $\alpha$ is a suffix of $w_d$. $|\alpha| \geq |\min(k_d)|$ because as there exists a transition from $k_d$ by $\sigma$ to $s$, $\min(k_d)\sigma$ is both a factor of $p$ and a suffix of $p\sigma$ and $\alpha\sigma$ is the longest of these factors. $\min(k_d)$ is a suffix of $\alpha$ and by lemma 4 $i \geq k_d$. From which $i = k_d$ and $\alpha\sigma$ leads to the same state as $w_d\sigma$: $s$.

Then the path $\alpha\sigma$ also leads to $s$ and therefore $s = S_{p\sigma}(m+1)$.
□

From these lemmas we can now deduce an algorithm **add_letter** to transform Oracle($p$) in Oracle($p\sigma$). It is given Figure 3.

**Lemma 12.** *The algorithm* **add-letter** *really builds Oracle($p\sigma$) from Oracle($p = p_1p_2\ldots p_m$) and updates the supply function of the new state $m+1$ of Oracle($p\sigma$).*

*Proof.* We go down the suffix path of $p$ in accordance with lemma 9. We stops in accordance with lemma 10 and we update the supply value of state $m+1$ according to lemma 11. □

The complete algorithm **Oracle-sequential** that builds Oracle($p = p_1p_2\ldots p_m$) just consists in upgrading the automaton by adding the letters $p_i$ one by one from left to right with the function **add_letter**.

**Theorem 1.** *The algorithm* Oracle-sequential *($p = p_1p_2\ldots p_m$) builds Oracle($p$).*

*Proof.* By induction on string $p$ using lemma 12. □

---

**Function add_letter**(Oracle$(p = p_1 p_2 \ldots p_m)$, $\sigma$)
1.    Create a new state $m + 1$
2.    Create a new transition from $m$ to $m + 1$ labeled by $\sigma$
3.    $k \leftarrow S_p(m)$
4.    **While** $k > -1$ and there is no transition from $k$ by $\sigma$ **Do**
5.            Create a new transition from $k$ to $m + 1$ by $\sigma$
6.            $k \leftarrow S_p(k)$
7.    **End While**
8.    **If** $(k = -1)$ **Then** $s \leftarrow 0$
9.    **Else** $s \leftarrow$ where leads the transition from $k$ by $\sigma$.
10.   $S_{p\sigma}(m + 1) \leftarrow s$
11.   **Return**  Oracle$(p = p_1 p_2 \ldots p_m \sigma)$

---

**Fig. 3.** Add a letter $\sigma$ to Oracle$(p = p_1 p_2 \ldots p_m)$ to get Oracle$(p\sigma)$.

**Theorem 2.** *The complexity of the algorithm* Oracle-sequential*$(p = p_1 p_2 \ldots p_m)$ is $O(m)$ in time and in space[1].*

*Proof.* The algorithm is in $O(m)$ in space. Indeed, all the transitions which are created by the algorithm are transitions of Oracle$(p)$. Exactly $m + 1$ states are created and a supply value associated to each of these states can be stored in constant space. So the algorithm requires linear space.

The algorithm is in $O(m)$ in time. As we only create the states and the transitions that are necessary, the only point to verify is that the total number of backward jumps on the supply path (lines 4-6, Figure 3) is linear.

In each stage $i$ of the construction, *i.e.* when letter $p_i$ is being added, the number $r_i$ of backward jumps on the supply path is bounded by $k_i = |\text{repet}_p(i - 1)|$. During the transition from stage $i$ to stage $i + 1$, we have $k_{i+1} \leq k_i - r_i + 2$ and $r_i \leq k_i - k_{i+1} + 2$. The sum $\sum_{i=1}^{n} r_i$ is therefore bound by $2n$ and the algorithm is linear in time. $\square$

*Example.* The sequential construction of Oracle(*abbbaab*) is given in Figure 4.

## 3    Suffix Oracle

The links between the suffix automaton and our factor oracle lead to a straightforward extension of the oracle: It is possible to mark some states as terminal on the factor oracle as on the suffix automaton in order to recognize suffixes of

---

[1] The constants involved in the asymptotic bound of the complexity of the sequential construction algorithm depend on the implementation and may involve the size of the alphabet $\Sigma$. If we implement the transitions in a way that they are accessible in $O(1)$ (use of tables), then the complexity is $O(m)$ in time and $O(|\Sigma| \cdot m)$ in space. If we implement the transitions in a way that they are accessible in $O(log|\Sigma|)$ (use of search trees), then the complexity is $O(log|\Sigma| \cdot m)$ in time and $O(m)$ in space.

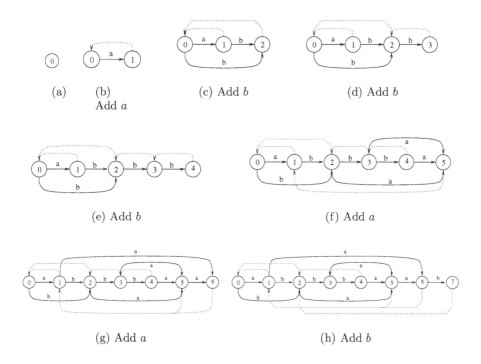

(a)    (b)           (c) Add $b$                      (d) Add $b$
       Add $a$

(e) Add $b$                        (f) Add $a$

(g) Add $a$                        (h) Add $b$

**Fig. 4.** Sequential construction of Oracle(abbaba). The dot-lined arrows represent the supply function.

$p$. This extension will allow us to use some properties of the suffix automaton. We call this new structure *suffix oracle* and we denote it by SOracle($p$).

**Definition 2.** *A state $q$ of the suffix oracle is terminal if and only if there is a path labeled by a suffix of $p$ leading from the initial state to $q$.*

The high-level construction algorithm of the factor oracle (see Figure 1) can not be easily modified in order to build the suffix oracle because it can not detect terminal states. Conversely, the sequential construction algorithm can because of the supply function. This is the point of the following lemma. Let us recall that for Oracle($p = p_1 p_2 \ldots p_m$), the sequence defined by $k_0 = m$, $k_i = S_p(k_{i-1})$ for $i >= 1$ is finite, strictly decreasing and ends in state 0. We denote $CS_p = \{k_0 = m, k_1 \ldots, k_t = 0\}$ the suffix path of $p$ in Oracle($p$).

**Lemma 13.** *The terminal states of SOracle($p$) are the states of Oracle($p$) that are on the suffix path $CS_p$.*

*Proof.*

(i) If $k \in CS_p$, then $k$ is a terminal state of SOracle($p$). We denote $w_0 = p$ and $w_i = repet_p(k_{i-1})$ for all $1 \leq i \leq t$. Corollary 4 sets that $w_l$ is a suffix of all the $w_i$ with $0 \leq i < l \leq t$, so $w_l$ is a suffix of $p$. As $w_i$ leads to $k_i$ with $0 \leq i \leq t$, the states $k_i$ are reachable by a suffix and so are terminal.

(ii) If $q$ is a state of Oracle($p$) such that there is a path from the initial state $0$ to $q$ labeled by a suffix $s$ of $p$, then $q \in CS_p$. If $q = m$, $q = k_0$ then the property holds. Assume now that $0 < q < m$. As $s$ is a suffix of $p$, there is an $i$ such that $s$ is a proper suffix of $w_{i-1}$ (which leads to $k_{i-1}$) and such that $w_i$ (which leads to $k_i$) is a proper suffix of $s$. According to lemma 5 we have $k_i \leq q \leq k_{i-1}$.

Assume that $k_i < q < k_{i-1}$. Consider $v = \min(q)$.

We first deal with the case where $w_i = \epsilon$ ($k_i = 0$). The path labeled by $s$ which leads to $q \neq 0$ ends by $v = \min(q)$ which is not the empty string. As $s$ is a proper suffix of $w_{i-1}$, $v$ is also a suffix of $w_{i-1}$ and $k_{i-1} \in \mathrm{endpos}_p(v)$. So there are occurrences of $v$ both in $q < k_i$ and in $k_i$ which contradicts $\epsilon = w_i = \mathrm{repet}_p(w_{i-1})$.

We can now assume that $|w_i| > 0$.

We first notice that $|v| < |w_i|$. Indeed, conversely assume that $|v| \geq |w_i|$. We can not have $|v| = |w_i|$ because in that case $v = w_i$ (they are both suffixes of $s$) and $q = \mathrm{poccur}(v, p)$. As we assume that $k_i < q$, this is a contradiction with lemma 4. So $|v| > |w_i|$ but in that case as (1) $q \in \mathrm{endpos}_p(v)$ (lemma 1) (2) $k_{i-1} \in \mathrm{endpos}_p(v)$ (lemma 2: $v$ is a suffix of $w_{i-1}$) (3) $q < k_i$ (by assumption), it follows that $v$ is a suffix of $w_{i-1}$, occurs strictly before $k_{i-1}$ and is greater than $w_i$. This contradicts the definition of $w_i = \mathrm{repet}_p(w_{i-1})$. As $|v| < |w_i|$ and as $w_i$ and $v$ (lemma 2) are suffixes of $s$, $v$ is a proper suffix of $w_i$. As $v = \min(q)$, lemma 4 contradicts the fact that $k_i < q$.

□

To transform the factor oracle of $p$ into a suffix oracle, we just go down the suffix path of the last state created by the sequential construction of Oracle($p$) marking each encountered state as terminal. The pseudo-code of the construction of the suffix oracle (using the sequential construction of the factor oracle) is given in Figure 5.

```
Suffix-oracle(p = p₁p₂ ... pₘ)
1.    Oracle-sequential(p)
2.    t ← m
3.    While Sₚ(t) ≠ −1 Do
4.        mark t as terminal
5.        t ← Sₚ(t)
6.    End While
```

**Fig. 5.** Construction algorithm of the suffix oracle SOracle($p = p_1 p_2 \ldots p_m$).

For instance, the construction of SOracle(*abbbaab*) is given in Figure 6.

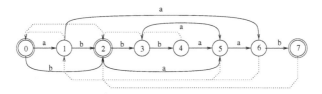

**Fig. 6.** Example of suffix oracle. Double-circled states are terminal. The supply function is represented by dot-lined arrows.

We mainly use the factor oracle rather than the suffix oracle because of the following reason. The power of the structure of factor oracle stands on its simplicity of construction (which becomes a little more complicated for the suffix oracle) but particularly on the few memory which is needed to implement it. This memory saving is based on the fact that the states of the automaton have not to be coded because we can consider a position in the string $p$ as a state. So we only have to code external transitions which are at most $m - 1$. It is more difficult to do the same with the suffix oracle because we need a way of marking the terminal states. This complicates the implementation and slows the terminality test if you want to keep a sharp implementation.

*Note.* For some strings, the suffix oracle matches the suffix automaton and therefore recognizes exactly the suffixes. On a binary alphabet $\Sigma = \{0, 1\}$, it is notably the case for Fibonacci words and more generally for any left special factor of an infinite sturmian word (a factor $u$ of a sturmian word is left special if and only if $0u$ and $1u$ are both factors of the same sturmian word). The interested reader can refer to [1] for more details on this point.

## 4   String Matching

The factor oracle of $p$ can be used in the same way as the suffix automaton in string matching in order to find the occurrences of a word $p = p_1 p_2 \ldots p_m$ in a text $T = t_1 t_2 \ldots t_n$ both on an alphabet $\Sigma$. The suffix automaton is used in [9,8] to get an algorithm called BDM (for *Backward Dawg matching*). Its average complexity is in $O(n \log_{|\Sigma|}(m)/m)$ under a Bernoulli model of probability where all the letters have the same probability. Yao proved in [11] that this bound is optimal. The BDM algorithm moves a window of size $m$ on the text. For each new position of this window, the suffix automaton of $p^r$ (the mirror image of $p$) is used to search for a factor of $p$ from the right to the left of the window. The basic idea of the BDM is that if this backward search failed on a letter $\sigma$ after the reading of a word $u$ then $\sigma u$ is not a factor of $p$ and moving the beginning of the window just after $\sigma$ is secure. This idea is then refined in the BDM using some properties of the suffix automaton.

However this idea is enough in order to get an efficient string matching algorithm. The most amazing is that the strict recognition of the factors (that the

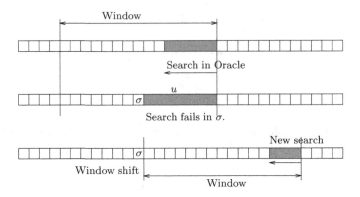

**Fig. 7.** Shift of the search window after the fail of the search by Oracle($p$). The word $\sigma u$ is not a factor of $p$.

factor and suffix automata allow) is not necessary. For the algorithm to work, it is enough to know that $\sigma u$ is not a factor of $p$. The oracle can be used to replace the suffix automaton as it is illustrated by Figure 7. We call this new algorithm **BOM** for *Backward Oracle Matching*. Its proof is given in lemma 14. We make the conjecture (according to the experimental results) that BOM is still optimal on average.

---

**BOM**($p = p_1 p_2 \ldots p_m$, $T = t_1 t_2 \ldots t_n$)
1.   Pre-processing
2.       Construction of the oracle of $p^r$
3.   Search
4.       $pos \leftarrow 0$
5.       **While** ($pos <= n - m$) **do**
6.           $state \leftarrow$ initial state of Oracle($p^r$)
7.           $j \leftarrow m$
8.           **While** $state$ exists **do**
9.               $state \leftarrow$ image state by $T[pos + j]$ in Oracle($p^r$)
10.              $j \leftarrow j - 1$
11.          **EndWhile**
12.          **If** $j = 0$ **do**
13.              **mark an occurrence at** $pos + 1$
14.              $j \leftarrow 1$
15.          **EndIf**
16.          $pos \leftarrow pos + j$
17.      **EndWhile**

**Fig. 8.** Pseudo-code of **BOM** algorithm.

**Lemma 14.** *The BOM algorithm marks all the occurrences of p in T and only them.*

*Proof.*

- BOM only marks valid occurrences because the only word of length $m$ recognized by Oracle($p^r$) is $p^r$ itself.
- BOM marks all the occurrences of $p$. Indeed, assume on the contrary that there is an occurrence of $p$ that no window matches. As the window shift is at most $m$, we have necessarily the situation described in Figure 9 ($u$ may be the empty word) where Window 1 and Window 2 are consecutive in the algorithm. The recognition failure should have occurred in $\sigma$, this is not possible because $\sigma u$ is a factor of $p$.

□

The worst-case complexity of BOM is $O(nm)$. However, in the average, we make the following conjecture based on experimental results (see Section 4.3):

**Conjecture 1.** *Under a model of independence and equiprobability of letters, the BOM algorithm has an optimal average complexity of $O(n \log_{|\Sigma|}(m)/m)$.*

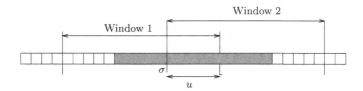

**Fig. 9.** Impossible situation in the BOM algorithm during the search phase.

## 4.1   Approach Using the Suffix Oracle

The use of suffix oracle instead of factor oracle allows a refinement of the preceding approach. This refinement comes directly from the use of suffix automaton in BDM. During the backward search phase in suffix oracle, if a terminal state (which does not correspond to the whole word) is encountered, the position in the window is save in a variable *last*. This enables us to give a bound on the longest read factor which is a suffix of $p^r$ ,*i.e.* a prefix of $p$. By saving the last state, we save a bound on the longest prefix and we can shift the window up to *last* (the string being or not being found in the current window). Figure 10 illustrates this improvement.

We call this algorithm **BSOM** for *Backward Suffix Oracle Matching*. Its complexity is still worst case $O(nm)$.

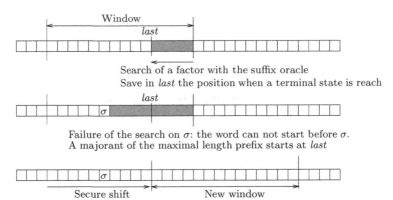

**Fig. 10.** Search with the suffix oracle. The terminal state marking allows an improvement of the search based on the factor oracle.

## 4.2   A Linear Worst Case Algorithm

Even if the preceding algorithms are very efficient in practice, they have a worst-case complexity in $O(mn)$. There are several techniques to make the BDM algorithm (using suffix automaton) linear in the worst case, and one of them can also be used to make our algorithms linear in the worst case. It uses the Knuth-Morris-Pratt (KMP) algorithm to make a forward reading of some characters in the text.

To explain the combined use of KMP and (factor or suffix) oracle, we consider the current position before the search with the oracle: a prefix $v$ of the string has already been read with KMP at the beginning of the search window and we start the backward search using the oracle from the right end of that current window. The end position of $v$ in the current window is called *critical position* and is denoted by *Critpos*.

The current position is schematized in Figure 11.

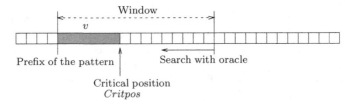

**Fig. 11.** Current position in the linear algorithm using both KMP and (factor or suffix) oracle.

We use the search with the oracle from right to left from the right end of the window. We consider two cases whether the critical position is reached or not.

1. The critical position is not reached. The failure of the recognition of a factor occurs on character $\sigma$ as in the general approach (Figure 12). We shift the window to the left until its beginning goes past character $\sigma$. We restart a KMP search on this new window rereading the characters already read by the oracle. This search stops in a new current position (with a new corresponding critical position) when the recognized prefix is small enough (less than $\alpha m$ with $0 < \alpha < 1$). The value of $\alpha$ is discussed with the experimental results (see section 4.3), typically $\alpha = 1/2$. This situation is schematized in Figure 12.

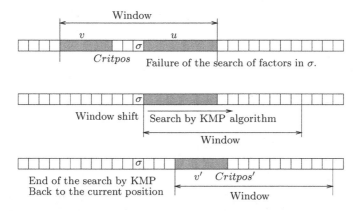

**Fig. 12.** First case: the critical position is not reached.

2. The critical position is reached. We resume the KMP search from the critical position, from the state we were before stopping, rereading at least the characters read by the oracle. We then go on reading the text until the longest recognized prefix is small enough (less than $\alpha m$). This situation is schematized in Figure 13.

This algorithm can be used with a backward search done with the factor oracle as well as with the suffix oracle (saving the last terminal state encountered). We call these two algorithms **Turbo-BOM** and **Turbo-BSOM**. Concerning the complexity in the worst case, we have the following result.

**Theorem 3.** *The two algorithms Turbo-BOM and Turbo-BSOM are: (i) linear considering the number of inspections of characters in the text, the number of these inspections is less than $2n$; (ii) linear considering the number of comparisons of characters, the number of these comparisons is less than $2n$ when the transitions of the oracle are available in $O(1)$ and less than $2n + n \log |\Sigma|$ when the transitions are available in $\log |\Sigma|$.*

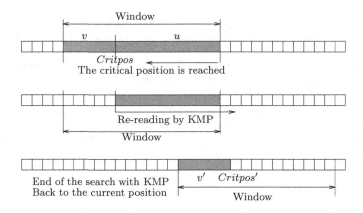

**Fig. 13.** Second case: the critical position is reached.

*Proof.*

(i) Each character in the text is read twice, once during the backward search using the oracle and once during the search with the KMP algorithm. The results follows.

(ii) This complexity comes directly from the fact that the number of comparisons done by the KMP algorithm is less than $2n$. If the transitions of the oracle are available in $O(1)$, the backward search using the oracle does not require any comparisons (it only requires inspections), and the total number of comparisons is bound by the number of comparisons in KMP: $2n$. If the transitions are available in $\log \Sigma$ (for instance with binary search trees), the number of comparisons done during the backward search is bounded by $n \log \Sigma$ and the total number by $2n + n \log \Sigma$.

□

### 4.3   Experimental Results

In this section, we present experimental results on the time complexity of our string matching algorithms, compared to the following algorithms: **Sunday**: the Sunday algorithm [10] is often considered as the fastest in practice; **BM**: the Boyer-Moore algorithm [6]; **BDM**: the classical Backward Dawg Matching with a suffix automaton [8]; **Suff**: the Backward Dawg Matching with a suffix automaton but without testing terminal states, this is equivalent to the basic approach with the factor automaton[2]; **BOM**: the Backward Oracle Matching with the

---

[2] The suffix automaton without taking in account the terminal states (i.e. considering every state as terminal) and the factor automaton recognize the same language. The difference is that the factor automaton is minimal, so its size is smaller or equal than the size of the suffix automaton. But the difference of size is not significant in practice, anyway not enough significant to justify the implementation of a factor

factor oracle; **BSOM**: the Backward Oracle Matching with the suffix oracle (testing terminal states); **Turbo-BOM**: the linear algorithm using BOM and KMP with $\alpha = 1/2$.

Our string matching experiments are done on DNA sequences (we took the Archaeoglobus Fulgidus sequence of 2 MB) and on natural language (English - we took a compilation of Wall Street Journal articles of 10 MB). We also performed experiments on random texts for alphabets of size 2, 4, 16 and 32. Results are obtain with an accuracy of +/- 2% with a confidence of 95% (which may require thousands of iterations). The machine used is a PC with a Pentium II processor at 350MHz running Linux 2.0.32 operating system. For all the algorithms, the transitions of the automata are implemented as tables which allow $O(1)$ branches.

Experimental results in string matching are always surprising because codes are smalls and the time taken by a character comparison is not much greater than the time taken by an integer incrementation. It is for instance the reason why Sunday algorithm is the fastest algorithm for small strings: a window shift is usually very small but require very few operations. It is also the reason why BDM is slower than Suff and BSOM slower than BOM whereas the window shifts in BSOM and BDM are greater. When searching in sequences of characters, it is obviously useless to mark and test terminal states in both suffix automaton and factor oracle.

The 4 sub-figures of Figure 14 show that BOM is as fast as Suff (except on a binary alphabet) which is much more complicated and requires much more memory. BOM reads more texts characters than Suff, but as the oracle automaton is much smaller than the suffix automaton, it performs less memory page breaks and the experimental search times are the sames.

Turbo-BOM algorithm is the slowest but it is the only one that can be used in real time and in that case its behavior is rather good. It has to be noticed that we arbitrarily set the value of $\alpha$ to $1/2$. However, according to the tests we performed for different values of $\alpha$, it turns out that $\alpha = 1/2$ is the more often the best value and that the variations of search times with other values of $\alpha$ (as far as they stay between $(2\log_{|\Sigma|} m)/m$ and $(m - 2\log_{|\Sigma|} m)/m$ ) are not very significant and anyway do not deserve by themselves an accurate study.

## 5    Conclusions

The two new structures we presented, the factor oracle and the suffix oracle, enable new string matching algorithms. These algorithms are very efficient in practice, as efficient as the ones which already existed, but are far more simple to implement and require less memory. According to the experimental results, we conjecture that they are optimal on the average (under a model of equiprobability of letters) but it remains to be shown.

About the structure of factor oracle itself, many questions stay open. Among others, it would be interesting to have a characterization of the language recog-

---

automaton which will complicate and slow the preprocessing phase of the string matching algorithm.

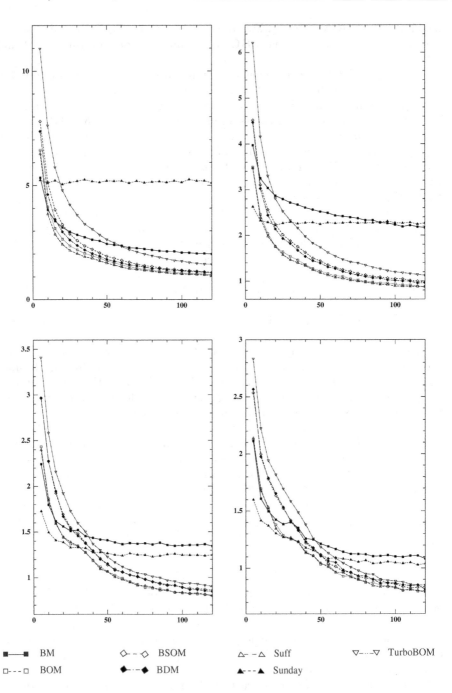

**Fig. 14.** Experimental results in time of the string matching algorithms on random texts of size 10 MB on alphabets of size 2, 4, 16 and 32. The X-axis represents the length of the string and the Y-axis the search time in 1/100th seconds per MBytes.

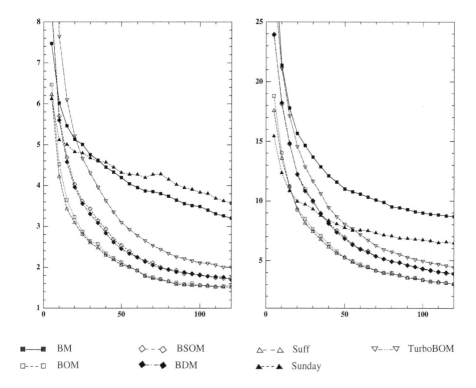

| ■——■ | BM | ◇ - ◇ | BSOM | △- -△ | Suff | ▽----▽ | TurboBOM |
| □---□ | BOM | ◆-·-◆ | BDM | ▲--▲ | Sunday | | |

**Fig. 15.** Experimental results in time of the string matching algorithms on DNA sequences (left plot - 2MB) and Natural language (right plot - 10 MB). The X-axis represents the length of the string and the Y-axis the search time in 1/100th seconds per MBytes.

nized by the oracle. It would also be of interest to study of the average number of external transitions in the oracle, to know the average memory space required by the string matching algorithms.

We notice that the factor oracle is not minimal considering the number of transitions among the automata of $m + 1$ states which recognize at least the factors. This reduced automaton may also be used in string matching provided that its construction can be done in linear time. This construction remains an open problem.

Finally, the factor oracle can be extended to a set of strings, and integrated in multi string matching algorithms. The experimental results are very promising, the new algorithms being by far the fastest in many practical cases [3].

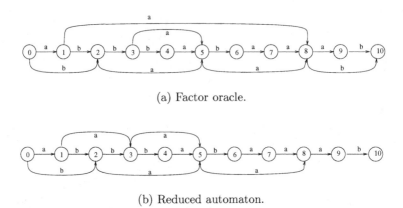

(a) Factor oracle.

(b) Reduced automaton.

**Fig. 16.** The factor oracle is not minimal considering the number of transitions among the automata of $m + 1$ states which recognize at least the factors.

# References

1. C. Allauzen. *Combinatoire sur les mots et recherche de motifs (Combinatorics on words and string matching)*. PhD thesis, Université de Marne-la-Vallée, 2001.
2. C. Allauzen, M. Crochemore, and M. Raffinot. Factor oracle: a new structure for pattern matching. In Miroslav Bartosek Jan Pavelka, Gerard Tel, editor, *SOF-SEM'99, Theory and Practice of Informatics (Brno, 1999)*, number 1725 in LNCS, pages 291–306. Springer-Verlag, 1999.
3. C. Allauzen and M. Raffinot. Factor oracle of a set of words. Techni-cal Report 99-11, Institut Gaspard-Monge, Université de Marne-la-Vallée, 1999. http://www-igm.univ-mlv.fr/~raffinot/ftp/IGM99-11-english.ps.gz.
4. R. A. Baeza-Yates. Searching subsequences. *Theor. Comput. Sci.*, 78(2):363–376, 1991.
5. A. Blumer, J. Blumer, A. Ehrenfeucht, D. Haussler, M. T. Chen, and J. Seiferas. The smallest automaton recognizing the subwords of a text. *Theor. Comput. Sci.*, 40(1):31–55, 1985.
6. R. S. Boyer and J. S. Moore. A fast string searching algorithm. *Commun. ACM*, 20(10):762–772, 1977.
7. M. Crochemore. Transducers and repetitions. *Theor. Comput. Sci.*, 45(1):63–86, 1986.
8. M. Crochemore and W. Rytter. *Text algorithms*. Oxford University Press, 1994.
9. A. Czumaj, M. Crochemore, L. Gasieniec, S. Jarominek, T. Lecroq, W. Plandowski, and W. Rytter. Speeding up two string-matching algorithms. *Algorithmica*, 12:247–267, 1994.
10. D. Sunday. A very fast substring search algorithm. *CACM*, 33(8):132–142, August 1990.
11. A. C. Yao. The complexity of pattern matching for a random string. *SIAM J. Comput.*, 8(3):368–387, 1979.

# Better Filtering with Gapped q-Grams

Stefan Burkhardt[1][*] and Juha Kärkkäinen[2][**]

[1] Center for Bioinformatics, Saarland University
Postfach 151150, 66041 Saarbrücken, Germany
stburk@mpi-sb.mpg.de
[2] Max-Planck-Institut für Informatik
Stuhlsatzenhausweg 85, 66123 Saarbrücken, Germany
juha@mpi-sb.mpg.de

**Abstract.** The $q$-gram filter is a popular filtering method for approximate string matching. It compares substrings of length $q$ (the $q$-grams) in the pattern and the text to identify the text areas that might contain a match. A generalization of the method is to use *gapped* $q$-grams, subsets of $q$ characters in some fixed non-contiguous shape, instead of contiguous substrings. Although mentioned a few times in the literature, this generalization has never been studied in any depth. In this paper, we report the first results from a study on gapped $q$-grams. We show that gapped $q$-grams can provide orders of magnitude faster and/or more efficient filtering than contiguous $q$-grams. The performance, however, depends on the shape of the $q$-grams. The best shapes are rare and often possess no apparent regularity. We show how to recognize good shapes and demonstrate with experiments their advantage over both contiguous and average shapes. We concentrate here on the $k$ mismatches problem, but also outline an approach for extending the results to the more common $k$ differences problem.

## 1  Introduction

Given a pattern string $P$ of length $m$, a text string $T$ of length $n$, and a distance $k$, the approximate string matching problem is to find all substrings of the text $T$ that are within a distance $k$ of the pattern $P$. The most commonly used distance measure, leading to the $k$ *differences problem*, is the Levenshtein distance, the minimum number of single character insertions, deletions and replacements needed to change one string into the other. A simpler variation, the $k$ *mismatches problem*, uses the Hamming distance, the minimum number of replacements needed to change one string into the other, i.e., the number of mismatching characters.

The fastest algorithm in practice for the $k$ differences problem is the bitparallel dynamic programming algorithm of Myers [9]. It works in time $O(nm/w)$,

---

[*] Supported by the DFG 'Initiative Bioinformatik' grant BIZ 4/1-1.
[**] Partially supported by the IST Programme of the EU under contract number IST-1999-14186 (ALCOM-FT).

A. Amir and G.M. Landau (Eds.): CPM 2001, LNCS 2089, pp. 73–85, 2001.

where $w$ is the word size of the machine. An extensive survey and comparison of algorithms is given in [10]. The $k$ mismatches problem is a simpler problem but we do not know of any asymptotically faster algorithms for it. As the pattern is usually short in comparison to the text, much faster string matching is often possible if the text has been preprocessed to build, e.g., the suffix tree of the text. This is called *indexed* or *offline* string matching. For the $k$ differences and the $k$ mismatches problems, dynamic programming over the suffix tree [1,17,4] can be fast, but only for short patterns and small distances $k$.

Filtering is a way to speed up approximate string matching. The idea is to narrow down the search to a small fraction of the text with some filtering method (the *filtering phase*) and search only those areas using a proper approximate string matching algorithm (the *verification phase*). A good filtering method is *fast* and *efficient*, i.e., leaves only a small area to be verified. A good survey of filtering methods is given in [10]. Among the most popular and studied filtering methods is the *q-gram method*.

A *q-gram* is a substring of length $q$. The basic $q$-gram method works as follows. First, find all matching $q$-grams between the pattern and the text. That is, find all pairs $(i, j)$ such that the $q$-gram at position $i$ in the pattern is identical to the $q$-gram at position $j$ in the text. We call such a pair a *hit*. Second, identify the text areas that have enough hits. These are the areas passed to the verification phase. There are different ways of defining the text areas and counting the hits in them (see, e.g., [6,5]). However, all of them have the same *threshold*, the significant number of $q$-grams. This number is given by the *q-gram lemma*.

**Lemma 1 (The $q$-Gram Lemma [6]).** *Let $P$ and $S$ be strings of length $m$ with (Levenshtein or Hamming) distance $k$. Then $P$ and $S$ have at least $t = m - q(k + 1) + 1$ common q-grams.*

The threshold given by the lemma is tight in the sense that using any lower value might miss an occurrence (see Lemmas 2 and 4). For example, strings ACAGCTTA and ACACCTTA have Hamming and Levenshtein distance 1 and have $8 - 3(1 + 1) + 1 = 3$ common 3-grams: ACA, CTT and TTA.

The above description of the $q$-gram method leaves many details open. Different realizations of the method are described in [6,16,5,2]. There are also many variations, e.g., not using all $q$-grams [15,14].

The $q$-gram method is particularly suitable for indexed string matching. An index of all text $q$-grams is simple to implement using table lookup, hashing or a trie. This makes the $q$-gram method very fast unless the number of hits is large. Thus, we would like $q$ to be large since the number of hits decreases exponentially as $q$ increases. On the other hand, as $q$ increases the threshold given by the $q$-gram lemma decreases, which reduces filtering efficiency. The best trade-off depends on the implementation and the application.

A generalization of the $q$-gram method uses *gapped* $q$-grams, subsets of $q$ characters of a fixed non-contiguous *shape*. For example, the 3-grams of shape ##-# in the string ACAGCT are AC.G, CA.C and AG.T. Gapped $q$-grams have been used in [3,11,8]. In [3,11], the motivation is to increase the filtration efficiency by considering multiple shapes. Pevzner and Waterman [11] use $q$-grams containing every $(k+1)$st character together with contiguous $q$-grams for the $k$ mismatches problem. The FLASH algorithm of Califano and Rigoutsos [3] uses as many as 40 different random shapes in a probabilistic manner, i.e., without a guarantee of finding all occurrences. Their approach is effective for high $k$ but they need a huge index (18GB for a 100 million nucleotide DNA database). The Grampse system of Lehtinen et al. [8] uses a shape containing every $h$th character for some $h$ (similar to [11]) for exact matching. Their motivation of using gapped $q$-grams is to reduce dependencies between the characters of a $q$-gram.

In this paper, we will show that gapped $q$-grams have advantages over contiguous $q$-grams even when using just one shape and not being concerned with dependencies between characters. Gapped $q$-grams of suitably chosen shape provide much faster and/or more efficient filtering. We have observed improvements of several orders of magnitude in our experiments.

The results in this paper apply only to the $k$ mismatches problem. The $k$ differences problem causes difficulties for gapped $q$-grams because they are affected by insertions and deletions in the gaps. However, the difficulties can be tackled by using multiple shapes. For example, we can use the shapes `#####-##` and `#####---##` to handle an insertion or a deletion in the gap of the shape `#####--##`. We are currently working on this approach to extend our results to the $k$ differences problem. As a preliminary result in this direction, we show that even $q$-grams with just one gap are better than contiguous ones. As the above example shows, $q$-grams with few gaps are of special interest for the $k$ differences problem.

## 2   Shapes

A *shape* $Q$ is a set of non-negative integers containing 0. The *size* of $Q$, denoted by $|Q|$, is the cardinality of the set. The *span* of $Q$ is $s(Q) = \max Q + 1$, i.e., the size of the minimum contiguous interval containing $Q$. A shape $Q$ with size $q$ and span $s$ is called a $q$-shape or a $(q, s)$-shape.

For any integer $i$ and shape $Q$, the positioned shape $Q_i$ is the set $\{i+j \mid j \in Q\}$. Let $Q_i = \{i_1, i_2, \ldots, i_q\}$, where $i = i_1 < i_2 < \cdots < i_q$, and let $S = s_1 s_2 \ldots s_m$ be a string. For $1 \le i \le m - s(Q) + 1$, the $Q$-*gram* at position $i$ in $S$, denoted by $S[Q_i]$, is the string $s_{i_1} s_{i_2} \ldots s_{i_q}$. Two strings $P$ and $S$ have a *common $Q$-gram* at position $i$ if $P[Q_i] = S[Q_i]$.

*Example 1.* Let $Q = \{0, 1, 3, 6\}$ be a shape. Using the notation from the introduction $Q$ is the shape `##-#--#`. Its size $|Q| = 4$ and its span $s(Q) = 7$. The string $S = \texttt{ACGGATTAC}$ has three $Q$-grams: $S[Q_1] = s_1 s_2 s_4 s_7 = \texttt{ACGT}$, $S[Q_2] = \texttt{CGAA}$ and $S[Q_3] = \texttt{GGTC}$.

## 3   Threshold

The $q$-gram lemma does not apply to gapped $q$-grams in the form of Lemma 1. A straightforward generalization would give a threshold of $t = m - s(Q) - |Q|k + 1$

(see Lemma 3 below). Pevzner and Waterman [11] give this threshold for shapes of form $\{0, h, 2h, \ldots, (q-1)h\}$.[1] This threshold for gapped shapes is strictly worse than for contiguous shapes of the same size. However, the threshold is not tight for gapped $q$-grams as shown by the following example.

*Example 2.* Let $m = 11$ and $k = 3$ and consider the 3-shapes ### and ##-#. The above threshold for the two shapes are 0 and $-1$, respectively. Thus, neither shape would seem to be useful for filtering in this case. However, the real threshold for the shape ##-# is 1. By full enumeration of all combinations of 3 mismatches it is possible to verify that at least one ##-#-gram is always unaffected by the mismatches. The following figure gives an example of a worst possible combination of 3 mismatches for both shapes.

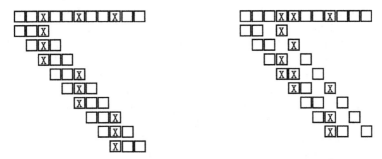

We will next define the tight threshold for arbitrary $q$-grams.

Let $P = p_1, \ldots p_m$ and $S = s_1 \ldots s_m$ be two strings of length $m$. Let $R(P, S)$ be the set of positions where $P$ and $S$ do not match, i.e., $R(P, S) = \{i \in \{1, \ldots, m\} \mid p_i \neq s_i\}$. Then $|R(P, S)|$ is the Hamming distance of $P$ and $S$.

To determine the common $Q$-grams of $P$ and $S$ only the mismatch set $R(P, S)$ is needed: $P[Q_i] = S[Q_i]$ if and only if $Q_i \cap R(P, S) = \emptyset$. The minimum number of common $q$-grams is the threshold value needed for the $q$-gram method.

**Definition 1.** *Let $m$ and $k$ be non-negative integers and $Q$ a shape. The threshold of $Q$ for pattern length $m$ and Hamming distance $k$ is*

$$t(Q, m, k) = \min_{R \subseteq \{1, \ldots, m\}, |R| = k} |\{i \in \{1, \ldots, m - s(Q) + 1\} \mid Q_i \cap R\}|.$$

From the above discussion we get the following tight form of the $q$-gram lemma for arbitrary shapes.

**Lemma 2 (The $Q$-Gram Lemma).** *Let $Q$ be a shape. For any two strings $P$ and $S$ of length $m$ with Hamming distance $k$, the number of common $Q$-grams of $P$ and $S$, i.e., the size of the set $\{i \in \{1, \ldots, m - s(Q) + 1\} \mid P[Q_i] = S[Q_i]\}$, is at least $t(Q, m, k)$. Furthermore, there exists two strings $P$ and $S$ of length $m$ and Hamming distance $k$, for which the number of common $Q$-grams is exactly $t(Q, m, k)$.*

---

[1] They also give a better threshold for the case when the span of the shape is close to the length of the pattern.

We have not found a closed form for the exact threshold in the general case. The following lower bound we already saw in the beginning of this section.

**Lemma 3.** $t(Q, m, k) \geq \max\{0, m - s(Q) - |Q|k + 1\}$.

*Proof.* Let $R$ be the set minimizing the expression in the definition of $t(Q, m, k)$. For each $j \in R$ there is exactly $|Q|$ integers $i$ such that $j \in Q_i$. Therefore, at most $k|Q|$ of the positioned shapes $Q_i$, $i \in \{1, \ldots, m - s(Q) + 1\}$, intersect with $R$, and at least $m - s(Q) - k|Q| + 1$ do not intersect with $R$.     □

Tighter bounds may be given for special cases. In particular, the old $q$-gram lemma for contiguous $q$-grams gives indeed the exact threshold as shown by the following lemma.

**Lemma 4.** *Let $Q$ be a contiguous shape, i.e., $Q = \{0, 1, \ldots, q - 1\}$. Then*
$$t(Q, m, k) = \max\{0, m - s(Q) - |Q|k + 1\} = \max\{0, m - q(k + 1) + 1\}.$$

*Proof.* The lower bound is shown by Lemma 3. Let $R = \{q, 2q, \ldots, kq\}$. Then $Q_i$ intersects with $R$ if and only if $i \in \{1, \ldots, kq\}$, and thus does not intersect with $R$ if $i \in \{kq + 1, \ldots, m - q + 1\}$. This shows the upper bound.     □

Using the lower bound of Lemma 3 as the threshold guarantees that all approximate occurrences are found, but it is very inefficient choice for gapped shapes. A difference of just one in the threshold value used makes a big difference in the efficiency of filtering. We have computed the exact thresholds for all shapes for $m = 50$ and $k \in \{4, 5\}$. Tables 1 and 2 give the highest threshold among $(q, s)$-shapes for all combinations of $q$ and $s$ that have shapes with positive thresholds (except $q = s = 1$).

The tables show that Example 2 was not an isolated case: in many cases, especially for higher values of $q$, best gapped shapes have much higher thresholds than contiguous shapes of the same or even smaller size. Thus, one can use a higher value of $q$ to get fewer hits, or have a higher threshold and better filtration, or even both.

However, it is not sufficient just to have gaps; the shape has to be chosen carefully. For instance, for the parameters of Example 2, $m = 11$, $k = 3$, $q = 3$, the shape ##-# and its mirror image #-## are the only ones that have a positive threshold. As a more impressive example, for the parameter values $m = 50$, $k = 5$ and $q = 12$, there are only two shapes, ###-#--###-#--###-# and #-#-#-#---#-----#-#-#-#---#-----#-#-#---#, (and their mirror images) with a positive threshold. In most cases shown in Tables 1 and 2, only a few shapes achieve the highest threshold. The distribution of threshold values for one typical case is shown in Figure 1.

## 4   Minimum Coverage

The filtering efficiency of a $Q$-gram clearly depends on the threshold $t(Q, m, k)$. However, the correlation is not direct. The following example shows an additional property of shapes that can affect the filtering efficiency.

*Example 3.* Let $m = 13$ and $k = 3$. Then both shapes ### and ##-# have a threshold of two. If two strings have four consecutive matching characters, they have two common 3-grams of shape ###. In contrast, to have two common 3-grams of shape ##-# two strings need to have at least 5 matching characters.

Motivated by the example, we define the following measure.

**Definition 2.** *Let* $Q$ *be a shape and* $t$ *a non-negative integer. The* minimum coverage *of* $Q$ *for threshold* $t$ *is*

$$c(Q, t) = \min_{C \subseteq N, |C| = t} | \cup_{i \in C} Q_i |.$$

The minimum coverage is, in essence, the minimum number of characters that need to match between a pattern and a text substring for there to be $t$ matching $Q$-grams. This gives a reasonable first order estimator for the probability of random strings having $t$ common $Q$-grams. We do not analyse this further in this paper, but our experiments support the conjecture that there is a strong correlation between the minimum coverage $c(Q, t(Q, m, k))$ and the filter efficiency (see Figure 2 in Section 5).

We have computed the minimum coverages of all $(q, s)$-shapes for $m = 50$ and $k \in \{4, 5\}$. Tables 1 and 2 give the highest minimum coverage among $(q, s)$-shapes for all combinations of $q$ and $s$ that have shapes with positive thresholds (except $q = s = 1$). The tables show that considering the minimum coverage further improves the advantage of the best gapped shapes over the contiguous shapes. There are even cases where the contiguous shape is the $q$-shape with the highest threshold but some gapped $q$-shapes have a higher minimum coverage.

As with threshold values, the shapes with the highest minimum coverage are rare. Figure 1 shows the distribution of minimum coverages in one typical case.

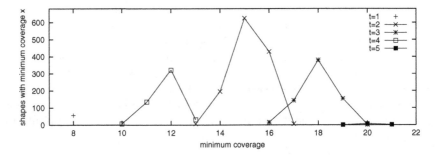

**Fig. 1.** Distribution of thresholds and minimum coverages of $(8, 17)$-grams for $m = 50$, $k = 5$. Mirror images and one shape with threshold 0 are not included.

**Table 1.** The best thresholds/minimum coverages for $m = 50$ and $k = 4$.

| $s$ | 2 | 3 | 4 | 5 | 6 | 7 | 8 | 9 | 10 | 11 | 12 | 13 | 14 |
|---|---|---|---|---|---|---|---|---|---|---|---|---|---|
| 2 | 41/42 | | | | | | | | | | | | |
| 3 | 40/41 | 36/38 | | | | | | | | | | | |
| 4 | 39/40 | 35/38 | 31/34 | | | | | | | | | | |
| 5 | 38/39 | 34/38 | 30/34 | 26/30 | | | | | | | | | |
| 6 | 37/38 | 33/38 | 29/34 | 25/30 | 21/26 | | | | | | | | |
| 7 | 36/37 | 32/38 | 28/34 | 24/30 | 20/26 | 16/22 | | | | | | | |
| 8 | 35/36 | 31/38 | 27/34 | 23/30 | 19/26 | 15/22 | 11/18 | | | | | | |
| 9 | 34/35 | 30/38 | 26/34 | 22/30 | 18/26 | 14/22 | 10/18 | 6/14 | | | | | |
| 10 | 33/34 | 29/38 | 25/34 | 21/30 | 17/26 | 13/22 | 9/18 | 5/14 | 1/10 | | | | |
| 11 | 32/33 | 28/36 | 24/34 | 20/30 | 17/27 | 14/24 | 10/20 | 7/17 | 4/14 | 0/0 | | | |
| 12 | 31/32 | 27/35 | 23/34 | 20/31 | 17/28 | 13/24 | 10/21 | 8/19 | 5/16 | 2/13 | 0/0 | | |
| 13 | 30/31 | 26/34 | 22/34 | 19/31 | 16/28 | 13/25 | 10/22 | 8/20 | 6/18 | 3/15 | 1/12 | 0/0 | |
| 14 | 29/30 | 25/33 | 21/34 | 18/31 | 15/28 | 12/25 | 10/23 | 8/21 | 5/18 | 4/17 | 2/15 | 1/13 | 0/0 |
| 15 | 28/29 | 24/32 | 20/34 | 17/31 | 14/28 | 12/26 | 9/23 | 7/21 | 5/19 | 3/17 | 2/16 | 1/13 | 0/0 |
| 16 | 27/28 | 23/31 | 19/34 | 16/31 | 14/29 | 11/26 | 9/24 | 7/22 | 5/20 | 3/18 | 2/16 | 1/13 | 0/0 |
| 17 | 26/27 | 22/30 | 18/33 | 16/31[a] | 13/29 | 11/27 | 9/25 | 7/23 | 5/21 | 4/20 | 2/17 | 1/13 | 0/0 |
| 18 | 25/26 | 21/28 | 17/32 | 15/31 | 12/29 | 10/27 | 8/25 | 6/23 | 5/22 | 3/20 | 2/18 | 1/13 | 0/0 |
| 19 | 24/25 | 20/27 | 17/30[a] | 14/30 | 12/29 | 9/27 | 7/25 | 6/24 | 4/22 | 3/20 | 2/18 | 1/13 | 1/14 |
| 20 | 23/24 | 19/26 | 16/29[a] | 13/30 | 11/29[a] | 9/28 | 7/26 | 5/24 | 4/23 | 3/21 | 2/18 | 1/13 | 0/0 |
| 21 | 22/23 | 18/25 | 15/29 | 12/29 | 10/28 | 8/27 | 6/26 | 5/24 | 3/20 | 2/18 | 2/17 | 1/13 | 0/0 |
| 22 | 21/22 | 17/24 | 15/28[a] | 12/29[a] | 9/29 | 7/26 | 6/24 | 4/22 | 3/21 | 2/18 | 1/12 | 1/13 | 0/0 |
| 23 | 20/21 | 17/24 | 14/28 | 12/28[a] | 9/28 | 7/26[a] | 5/25 | 4/22 | 2/17 | 2/17 | 1/12 | 0/0 | 0/0 |
| 24 | 19/20 | 17/24 | 15/26 | 12/29[a] | 10/27[a] | 7/27 | 5/25 | 4/20[a] | 2/17 | 1/11 | 1/12 | 0/0 | 0/0 |
| 25 | 20/21 | 17/24 | 15/28[a] | 12/29[a] | 10/29[a] | 7/26 | 5/25 | 4/23 | 3/20 | 2/17 | 1/12 | 0/0 | 0/0 |
| 26 | 21/22 | 17/24 | 16/29[a] | 12/30 | 11/31[a] | 8/28 | 6/27 | 4/24 | 3/22 | 2/18 | 1/12 | 1/13 | 0/0 |
| 27 | 20/21 | 16/23 | 16/29[a] | 12/30 | 12/30[a] | 8/27 | 7/28[a] | 5/26 | 4/24 | 3/21 | 2/18 | 1/13 | 0/0 |
| 28 | 19/20 | 15/21 | 15/29 | 11/30 | 11/30[a] | 8/29 | 7/29[a] | 6/25[a] | 4/25 | 3/22 | 2/19 | 1/13 | 1/14 |
| 29 | 18/19 | 14/20 | 14/28 | 10/28 | 10/30 | 8/27[a] | 7/27[a] | 5/27 | 4/25 | 3/22 | 2/19 | 1/13 | 1/14 |
| 30 | 17/18 | 13/19 | 13/26 | 9/25 | 9/30 | 7/27 | 6/27 | 4/25 | 4/24 | 3/23 | 2/19 | 1/13 | 1/14 |
| 31 | 16/17 | 12/18 | 12/25 | 9/24 | 9/28[a] | 6/27 | 6/27[a] | 4/25 | 4/24 | 2/19 | 2/19 | 1/13 | 1/14 |
| 32 | 15/16 | 11/17 | 11/23 | 9/24[a] | 7/25 | 5/25 | 5/25 | 3/22 | 3/22 | 2/18 | 2/18 | 1/13 | 1/14 |
| 33 | 14/15 | 12/17[a] | 10/20 | 8/24 | 8/24[a] | 5/25 | 4/24 | 3/22 | 2/18 | 2/18 | 1/12 | 1/13 | 0/0 |
| 34 | 13/14 | 12/18 | 9/19 | 8/24 | 8/24[a] | 5/24 | 4/23 | 4/21[a] | 2/18 | 2/17 | 1/12 | 1/13 | 0/0 |
| 35 | 12/13 | 12/18 | 8/18 | 8/24 | 8/25[a] | 4/22 | 4/23 | 4/21 | 2/17 | 2/17 | 2/16 | 1/13 | 1/14 |
| 36 | 11/12 | 11/17 | 7/16 | 7/22 | 7/24 | 4/22 | 4/21 | 4/21 | 2/17 | 2/17 | 2/16 | 1/13 | 1/14 |
| 37 | 10/11 | 10/15 | 8/16[a] | 6/19 | 6/21 | 4/22 | 4/21 | 3/18 | 2/16 | 2/17 | 2/16 | 0/0 | 0/0 |
| 38 | 9/10 | 9/14 | 8/16[a] | 5/15 | 5/20 | 4/22 | 4/19[a] | 3/18 | 2/17 | 2/14 | 1/12 | 0/0 | 0/0 |
| 39 | 8/9 | 8/13 | 8/16[a] | 5/14[a] | 4/18 | 4/21 | 4/19[a] | 2/16 | 2/17 | 2/14 | 1/12 | 0/0 | 0/0 |
| 40 | 7/8 | 7/12 | 7/16 | 5/14 | 3/15 | 3/18 | 3/19 | 2/15 | 2/12 | 1/11 | 1/12 | 0/0 | 0/0 |
| 41 | 6/7 | 6/10 | 6/15 | 6/14[a] | 4/14[a] | 2/13 | 2/14 | 2/14 | 1/10 | 1/11 | 1/12 | 1/13 | 0/0 |
| 42 | 5/6 | 5/9 | 5/13 | 5/14 | 4/14[a] | 2/12 | 2/13 | 2/14 | 1/10 | 0/0 | 0/0 | 0/0 | 0/0 |
| 43 | 4/5 | 4/8 | 4/10 | 4/14 | 4/14 | 3/12[a] | 2/9 | 1/9 | 1/10 | 0/0 | 0/0 | 0/0 | 0/0 |
| 44 | 3/4 | 3/6 | 3/9 | 3/12 | 3/14 | 3/12[a] | 2/10 | 1/9 | 1/10 | 0/0 | 0/0 | 0/0 | 0/0 |
| 45 | 2/3 | 2/5 | 2/7 | 2/9 | 2/11 | 2/12 | 2/11 | 1/9 | 0/0 | 0/0 | 0/0 | 0/0 | 0/0 |
| 46 | 1/2 | 1/3 | 1/4 | 1/5 | 1/6 | 1/7 | 1/8 | 1/9 | 1/10 | 0/0 | 0/0 | 0/0 | 0/0 |

[a] The shapes with the highest threshold and the highest coverage are different.

**Table 2.** The best thresholds/minimum coverages for $m = 50$ and $k = 5$.

| $s$ | 2 | 3 | 4 | 5 | 6 | 7 | 8 | 9 | 10 | 11 | 12 |
|---|---|---|---|---|---|---|---|---|---|---|---|
| 2 | 39/40 | | | | | | | | | | |
| 3 | 38/39 | 33/35 | | | | | | | | | |
| 4 | 37/38 | 32/35 | 27/30 | | | | | | | | |
| 5 | 36/37 | 31/35 | 26/30 | 21/25 | | | | | | | |
| 6 | 35/36 | 30/35 | 25/30 | 20/25 | 15/20 | | | | | | |
| 7 | 34/35 | 29/35 | 24/30 | 19/25 | 14/20 | 9/15 | | | | | |
| 8 | 33/34 | 28/35 | 23/30 | 18/25 | 13/20 | 8/15 | 3/10 | | | | |
| 9 | 32/33 | 27/35 | 22/30 | 18/26 | 14/22 | 9/17 | 5/13 | 0/0 | | | |
| 10 | 31/32 | 26/34 | 21/30 | 18/27 | 13/22 | 10/19 | 6/15 | 3/12 | 0/0 | | |
| 11 | 30/31 | 25/33 | 20/30 | 16/26 | 13/23 | 10/20 | 7/17 | 4/14 | 2/12 | 0/0 | |
| 12 | 29/30 | 24/32 | 19/30 | 16/27 | 12/23 | 9/20 | 7/18 | 4/15 | 2/13 | 0/0 | 0/0 |
| 13 | 28/29 | 23/31 | 19/30[a] | 15/27 | 12/24 | 9/21 | 6/18 | 4/16 | 2/14 | 1/11 | 0/0 |
| 14 | 27/28 | 22/30 | 17/30 | 14/27 | 11/24 | 8/21 | 6/19 | 4/17 | 2/14 | 1/11 | 0/0 |
| 15 | 26/27 | 21/28 | 17/29[a] | 13/27 | 10/24 | 8/22 | 5/19 | 3/17 | 2/15 | 1/11 | 0/0 |
| 16 | 25/26 | 20/27 | 16/29[a] | 13/27 | 10/24 | 7/22 | 5/20 | 3/17 | 2/15 | 1/11 | 0/0 |
| 17 | 24/25 | 19/26 | 15/28 | 12/27 | 9/25 | 7/23 | 5/21 | 3/18 | 2/15 | 1/11 | 0/0 |
| 18 | 23/24 | 18/25 | 14/27 | 11/27 | 8/25 | 6/23 | 4/20 | 3/18 | 2/16 | 1/11 | 0/0 |
| 19 | 22/23 | 17/24 | 14/26[a] | 11/26[a] | 8/25 | 6/22 | 4/20 | 2/15 | 1/10 | 1/11 | 1/12 |
| 20 | 21/22 | 16/23 | 13/25[a] | 10/24 | 7/24 | 5/22 | 3/18 | 2/15 | 1/10 | 1/11 | 0/0 |
| 21 | 20/21 | 15/21 | 12/25 | 9/23 | 7/22 | 5/21 | 3/18 | 2/15 | 1/10 | 0/0 | 0/0 |
| 22 | 19/20 | 15/21 | 12/23[a] | 9/23 | 6/21 | 4/20 | 2/14 | 1/9 | 1/10 | 0/0 | 0/0 |
| 23 | 18/19 | 15/21 | 12/23[a] | 9/23 | 6/21 | 4/19 | 2/14 | 1/9 | 0/0 | 0/0 | 0/0 |
| 24 | 18/19 | 15/21 | 13/24 | 9/23 | 7/20[a] | 4/19 | 2/14 | 1/9 | 0/0 | 0/0 | 0/0 |
| 25 | 19/20 | 15/21 | 13/25[a] | 9/23[a] | 7/22 | 4/20 | 3/16 | 1/9 | 1/10 | 0/0 | 0/0 |
| 26 | 20/21 | 15/21 | 14/26[a] | 9/24 | 8/24 | 5/23 | 3/19 | 2/15 | 1/10 | 0/0 | 0/0 |
| 27 | 19/20 | 14/20 | 14/26[a] | 9/24 | 9/24[a] | 6/23[a] | 4/19 | 2/15 | 1/10 | 1/11 | 0/0 |
| 28 | 18/19 | 13/19 | 13/26 | 8/24 | 8/24 | 5/22 | 4/20 | 3/17 | 2/15 | 1/11 | 0/0 |
| 29 | 17/18 | 12/18 | 12/25 | 8/23 | 8/24[a] | 5/22 | 5/22 | 2/15 | 2/15 | 1/11 | 0/0 |
| 30 | 16/17 | 11/17 | 11/23 | 7/21 | 6/21 | 4/20 | 4/20 | 2/15 | 2/15 | 1/11 | 0/0 |
| 31 | 15/16 | 10/15 | 10/20 | 7/21 | 7/20[a] | 4/18 | 4/18[a] | 2/14 | 2/14 | 1/11 | 0/0 |
| 32 | 14/15 | 10/15 | 9/19 | 7/19[a] | 5/19 | 3/17 | 3/16 | 1/9 | 1/10 | 0/0 | 0/0 |
| 33 | 13/14 | 10/15[a] | 8/18 | 6/19 | 6/19[a] | 3/17 | 2/14 | 1/9 | 1/10 | 1/11 | 0/0 |
| 34 | 12/13 | 11/17 | 7/16 | 6/19 | 6/20[a] | 3/17 | 3/14 | 2/12 | 1/10 | 1/11 | 0/0 |
| 35 | 11/12 | 11/17 | 6/15 | 6/19 | 6/20[a] | 3/16 | 3/14 | 2/14 | 1/10 | 1/11 | 0/0 |
| 36 | 10/11 | 10/15 | 6/13[a] | 5/15 | 5/20 | 2/13 | 2/13 | 2/14 | 1/10 | 0/0 | 0/0 |
| 37 | 9/10 | 9/14 | 7/15[a] | 4/14 | 4/18 | 3/13[a] | 2/13 | 2/13 | 1/10 | 1/11 | 1/12 |
| 38 | 8/9 | 8/13 | 7/15[a] | 4/12[a] | 3/15 | 2/13 | 2/12 | 1/9 | 1/10 | 1/11 | 0/0 |
| 39 | 7/8 | 7/12 | 7/15[a] | 4/12 | 3/12 | 2/12 | 2/12 | 1/9 | 1/10 | 1/11 | 0/0 |
| 40 | 6/7 | 6/10 | 6/15 | 4/13 | 2/10 | 2/11 | 1/8 | 1/9 | 1/10 | 0/0 | 0/0 |
| 41 | 5/6 | 5/9 | 5/13 | 5/13[a] | 3/10[a] | 1/7 | 1/8 | 1/9 | 0/0 | 0/0 | 0/0 |
| 42 | 4/5 | 4/8 | 4/10 | 4/13 | 3/10 | 1/7 | 1/8 | 1/9 | 0/0 | 0/0 | 0/0 |
| 43 | 3/4 | 3/6 | 3/9 | 3/12 | 3/12 | 2/8 | 1/8 | 0/0 | 0/0 | 0/0 | 0/0 |
| 44 | 2/3 | 2/5 | 2/7 | 2/9 | 2/10 | 2/9 | 1/8 | 0/0 | 0/0 | 0/0 | 0/0 |
| 45 | 1/2 | 1/3 | 1/4 | 1/5 | 1/6 | 1/7 | 1/8 | 0/0 | 0/0 | 0/0 | 0/0 |

[a] The shapes with the highest threshold and the highest coverage are different.

# 5   Experiments

To test the $q$-grams in practice, we performed some experiments on DNA data. We used two different databases of size 50 million, one randomly generated (with even and independent distribution of characters), the other containing the first 50 million basepairs of the GenBank Mouse EST database. The queries we used were random strings of length 500. However, the threshold used in filtering was computed for $m = 50$. The effect is that the filtering is guaranteed to report all positions where there is an approximate occurrence of a substring of length 50 of the query.[2] The distance $k$ varied between 3 and 6. The experimental setting corresponds to the high similarity local alignment problems in shotgun sequencing [18] and EST clustering [7,2]. No actual matches were found in the databases, i.e., all potential matches reported by the filtering were false positives.

For a filter algorithm the two main properties of interest are speed and filtering efficiency. As a measure of these properties we use the number of *hits* and the number of *matches*, respectively. A hit is a pair $(i, j)$ such that the query $q$-gram at position $i$ matches the database $q$-gram at position $j$. The time to process the hits usually dominates the running time of the filtering phase. A hit $(i, j)$ is counted for the position $j - i$. A match is a position that has at least $t$ hits. The number of matches reflects the amount of work that the verification phase must do.

The expected number of hits is proportional to $4^{-q}$. Our conjecture is that there is a similar dependence between the number of matches and the minimum coverage of the shape. We tested a large number of shapes using different values of $k$ and the two databases described above. In figure 2 the relation between the minimum coverage and the number of matches per billion characters are shown. In most cases, the number of matches was computed from an average over 100 queries, although for some of the shapes, 1000 queries was used. It is clear that there is a strong but not stringent correlation of the form we expected.

For a more detailed comparison of shapes, we chose four classes of shapes for different values of $q$ and $k$:

- *Best.* This is the shape with the highest minimum coverage. To choose between multiple shapes with the same coverage we used the number of distinct covers of the minimum size, the number of distinct covers of the minimum size plus one, and the threshold as secondary keys (in this order).
- *Median.* For each span $s$, all $(q, s)$-shapes (without mirror images) were ordered by the minimum coverage, and by the same secondary keys as for the best shape, and the median shape in this order was identified. Of these shapes (one for each $s$) the best one was used in the experiments. The chance that a randomly chosen shape is better than this shape is at most half.

---

[2] A better filter efficiency could be achieved by counting the hits separately for each substring of length 50, but we chose the simpler approach. This should not have a significant effect on the relative performance of different shapes. However, looking at the results of the experiments one should keep in mind that we do not achieve maximal filter efficiency.

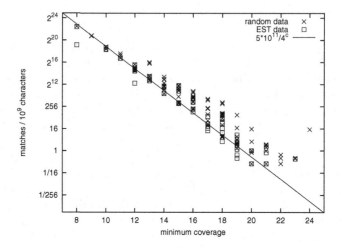

**Fig. 2.** Correlation of minimum coverage and filter efficiency.

- *1-gap.* The best shape with exactly one gap.
- *Contiguous.*

The chosen shapes except the contiguous ones are shown in Table 3. A missing shape means that the shape in that category had a threshold of 0. Figure 3 compares the chosen shapes both in theory ($q$ vs. minimum coverage) and in practice (hits vs. matches). The experimental results are the averages from 1000 queries against the random database. The missing datapoints in the experimental graph either had a threshold of 0 (bottom of the graph) or had no matches at all (top of the graph).

There are several things of interest in Figure 3. First, there is a high correlation between theoretical and experimental behavior. Second, the performance of the median shapes shows that while a randomly chosen shape is likely to be better than a contiguous one, still much better results can be achieved by a careful choice of the shape. Third, the shapes with one gap, which are of particular interest for the $k$ differences problem, are not as good as the best ones overall but still much better than contiguous ones. Finally, the best shapes have several orders of magnitude better performance than the contiguous shapes.

## 6    Concluding Remarks

We have shown that suitably chosen gapped $q$-grams can significantly improve the performance of the basic $q$-gram filtering for the $k$ mismatches problem. While interesting in itself, it also opens the door to the possibilities of gapped $q$-grams in the numerous other algorithms and applications, where contiguous $q$-grams and related methods have been found to be useful. In fact, most filtering

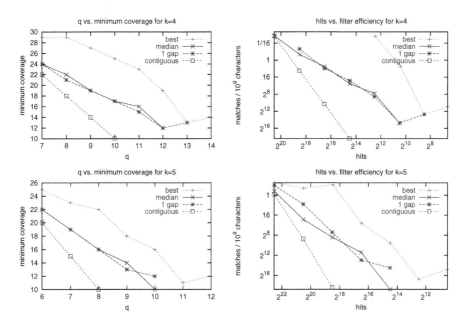

**Fig. 3.** Comparison of four classes of shapes.

| | | $k = 4$ | |
|---|---|---|---|
| $q$ | best | median | 1 gap |
| 7 | ##--#---------#---------##--# | ###-#-----#--#--# | ######-----# |
| 8 | ##--#------#-------#----##-# | #-#-#-#-#--##---#-# | ######----## |
| 9 | ##--#-#--------#------##-#-# | ####-#-#------### | #######----## |
| 10 | ##--#-#-----#------#--##-#-# | ###-###--#-----### | #######---### |
| 11 | ###-#--#------#------#-#-#-## | ######-###-#-# | ########--### |
| 12 | ##-#-#---##---------#-##-#---## | ##-##-####-###-# | #########--### |
| 13 | #########-#### | | #########-#### |
| 14 | ####-##---####-## | | |
| | | $k = 5$ | |
| $q$ | best | median | 1 gap |
| 6 | ##------#--#--#-# | #-###-----#-# | #####----# |
| 7 | #-##------##-#-# | ##--#--#--##-# | #####---## |
| 8 | ##--#-#------------##--#-# | #-#--####---#-# | ######---## |
| 9 | ###--#--#-#---#-## | ######--#-#-# | #######---## |
| 10 | ###--#--#-#-#--###-# | ##-##--#-#-###-# | #######-### |
| 11 | #######-##-## | | |
| 12 | ###-#--###-#--###-# | | |

**Table 3.** The shapes used in Figure 3.

methods for string matching use short substrings one way or another. We are currently working on extending the use of gapped $q$-grams to the $k$ differences problem.

There may be applications even beyond string matching. For example, in DNA sequencing by hybridization (SBH) the problem is, in essence, to construct a string given its $q$-grams. Preparata et al. [12,13] have recently shown that SBH can be significantly improved by using gapped probes ($q$-grams) instead of contiguous ones.

## Acknowledgements

We thank Kurt Melhorn for useful remarks.

## References

1. R. Baeza-Yates and G. Gonnet. All-against-all sequence matching. Technical report, Dept. of Computer Science, University of Chile, 1990.
2. S. Burkhardt, A. Crauser, P. Ferragina, H.-P. Lenhof, E. Rivals, and M. Vingron. $q$-gram based database searching using a suffix array. In S. Istrail, P. Pevzner, and M. Waterman, editors, *Proceedings of the 3rd Annual International Conference on Computational Molecular Biology (RECOMB-99)*, pages 77–83, Lyon, France, 1999. ACM Press.
3. A. Califano and I. Rigoutsos. FLASH: A fast look-up algorithm for string homology. In L. Hunter, D. Searls, and J. Shavlik, editors, *Proceedings of the First International Conference on Intelligent Systems for Molecular Biology*, pages 56–64, Bethesda, MD, 1993. AAAI Press.
4. A. L. Cobbs. Fast approximate matching using suffix trees. In Z. Galil and E. Ukkonen, editors, *Proceedings of the 6th Annual Symposium on Combinatorial Pattern Matching*, number 937 in Lecture Notes in Computer Science, pages 41–54, Espoo, Finland, 1995. Springer-Verlag, Berlin.
5. N. Holsti and E. Sutinen. Approximate string matching using $q$-gram places. In *Proceedings of the 7th Finnish Symposium on Computer Science*, pages 23–32, 1994.
6. P. Jokinen and E. Ukkonen. Two algorithms for approximate string matching in static texts. In A. Tarlecki, editor, *Proceedings of the 16th Symposium on Mathematical Foundations of Computer Science*, number 520 in Lecture Notes in Computer Science, pages 240–248, Kazimierz Dolny, Poland, 1991. Springer-Verlag, Berlin.
7. A. Krause and M. Vingron. A set-theoretic approach to database searching and clustering. *Bioinformatics*, 14:430–438, 1998.
8. O. Lehtinen, E. Sutinen, and J. Tarhio. Experiments on block indexing. In R. Baeza-Yates N. Ziviani and K. Guimarães, editors, *Proceedings of the 3rd South American Workshop on String Processing (WSP'96)*, pages 183–193, Recife, Brazil, 1996. Carleton University Press.
9. G. Myers. A fast bit-vector algorithm for approximate string matching based on dynamic programming. *J. Assoc. Comput. Mach.*, 46(3):395–415, 1999.
10. G. Navarro. *Approximate Text Searching*. PhD thesis, Dept. of Computer Science, University of Chile, 1998.

11. P. A. Pevzner and M. S. Waterman. Multiple filtration and approximate pattern matching. *Algorithmica*, 13(1/2):135–154, 1995.
12. F. P. Preparata, A. M. Fieze, and E. Upfal. On the power of universal bases in sequencing by hybridization. In *Proceedings of the 3rd Annual International Conference on Computational Molecular Biology (RECOMB-99)*, pages 295–301, Lyon, France, 1999. ACM Press.
13. F. P. Preparata and E. Upfal. Sequencing-by-hybridization at the information-theory bound: An optimal algorithm. In R. Shamir, S. Miyano, S. Istrail, P. Pevzner, and M. Waterman, editors, *Proceedings of the 4th Annual International Conference on Computational Molecular Biology (RECOMB-00)*, pages 245–253, Tokio, 2000. ACM Press.
14. E. Sutinen and J. Tarhio. On using q-gram locations in approximate string matching. In P. G. Spirakis, editor, *Proceedings of the 3rd Annual European Symposium on Algorithms*, number 979 in Lecture Notes in Computer Science, pages 327–340, Corfu, Greece, 1995. Springer-Verlag, Berlin.
15. T. Takaoka. Approximate pattern matching with samples. In Ding-Zhu Du and Xiang sun Zhang, editors, *Proceedings of the 5th International Symposium on Algorithms and Computation*, number 834 in Lecture Notes in Computer Science, pages 236–242, Beijing, P.R. China, 1994. Springer-Verlag, Berlin.
16. E. Ukkonen. Approximate string matching with q-grams and maximal matches. *Theor. Comput. Sci.*, 92(1):191–212, 1992.
17. E. Ukkonen. Approximate string matching over suffix trees. In A. Apostolico, M. Crochemore, Z. Galil, and U. Manber, editors, *Proceedings of the 4th Annual Symposium on Combinatorial Pattern Matching*, number 684 in Lecture Notes in Computer Science, pages 228–242, Padova, Italy, 1993. Springer-Verlag, Berlin.
18. J. Weber and H. Myers. Human whole genome shotgun sequencing. *Genome Research*, 7:401–409, 1997.

# Fuzzy Hamming Distance:
# A New Dissimilarity Measure
## (Extended Abstract)

Abraham Bookstein[1], Shmuel Tomi Klein[2], and Timo Raita[3]

[1] Center for Information and Language Studies
University of Chicago, Chicago, IL 60637
a-bookstein@uchicago.edu
[2] Dept. of Math. & CS, Bar Ilan University
Ramat-Gan 52900, Israel
tomi@cs.biu.ac.il
[3] Comp. Sci. Dept., University of Turku
20520 Turku, Finland
raita@cs.utu.fi

**Abstract.** Many problems depend on a reliable measure of the distance or similarity between objects that, frequently, are represented as vectors. We consider here vectors that can be expressed as bit sequences. For such problems, the most heavily used measure is the Hamming distance, perhaps normalized. The value of Hamming distances is limited by the fact that it counts only exact matches, whereas in various applications, corresponding bits that are close by, but not exactly matched, can still be considered to be almost identical. We here define a "fuzzy Hamming distance" that extends the Hamming concept to give partial credit for near misses, and suggest a dynamic programming algorithm that permits it to be computed efficiently. We envision many uses for such a measure.

## 1 Introduction

The Hamming Distance has long been used to quantify the extent to which two bit sequences, or *bitmaps*, of the same dimension, differ. An early application was in the theory of error-correcting codes (see, e.g., [9]), where the Hamming distance measured the error introduced by noise over a channel when a message is sent between its source and destination. Within an Information Retrieval environment, bitmaps may indicate the documents a term occurs in; in such applications, the Hamming distance quantifies differences in the occurrence patterns of terms.

In a traditional application of the Hamming distance, the only concern is whether the corresponding bits in two strings agree; the distance doesn't distinguish whether a discrepancy between a target and source 1-bit are separated by one or many positions. Consider the following target bitmap (a), and the

A. Amir and G.M. Landau (Eds.): CPM 2001, LNCS 2089, pp. 86–97, 2001.
© Springer-Verlag Berlin Heidelberg 2001

candidate source bitmaps (b) and (c):

$$(a) \ 1100100000$$
$$(b) \ 1100010000$$
$$(c) \ 1100000001$$

As assessed by the traditional Hamming distance, both (b) and (c) are equally good efforts to match the target: both differ from it by a Hamming distance of 2. But intuitively, one is inclined to regard (b) as a better match than (c): while both (b) and (c) fail at matching the last 1-bit, (b) misses by only one unit, which may be quite acceptable for several applications. If so, we would like, for such a distance measure, that (b) be assessed as closer to (a) than is (c). In general, there is a great deal of arbitrariness in defining a measure of goodness. But a minimal desideratum of a measure of quality is that it at least satisfy such intuitive criteria as suggested by the applications for which it is intended. The *fuzzy* Hamming distance we define below introduces this flexibility.

As a possible application, consider the problem of automatically *segmenting* a body of text [10,11]. In this context, it is useful to have a measure of how well an algorithmic text segmentor agrees with a target partition, as defined, for example, by a judge. In this case, both the target and the source can be represented by bitmaps, with each sentence (or other text unit) being represented by a bit position, and 1-bits indicating segment boundaries. Here, ideally, we should have an exact matching of 1-bits, and the number of discrepancies is the most obvious measure of algorithm failure. However, unlike the many coding applications in which the Hamming distance is used, in the text segmentation context we have a notion of bit-site proximity, and the fuzzy measure takes this into account.

A further application is *term clustering* [3]: bitmaps may be used to indicate the units (sentences, paragraphs, etc.) within a single document that a given term occurs in; in this context, it is troubling that two terms that tend to occur "close" to one another, even if not in exactly the same units, are assessed by the Hamming distance in the same way as terms that are completely unrelated to one another. That is, the original Hamming distance does not recognize the idea of neighboring units.

Furthermore, it is evident that the new measure could be used in image processing, for example in the *compression* of black and white images and edge-detection, with possible applications to vector quantization, computer vision, robotics and fax transmisions.

An example directly parallel to the segmentation problem is measuring the effectiveness of techniques for *parsing* Chinese text into words [7]. A more flexible Hamming distance can also be of use in situations which require assessing the closeness of pairs of event sequences, in which nearby events are considered to be associated; such a requirement occurs in some *data-mining* applications [1].

It is our intention in this paper to extend the classic Hamming distance to be sensitive to situations in which the concept of neighboring locations is important. This extension will be defined in the next section in terms of the operations necessary to transform the source bitmap to target bitmap. Conditions sufficient for this measure to be a true distance will be derived. One important motivation for constructing the new measure is that we believe it to be more reliable than the crisp measure when the exact placement of 1-bits is influenced by random effects. We will test this by comparing the two measures on simulated sets of bit string.

## 2   Fuzzy Hamming Distance

The traditional definition of a Hamming distance is simple and intuitively appealing: given two bit-strings (more generally, strings of symbols) of the same dimension, the Hamming distance is the minimum number of symbol changes needed to change one bit-map into the other. One can view this as a type of edit-distance [13,6], but with a highly restricted set of edit operations.

The Fuzzy Hamming distance we are introducing is also a type of edit distance. But by extending the edit-operation set, we are able to recognize a notion of neighborhood that gives credit for near misses. Unlike most edit distances, ours compares pairs of fixed size bitmaps instead of general strings. This simplifies our task in that it fixes the size of the strings, and we need concentrate only on the 1-bits. However, it relies critically on a "shift" operation that while important for us, has not gotten much, if any, attention in the string literature.

Suppose then that we wish to measure the distance between two bitmaps. For many applications, these enter symmetrically, but this need not be the case. For example, one bitmap may represent a target bitmap $(B_T)$; the other may be a source bitmap $(B_S)$ that is the output of an algorithm that is trying to match that target. In both cases, we wish to measure how similar the two bitmaps are. Ideally, $B_S$ and $B_T$ should be identical.

Notationally, we let $M$ denote the dimension of our bitmaps. For example, we might have a bit-site for each of $M$ sentences, with a 1-bit indicating that the corresponding sentence ends a document segment. We shall use the notation $N(B)$ to denote the number of 1-bits in the bitmap $B$. We next define the fuzzy Hamming distance as an edit distance that measures the difficulty of transforming one such bitmap (say $B_S$) into the other $(B_T)$.

### 2.1   Edit Distance: Overview

To compute an edit distance we first define a set of elementary edit-operations, and associate with each a cost. The edit operations we define must certainly include the *insert/delete* operations of the crisp Hamming distance. In addition, we introduce a *shift* operation that allows us to transfer a 1-bit in $B_S$ to a nearby 1-bit in $B_T$ at less cost than deleting the 1-bit in $B_S$ and inserting it in $B_T$. The shift operation is an abstraction of the concrete task of attempting to match a 1-bit in a target and missing, but getting close — it thus captures, for our measure, the notion of neighboring bit-sites.

Given a set of elementary operations, with a cost assigned to each, a measure of the difference between two bitmaps can be computed by using these operations to transform one to the other, and adding up the costs of the operations used. To eliminate the ambiguity attached to the multiplicity of possible transformation sequences, we quantify how close one bitmap is to another by the *minimum* cost over sequences of elementary operations that transform the source bitmap into the target bitmap.

Computing edit distances typically requires processing whole strings. However, the special nature of our problem will allow us to focus on only the 1-bits of the bitmap, which greatly reduces the computational cost of our algorithm. Thus, to develop our algorithm, it is convenient to describe a participating bitmap by a list of the index values, in ascending order, of its 1-bits. For example, we might represent the bitmap $B_S$ by $S = \langle s_1, s_2, \ldots, s_{N(S)} \rangle$, where $s_i$ is an integer denoting the position of the $i$-th 1-bit of the bitmap $B_S$, and $N(S) = N(B_S)$. Below we shall refer to both representations, $B_S$ and $S$, as "the source bitmap".

## 2.2   Edit Distance: Details

In general, let $c(i, j)$ denote the minimum cost of transforming the first $i$ 1-bits of a source bitmap, $S$, into the first $j$ 1-bits of a target bitmap, $T$, where $1 \leq i \leq N(S)$ and $1 \leq j \leq N(T)$. Our objective, then, is to compute the distance $d(S, T)$ between the bitmaps $S$ and $T$ as $c(N(S), N(T))$, the minimum cost of transforming the source bitmap $S$ into the target bitmap $T$. We will represent the function $c$ in a tableau, whose values will be computed by means of a dynamic programming technique, as is standard in the string processing literature [13,6]. The dynamic programming technique further assures that we satisfy the desired constraint that any two shift operations do not cross: i.e., if $s_i$ is shifted to $t_j$ and $s_{i'}$ is shifted to $t_{j'}$, then the dynamic programming technique assures that if $i < i'$, then $j < j'$.

To define our edit distance, we use the following elementary operations, motivated by reference to a problem in which a source bitmap is created with the intention of matching a target bitmap:

- *Insertion:* Here the algorithm generating the source is considered to have missed the $j$-th 1-bit of the target. To correct, a 1-bit is inserted into the source at location $s_j$, incurring a cost $c_I > 0$. If this operator is applied in an optimal sequence of operations taking $\langle s_1, \ldots, s_i \rangle$ to $\langle t_1, \ldots, t_j \rangle$, then

$$c(i, j) = c_I + c(i, j - 1),$$

   and the $j$-th target bit is considered disposed of. Note that this insertion doesn't preclude the possibility of a 1-bit already being present in $B_S$, as might be the case if $i > j$. But permitting this greatly simplifies our evaluating this measure. Also note that in our problem, we are changing 1-bits. But the lengths of the bitmaps are fixed — only the number of 1-bits is changed. Thus when defined as operations on bitmaps, our terminology is somewhat non-standard, with insertion denoting a change of bitmap value rather than an operation that increases the size of the bitmap. Similar considerations apply below, when we define deletions.

- *Deletion:* Here the $i$-th 1-bit of the source is assessed as spurious. The 1-bit is changed to a 0-bit, incurring a cost $c_D > 0$. If this is optimal, then

$$c(i,j) = c_D + c(i-1,j),$$

and the $i$-th source bit is considered disposed of.

- *Shift:* Here the $j$-th 1-bit of the target and the $i$-th 1-bit of the source are considered to represent the same bit value, but misaligned by a small amount. That is, the source generator correctly sensed the need for a 1-bit, but its exact location may have been in error. Now the source 1-bit is shifted $\Delta$ locations to align it with the target 1-bit; here, $\Delta$ is a non-negative value, insensitive to the direction of the shift. The cost incurred by this operation is given by $c_S(\Delta)$, a non-negative function, monotonically increasing with $\Delta$. If the match was accurate, then $\Delta = 0$, with $c_S(0) = 0$, will denote the null operation. If a shift operation is optimal, then, for $\Delta = |j - i|$,

$$c(i,j) = c_S(\Delta) + c(i-1,j-1),$$

and the $i$-th source bit and $j$-th target bit are considered disposed of. Most simply, we can assess $c_S(\Delta) = A\,\Delta$, for some non-negative constant $A$. We would want to adjust $A$ relative to $c_I$ and $c_D$ so that for $\Delta$ "large," it is cheaper to *delete* and *insert* rather than shift, while the opposite is true for small $\Delta$. Although the shift operation is unconventional in the string processing literature, it is easily accommodated within the dynamic programming framework.

To implement the dynamic programming method, we initialize the tableau by inserting the following boundary values:

$$c(0,j) = j\,c_I \qquad \text{and} \qquad c(i,0) = i\,c_D.$$

The optimal costs are then developed recursively as indicated in the definition of the operations. Note that the size of the tableau is $N(\mathcal{S})N(\mathcal{T})$, rather than $M^2$, which may dramatically reduce the complexity of the computation. Further efficiencies are possible if the bitmaps are sparse or the maximal shift distance is sufficiently small.

Below, when we wish to reveal all parameters in the edit distance between two arbitrary source and target bitmaps of the same dimension, represented as above by $\mathcal{S}$ and $\mathcal{T}$, we adopt the notation $d(\mathcal{S}, \mathcal{T}; c_I, c_D, A)$.

## 3   Properties

The fuzzy Hamming distance has some very interesting properties and relations to other functions. We first examine the conditions under which the fuzzy Hamming distance can indeed be shown to be a distance. To do this it is convenient to define a function on the integers, as *concave* as follows: given integers $r, s, t, u$, with $r < s \leq t < u$, then a function $f$ defined on the integers is concave provided

$$\frac{f(u) - f(t)}{u - t} \leq \frac{f(s) - f(r)}{s - r}.$$

THEOREM. *The fuzzy Hamming distance, $d(S, T)$, is a true distance function, if, when $x \geq 0$ denotes the absolute size of a shift:*
   (a) $c_S(x) \geq 0$, taking the value 0 if and only if $x = 0$;
   (b) $c_S(x)$ increases monotonically;
   (c) $c_S(x)$ is concave on the integers; and
   (d) $c_D = c_I > 0$.

*Proof.* For the fuzzy Hamming distance to be a metric, three conditions must be satisfied.

*Positivity:* $d(S, T) \geq 0$. Clearly, if $S = T$, no operations (technically, only shifts of length zero) are required to transform one to the other, so for this case $d(S, T) = 0$. On the other hand, if the bitmaps are not identical, at least one non-trivial operation must be applied, incurring a positive cost. Thus, $d(S, T) \geq 0$, taking the value zero only for identical bitmaps.

*Symmetry:* $d(S, T) = d(T, S)$. For any sequence of operations, $o_1, o_2, \ldots, o_n$, taking $S$ to $T$, a complementary sequence of operations can be defined: $o'_n, o'_{n-1}, \ldots, o'_1$, where

$$o'_i = \begin{cases} delete & \text{if } o_i = insert \\ insert & \text{if } o_i = delete \\ shift(-j) & \text{if } o_i = shift(j). \end{cases}$$

Clearly, the complementary sequence systematically undoes the effect of the original sequence, and acting on $T$, transforms it to $S$. Under the conditions of the theorem, it incurs the identical cost. Since this is true as well for an optimal sequence, we see the symmetry condition is satisfied.

*Triangle Inequality:* For bitmaps $S$, $T$, and $U$, $d(S, T) \leq d(S, U) + d(U, T)$. Consider an optimal sequence of operations, $t_1$, that transforms $S$ to $T$. Each operation either processes a 1-bit of $S$ (by deleting it); processes a 1-bit of $T$ (by inserting it); or both (by shifting the bit from $S$ onto the bit in $T$). It is convenient for our proof to include a *null* operator (a shift of zero distance), so that below, a shift denotes a non-trivial movement.

To prove the triangle inequality, the costs of operations that transform $S$ to $T$ via the intermediary $U$ must evaluated, and compared to the cost of $t_1$. So consider the sequence of operations, $t_2$, first taking $S$ to $U$; then the sequence, $t_3$, taking $U$ to $T$.

Consider, then, a 1-bit in $S$. It could be considered disposed of in any of the following cases: if it is

i) deleted in $t_2$;

ii) left alone in both $t_2$ and $t_3$;

iii) left alone in $t_2$ and deleted in $t_3$;

iv) left alone in $t_2$ and shifted in $t_3$;

v) shifted in $t_2$ and left alone in $t_3$;

vi) deleted in $t_2$, then reinserted in $t_3$;

vii) shifted in $t_2$ then deleted in $t_3$; or

viii) shifted in both $t_2$ and $t_3$.

A similar breakdown is possible to describe how a 1-bit in $\mathcal{T}$ could be disposed of. In each case, the set of operations similar to i)–v) in effect describes operations legitimate in transforming $\mathcal{S}$ to $\mathcal{T}$ directly. But since $t_1$ is optimal, these operations, considered as a description of a transformation of $\mathcal{S}$ to $\mathcal{T}$, collectively can only increase the cost or leave it unchanged; this is consistent with the triangle inequality.

Any disposition involving two operations that cancel, as in vi), or a single shift combined with a non-null operation, as in vii), can only increase the cost relative to the disposition of the bit in $t_1$, again in accordance with the triangle inequality. By systematically examining all possibilities in this manner, it is straightforward to conclude that the only way the triangle inequality can break down is if the concatenated sequence $t_2\, t_3$ shifts a bit twice: in effect shifting the bit from its initial location in $\mathcal{S}$ to its final destination in $\mathcal{T}$ by means of two shifts (a disposition illegal when directly transforming $\mathcal{S}$ to $\mathcal{T}$).

Thus, to prove the triangle inequality we need to examine only the case when two successive shifts, say of lengths A and B, are encountered. Consider two cases: (a) both shifts are to the same direction, resulting in the combined shift length A+B and (b) one shift, say the first, requires a reverse shift for the second, resulting in the combined shift length A-B. These conditions imply that for the triangle inequality to be valid, the following inequalities must hold for A,B $\geq 0$ and A $\geq$ B:

$$c_S(A) + c_S(B) \geq c_S(A + B)$$
$$c_S(A) + c_S(B) \geq c_S(A - B).$$

However, the second inequality is a trivial consequence of the monotonicity property, since $c_S(A)$ alone is greater than $c_S(A - B)$, and gives nothing new. So we need focus only on the consequences of the first inequality.

First note that the triangle inequality is trivially valid if one term, say $B$, is zero. So consider two integers, $A \geq B > 0$. If $c_S(.)$ is concave on the integers, then certainly,

$$\frac{c_S(A + B) - c_S(A)}{B} \leq \frac{c_S(B) - c_S(0)}{B} = \frac{c_S(B)}{B}.$$

Thus $c_S(A + B) \leq c_S(A) + c_S(B)$. That is, the concavity property is sufficient to assure the triangle inequality, as was to be proved.  ∎

The class of integer concave functions is quite broad. In particular, if $c_S(.)$ is linear, or the integer restriction of a traditionally concave function defined over the reals, then we can assume that our measure is a distance, and benefit from the intuition that this provides.

The edit distance we defined above is very general, and it is interesting to relate it to other distance-measures between bitmaps.

**Hamming Distance:** An obvious measure for the distance between two bitmaps is the simple Hamming distance: the minimum number of bits that must be changed so the source and target agree. The Hamming distance between the bitmaps represented by $S$ and $T$ may be expressed as a special case of the edit distance: $d(S, T; 1, 1, \infty)$. That is, our measure is a genuine generalization of the traditional Hamming distance.

**Recall/Precision Type Measures:** In Information Retrieval, one customarily uses two measures to evaluate performance. *Recall* indicates the fraction of all relevant documents that appear in a retrieval set, while *precision* indicates the fraction of documents in a retrieval set that are relevant. These measures can be adapted to evaluating the distance between a source and target bitmap [10]. To do this, we define two functions:

$$p(S, T) = \frac{N(S \text{ AND } T)}{N(S)}, \qquad r(S, T) = \frac{N(S \text{ AND } T)}{N(T)},$$

where the AND operator acts on the bitmaps indicated by $S$ and $T$.

Thus $p(S, T)$ is the fraction of the 1-bits of the bitmap represented by $S$ to be evaluated which indeed match the 1-bits of the target bitmap represented by $T$, e.g., the percentage of segment boundaries produced by our algorithm, which are "correct" in the sense that they are also boundaries in the given reference partition; similarly $r(S, T)$ is the fraction of the 1-bits of the target bitmap that have corresponding 1-bits in the source bitmap (e.g. the percentage of segment boundaries of the reference partition that are detected by our algorithm).

We thus assign, after having fixed $T$, a pair of numbers $(r, p)$ to each partition. Relative to $T$, we consider a partition represented by $S_1$ to be better than another partition, represented by $S_2$, if

$$p(S_1, T) \geq p(S_2, T) \qquad \text{and} \qquad r(S_1, T) \geq r(S_2, T).$$

Generally, however, the values $r$ and $p$ tend to be inversely related, and trying to raise one usually lowers the other. By changing the parameters in a segmentation algorithm, one could get a series of $(r, p)$-pairs, and produce curves similar to the recall/precision curves in Information Retrieval.

While these measures are interesting because of their relation to the tradition of research in information retrieval, in fact, they can be simply expressed in terms of our edit distance. Letting $\mathbf{0}$ denote the zero-bitmap, we find

$$p(S, T) = \frac{d(S, \mathbf{0}; 0, 1, \infty) - d(S, T; 0, 1, \infty)}{d(S, \mathbf{0}; 0, 1, \infty)}$$

$$r(\mathcal{S}, \mathcal{T}) = \frac{d(\mathcal{S}, \mathbf{0}; 0, 1, \infty) - d(\mathcal{S}, \mathcal{T}; 0, 1, \infty)}{d(\mathcal{T}, \mathbf{0}; 0, 1, \infty)}.$$

Since recall and precision can be expressed in terms of fuzzy costs, this representation immediately suggests an interesting generalization. By relaxing the infinity value in the cost, we can define fuzzy versions of the classic recall and precision measures as used in this context.

## 4   Preliminary Tests on Text Collections

Our primary motivation is to develop a distance measure that is suitable for bitmaps, and also takes into account that bit locations have a proximity property. Such a distance would have face validity for many problems, such as those noted in our introduction, in which bitmaps whose corresponding one-bit locations are nearby should be considered closer than pairs of bitmaps, with the same numbers of one-bits, but in which the locations of corresponding one-bits are well separated. The measure described above has this quality, and in addition reduces to the conventional Hamming distance when our shift cost is made large enough.

But while the fuzzy Hamming distance's face validity justifies its development, we expect that such a measure would also offer performance advantages in situations in which bitmaps are generated in a noisy environment. Specifically, we speculate that an important distinction between the Hamming distance and our fuzzy generalization will be differences in the response of these measures to random influences. One way to test this in a controlled manner is to specifically introduce random effects by means of a simulation. We have devised a variety of such simulations, but the details are deferred to a follow-up paper.

In this section we shortly report on some simple tests which empirically assess the usefulness of the fuzzy Hamming distance.

The first test considers the *segmentation* problem mentioned above. The chosen text were the hundred first sentences of Ernest Wright's novel *Gadsby*, in which the letter **E** never appears. The text was prepared as a numbered sequence of sentences, but without any mention of the original paragraph breaks by the author. Two independent assessors, A and B, were then asked to partition the text into paragraphs. They had no knowledge of the "true" partition, but were given as guideline that the average paragraph consisted of 4 to 5 sentences.

The true partition, as well as those produced by A and B were then translated into bitmaps, each consisting of one bit-site per sentence, with a 1-bit representing a sentence which starts a new paragraph. Figure 1 shows two pairs of such bitmaps, the upper pair corresponding to A and the lower to B; in each pair, the upper map is the one corresponding to the original partition and the lower map is the one produced by the assessor.

These bitmap pairs were then presented to a group of 102 evaluators, which were asked to "grade" the similitude of the maps in each of the pairs. It has

101000010010000010010010100010000001010000010000001100000010010010010010000010010000000000000000100000
100001000010000100001001001001001000010000100000010010001100100010001000010010000010010000100001000001010

101000010010000010010010100010000001010000010000001100000010010010010010000010010000000000000000100000
101000100001000001001000001001001000101000000101000101001000001000001001001001000100000001001001000010010

FIGURE 1: *Partitions of the 100 first sentences of* Gadsby *into paragraphs*

been emphasized that the grading should be done on intuitive grounds, and that there is no *correct* grade which they ought to match. Their challenge was rather to try to quantize their overall feeling of how close the bitmaps in each of the pairs look to them by assigning to each pair a number between 0 and 10, with 0 standing for a perfect match and 10 denoting that there is no connection whatsoever between the maps within the pair. Table 1 summarizes the results.

TABLE 1: *Evaluation of closeness of bitmap pairs*

| Bitmap pair | Hamming distance | Fuzzy distance | Average grade |
|---|---|---|---|
| A | 19 | 6.82 | 3.73 |
| B | 22 | 6.61 | 3.55 |

We see that when closeness is measured by the strict Hamming distance, the guess of assessor A was closer to the original than that of assessor B. Nevertheless, most of the evaluators gave a better grade to the latter pair, and so did the fuzzy Hamming distance, which has been applied with parameters $c_I = c_D = 1$ and $A = 0.15$.

The second test relates to term clustering. The experiment was run on the King James version of the English Bible, which contains 10,644 different terms. The Bible consists of $M = 929$ chapters, each of which was taken as a textual unit. However, most of the terms occur only rarely, many in just one or two chapters, and these had to be excluded from our clustering tests. On the other end of the spectrum, some words are so frequent that they appear in practically every chapter; these too are not interesting from the clustering point of view. We thus restricted our attention to those terms appearing in $N$ chapters, where $20 \le N \le 500$. The number of terms satisfying these constraints was $n = 1462$. For each of these, a bitmap of $M$ bits was generated, with bit $i$ of map $j$ being set to 1 if and only if term $j$ appeared in chapter $i$ of the Bible.

In preparation for the clustering, the crisp Hamming distance was evaluated for all the possible $n(n-1)/2 = 1,067,991$ pairs, and arranged in an upper-triangular matrix; their values ranged from 7 (for the terms Asher and Issachar, which appeared in 28 and 29 chapters respectively) to 550 (for the terms soul and came).

The clustering was performed by using a traditional agglomerative method and the stop condition we chose for the clustering process was to reduce the number of clusters by half. Since each iteration decreases this number by one, we had 731 iterations. The test was then repeated with the fuzzy instead of the crisp distance, and some striking differences in the contents of the formed clusters were found.

For example, one of the clusters that emerged in the process using the crisp measure was

```
(joseph (levi ((gad reuben) simeon))),
```

where parentheses are used to indicate the hierarchical structure of the cluster. Note that all the terms here are names of tribes, and therefore clearly related. But there are still several tribe names missing. On the other hand, there was a similar, but larger, cluster when the process was based on the fuzzy distance:

```
(manasseh ((gad (reuben simeon)) (naphtali
            ((asher issachar) zebulun)))).
```

A tribes cluster is natural, since the tribe names are often mentioned in the same contexts. However, due to the chronological nature of the biblical description, several consecutive chapters (chosen as textual units in the test) often share the same topic. The fuzzy Hamming distance overcomes such – from the content point of view, possibly erroneous – chapter boundaries, which is why it succeeds in detecting more hidden connections.

Another noteworthy example is a cluster with terms connected to sacrifices. The crisp measure gave:

```
(((blemish bullock) (kid lamb))
 ((flour mingled) (atonement ram))),
```

and a similar cluster, albeit with different internal structure, was produced by the fuzzy distance:

```
(((atonement (blemish bullock)) (lamb ram))
((ordinance plague) ((flour mingled) (kid unleavened)))).
```

What is remarkable here is the appearance of the term unleavened. While most of its occurrences in the Bible are connected to the unleavened bread eaten on Passover, there is also a secondary connection to sacrifices which are accompanied by unleavened bread. But the term bread appears in many other contexts and does therefore not form a cluster with unleavened. Indeed, bread appears as a singleton in both clustering processes based on the crisp or fuzzy measures. Nevertheless, the fuzzy distance was able to detect the secondary connection.

## 5    Conclusions

The Hamming distance is a highly respected measure of the distance between equi-dimensional strings. But when the strings have a natural neighborhood property, as is typical of the bitmaps used in Information Retrieval, then the rigidity of the Hamming distance diminishes its value. We defined here a generalization of the Hamming distance that relaxes the condition for elements in a bitmap to match, by assigning partial credit to pairs of bits that don't exactly match, but are nonetheless close. We expect this measure to more accurately represent the distance between bitmaps subject to "noise." In this paper, we introduced the measure and explored some of its properties; in particular we found conditions for its defining a metric. The resistance of the fuzzy Hamming distance to random fluctuation was then compared to that of the classical crisp Hamming distance by means of controlled simulations; both the simulations and the evaluation measures were guided by a clustering metaphor.

The fundamental reason for defining the new measure is that, on its face, it more clearly captures the quality we are trying to evaluate. Our preliminary tests suggest that it may offer performance advantages as well in contexts in which the effects of randomness are a concern.

# References

1. ADRIAANS P., ZANTINGE D., *Data Mining*, Addison Wesley Longman Ltd, Harlow, England (1996).
2. BOOKSTEIN A., KLEIN S.T., Information Retrieval Tools for Literary Analysis, in *Database and Expert Systems Applications*, edited by A M. Tjoa, Springer Verlag, Vienna (1990) 1–7.
3. BOOKSTEIN A., KLEIN S.T., Compression of Correlated Bit-Vectors, *Information Systems* **16** (1991) 387–400.
4. BOOKSTEIN A., KLEIN S.T., RAITA T., Clumping properties of content-bearing words, *Journal of the American Society for Information Science* **49** (1998) 102–114.
5. CORMEN T.H., LEISERSON C.E., RIVEST R.L., *Introduction to Algorithms,* MIT Press, Cambridge, MA (1990).
6. CROCHEMORE M., RYTTER W., *Text algorithms*, New York, Oxford University Press (1994).
7. FAN, C.K., TSAI, W.H., Automatic Word Identification in Chinese Sentences by the Relaxation Technique. *Computer Processing of Chinese & Oriental Languages* **4** (1988) 33–56.
8. DOYLE L., Semantic Road Maps for Literature Searchers, *Journal of the ACM,* **8**(4) (1961) 553–578.
9. HAMMING R.W., *Coding and Information Theory,* Englewood Cliffs, NJ, Prentice-Hall (1980).
10. HEARST M.A., Multi-Paragraph Segmentation of Expository Text, *Proc. ACL Conf.,* Las Cruces (1994).
11. HEARST M.A., PLAUNT C,. Subtopic Structuring for Full-Length Document Access, *Proc. 16-th ACM-SIGIR Conf.,* Pittsburgh (1993) 59–68.
12. KNUTH D.E., *The Art of Computer Programming, Vol* **I**, *Fundamental Algorithms*, Addison-Wesley, Reading, Mass. (1973).
13. SANKOFF D., KRUSKAL J.B., *Time Warps, String Edits, and Macromolecules: the Theory and Practice of Sequence Comparison*, Reading, Mass., Addison-Wesley Pub. Co. (1983).

# An Extension of the Periodicity Lemma to Longer Periods

## (Invited Lecture)

Aviezri S. Fraenkel[1] and Jamie Simpson[2]

[1] Department of Computer Science and Applied Mathematics
The Weizmann Institute of Science
Rehovot 76100, Israel
fraenkel@wisdom.weizmann.ac.il
http://www.wisdom.weizmann.ac.il/~fraenkel
[2] School of Mathematics, Curtin University
Perth WA 6001, Australia
simpson@cs.curtin.edu.au
http://www.cs.curtin.edu.au/~simpson

**Abstract.** The well-known periodicity lemma of Fine and Wilf states that if the word $x$ of length $n$ has periods $p, q$ satisfying $p + q - d \leq n$, then $x$ has also period $d$, where $d = \gcd(p, q)$. Here we study the case of *long* periods, namely $p + q - d > n$, for which we construct recursively a sequence of integers $p = p_1 > p_2 > \ldots > p_{j-1} \geq 2$, such that $x_1$, up to a certain prefix of $x_1$, has these numbers as periods. We further compute the maximum alphabet size $|A| = p + q - n$ of $A$ over which a word with long periods can exist, and compute the subword complexity of $x$ over $A$.

## 1    Introduction

We consider words over a finite alphabet, not necessarily binary.

**Definition 1.** Let $x$ be a word of length $|x| = n$ having integer periods $p$ and $q$. Throughout we put $d = \gcd(p, q)$, $k = \lfloor q/p \rfloor$, and assume

$$0 < p < q < n. \tag{1}$$

In particular, we disregard the period of length $n$ of $x$. If

$$p + q - d > n, \tag{2}$$

then $p$ and $q$ are called *long periods* of $x$. If $p$, $q$, $d$ satisfy $p + q - d \leq n$, then $p$ and $q$ are called *short periods* of $x$.

Note that in view of the natural requirement (1), long periods aren't all that long compared to short periods!

The following *Periodicity Lemma* of Fine and Wilf applies to words with short periods.

A. Amir and G.M. Landau (Eds.): CPM 2001, LNCS 2089, pp. 98–105, 2001.

**Lemma 1.** *If $x$ has short periods $p$, $q$, then $x$ has also period $d$.*

Many words have both short and long periods.

*Example 1.* The periods $< n = 9$ of $x_1 = 110111011$ are $4, 7, 8$. Of these, $(p, q) = (4, 8)$ are short (Lemma 1 gives no new information for this case); and $(4, 7)$ and $(7, 8)$ are long.

We wish to describe the structure of a word $x$ in terms of its long periods. Part of our discussion involves an inductive argument relating to two types of long periods which we call Type I and Type II. We'll see that if $x$ has Type II periods, then it can be described in terms of a prefix of $x$ with long periods, and that prefix, possibly, in terms of a still shorter prefix of $x$ with long periods. The process continues while we deal with prefixes with Type II periods, and terminates when we get a prefix with Type I periods.

## 2   Structure of Words with Long Periods

For a word $x$ of length $n$ with long periods $p < q$, put

$$p_1 = p, \; q_1 = q, \; d_1 = d, \; k_1 = \lfloor q_1/p_1 \rfloor, \; n_1 = n, \; x_1 = x. \tag{3}$$

For $i \geq 2$ apply Euclid's algorithm for creating a simple continued fraction expansion of $q_j/p_j$, to construct recursively the following items,

$$p_i = q_{i-1} - k_{i-1}p_{i-1}, \; q_i = p_{i-1}, \; d_i = \gcd(p_i, q_i), \; k_i = \lfloor q_i/p_i \rfloor, \tag{4}$$
$$n_i = n_{i-1} - k_{i-1}p_{i-1}, \; x_i = x_1[1 \ldots n_i].$$

**Definition 2.** For $i \geq 1$, the long periods $p_i$ and $q_i$ of the word $x_i$ are Type I if

$$n_i \leq (k_i + 1)p_i, \quad \text{i.e.,} \quad n_{i+1} \leq q_{i+1}, \tag{5}$$

and Type II if

$$n_i > (k_i + 1)p_i, \quad \text{i.e.,} \quad n_{i+1} > q_{i+1}. \tag{6}$$

*Example 2.* For Example 1, the periods $(p_1, q_1) = (4, 7)$ are Type II, and $(7, 8)$ are Type I.

**Lemma 2.** (i)  *We have $d_i = d_1$ for all $i \geq 2$. Let $i \geq 1$. If $p_i$, $q_i$, $n_i$ satisfy (1) and (2), then $p_i \nmid q_i$ ($p_i$ does not divide $q_i$), $p_i \geq 3$, and*

$$q_i - p_i + 1 \leq k_i p_i \leq q_i - 1. \tag{7}$$

(ii)  *For all $i \geq 1$, $p_i + q_i - n_i$ is a constant.*
(iii)  *Let $i \geq 2$. If $p_{i-1}$, $q_{i-1}$ are Type II periods of $x_{i-1}$, then $p_i$, $q_i$, $n_i$ satisfy (1) and (2).*
(iv)  *If $p_1, q_1$ are long periods of $x_1$, then there exists a smallest $j \in \mathbb{Z}_{>0}$ for which $n_j \leq (k_j + 1)p_j$, i.e., $p_j, q_j$ are Type I periods of $x_j$.*
(v)  *For this smallest value of $j$ we have $p_{j+1} < n_{j+1} \leq q_{j+1}$.*

**Proof.** (i) $d_i = \gcd(p_i, q_i) = \gcd(q_{i-1} - k_{i-1}p_{i-1}, p_{i-1}) = \gcd(p_{i-1}, q_{i-1})$. This implies $d_i = d_1$. We have (1) and (2) $\Longrightarrow q_i < n_i < p_i + q_i - d_i \Longrightarrow d_i < p_i \Longrightarrow p_i \nmid q_i$. Then (7) follows from the fact that $q_i/p_i$ is not an integer. Also $p_i > 1$. If $p_i = 2$, then $d_i = 1$, so (1) and (2) imply $q_i < n_i < q_i + 1$, a contradiction. Hence $p_i \geq 3$.

(ii)  By (3), (4), $p_{i+1} + q_{i+1} - n_{i+1} = p_i + q_i - n_i$.

(iii)  From (4) and (6), $n = n_{i-1} - k_{i-1}p_{i-1} > p_{i-1} = q_i$, which is one part of (1). The hypothesis of (ii) implies that $p_{i-1}, q_{i-1}$ satisfy (1) and (2), so (7) holds with $i$ replaced by $i - 1$. Hence $1 \leq p_i = q_{i-1} - k_{i-1}p_{i-1} < p_{i-1} = q_i$, which are the other two parts of (1).

As we just remarked, $p_{i-1}, q_{i-1}$ satisfy (2). Hence we have by (ii) and (i), $p_{i-1} - q_{i-1} - n_{i-1} = p_i + q_i - n_i > d_{i-1} = d_i$, which is (2).

(iv)  By (4), $0 < n_i < n_{i-1}$ and $(k_i + 1)p_i \geq 2$ for all $i$. The well-ordering principle then implies that $j$ with the desired property exists.

(v)  The right side is the right side of (5) and the left side follows from (1), (4).  ∎

**Lemma 3.** *Definitions 1 and 2 are consistent.*

**Proof.** For Type I periods, (2) and (5) do not imply one another in either direction. For Type II periods, (2) and (6) imply $kp \leq q - d - 2 \Longrightarrow q + 1 - p \leq q - d - 2$ (by(7)) $\Longrightarrow d \leq p - 3$. Now $d = p - 2 \Longrightarrow d|2 \Longrightarrow d \in \{1, 2\} \Longrightarrow p \in \{3, 4\}$. Similarly, $d = p - 1 \Longrightarrow p = 2$. Also $p = 1$ is excluded since $p \nmid q$. Hence for a Type II period we need $p \geq 5$ or $p = 4$, $d = 1$. We show now that this requirement is satisfied "automatically".

Let there be a word with long periods $p = 4 < q$ and $d = 2$. Then $q = 4t + 2$, $t \in \mathbb{Z}_{>0}$. By (1), (2), $q < n < q + 2$. Thus $n = q + 1$. Hence $n = 4t + 3 < (\lfloor q/4 \rfloor + 1)4 = 4t + 4$, so $(p, q)$ are Type I periods. We can see similarly that $p \in \{2, 3\}$ (whence $d = 3$) implies that $(p, q)$ are Type I periods.

Note that (1) and (6) do not imply one another in either direction.

In conclusion, Definition 1 is consistent with a word with long periods to be either Type I or Type II; and Type I or Type II words are consistent with Definition 1.  ∎

**Lemma 4.** *Let $x_1$ be a word of length $n_1$ with long periods $q_1 > p_1$, $k_1 = \lfloor q_1/p_1 \rfloor$. Then for $g_1 \in \{0, \ldots, k_1\}$ we have,*

$$x_1[i] = x_1[i + g_1 p_1] \tag{8}$$

*for $i = 1, \ldots, n_2 = n_1 - k_1 p_1$. Moreover, the prefix $x_2 = x_1[1 \ldots n_2]$ of $x_1$ has period $p_2 = q_1 - k_1 p_1$.*

**Proof.** Note that (1) and the right hand side of (7) imply the inequalities:

$$n_2 = n_1 - k_1 p_1 \geq 2 \tag{9}$$

and $q_1 - k_1 p_1 \geq 1$. Since $x_1$ has period $p_1$, (8) follows. For verifying the second statement we have to show: $x_1[i] = x_1[i + p_2]$ for $i = 1, \ldots, n_1 - k_1 p_1 - (q_1 - k_1 p_1) = n_1 - q_1 \geq 1$ (by (1)). Indeed, since $x_1$ has period $q_1$ we have for $i \in [1, n_1 - q_1]$,

$$x_1[i] = x_1[i + q_1] = x[i + q_1 - k_1 p_1] = x[i + p_2],$$

where the second equality follows from the $p_1$-periodicity of $x_1$.     ∎

**Corollary 1.** (i) *If $x_j$ has Type I periods $p_j < q_j$, then $x_j = (x_{j+1}z)^{k_j}x_{j+1}$, where the border $x_{j+1}$ has period $p_{j+1}$, and $z = x_j[n_{j+1} + 1 \ldots p_j]$.*
(ii) *If $x_1$ has Type II periods $p_1 < q_1$, then $x_2$ is a word with long periods $p_2$ and $q_2$ which satisfy (1), namely, $n_2 > q_2 > p_2 > 0$.*

**Proof.** (i) Inequalities (9) and (5) imply $|x_{j+1}| > 0$, $|z| \geq 0$. Evidently $|x_{j+1}z| = p_j$. The result of the first part of (i) now follows from the structure of the word (8), which has period $p_j$.
   (ii) Follows directly from Lemma 2(iii).     ∎

*Example 3.* Let $x_1 = 1101110111$, with $n_1 = 10$. The periods $< 10$ are $4, 8, 9$, where $(4, 8)$ is short; $(4, 9)$, $(8, 9)$ are long, of Type I. Corollary 1(i) for $(4, 9)$ (with $k = 2$) states that $x_2 = 11$, $z = 01$, so $x_1 = (1101)^2 11$, where $x_2$ has period $p_2 = 1$. Corollary 1(ii) is illustrated, say, by $x_1 = (1102)^2 11$, which has the same parameters, but a larger alphabet. With respect to $(p_1, q_1) = (8, 9)$, Corollary 1(ii) implies that we can take $x_1 = (11023456)11$.

We give two more examples for illustrating Corollary 1(ii).

*Example 4.* $x_1 = 11011101$ ($n_1 = 8$) has Type I periods $p_1 = 4$, $q_1 = 7$, so $k_1 = 1$, $p_2 = 3$, $n_2 = 4$ and $z$ is empty. By Corollary 1(i), $x_1 = (1101)^1 1101$. By Corollary 1(ii) also $x_1 = (1201)^1 1201$ has the same parameters.

*Example 5.* $x_1 = 101101110110$ ($n_1 = 12$) has Type I periods $p_1 = 7$, $q_1 = 10$, so $k_1 = 1$, $p_2 = 3$, $n_2 = 5$, $x_2 = 10110$ and $z = 11$. By Corollary 1(i), $x_1 = ((10110)(11))^1 10110$. By Corollary 1(ii), $x_1 = ((10210)(34))^1 10210$ has the same parameters.

*Example 6.* Referring to Example 2, Corollary 1(iii) (for $(p_1, q_1) = (4, 7)$) states that $x_2 = 11011$ of length $n_2 = 5$ has long periods $q_2 = 4$ and $p_2 = 3$, which are, in fact, Type I periods.

Corollary 1(i) and (ii) describe the structure of words with Type 1 periods.

Corollary 1(iii) suggests an iterative procedure for expressing the structure of $x_1$ in terms of the shorter of a sequence of long periods $p_i$ of its prefixes $x_i$. The recursion terminates when the smallest $j$ is reached for which $n_j \leq (k_j + 1)p_j$. That is, the recursion terminates when the triple $(p_j, q_j, n_j)$ corresponds for the first time to Type I periods.

**Lemma 5.** *Let $s \in \{1, \ldots, j-1\}$, and let $j \geq 2$ be fixed. For $g_\ell = 0, \ldots, k_\ell$ ($1 \leq \ell < j$), put $E_s = g_s p_s + \ldots + g_{j-1} p_{j-1}$. Then for $i = 1, \ldots, n_j$, $E_s + i$ assumes all the integer values in the integer interval $[1, n_s]$.*

**Proof.** Descent on $s$. For $s = j - 1$, $E_{j-1} + i = i + g_{j-1} p_{j-1}$. For fixed values of $g_{j-1} \in \{0, \ldots, k_{j-1}\}$, we let $i$ range over $[1, n_j]$. This produces the intervals $I_0 = [1, n_j], I_1 = [1 + p_{j-1}, n_j + p_{j-1}], \ldots, I_{k_{j-1}} = [1 + k_{j-1} p_{j-1}, n_j + k_{j-1} p_{j-1}]$. Since $n_j + k_{j-1} p_{j-1} = n_{j-1}$, the first interval begins with 1 and the last ends with $n_{j-1}$. Moreover, every two consecutive intervals overlap: $I_t$ ends with $n_j + t p_{j-1}$, and $I_{t+1}$ begins with $1 + (t+1) p_{j-1}$ and we have $1 + (t+1) p_{j-1} \leq n_j + t p_{j-1} = n_{j-1} - k_{j-1} p_{j-1} + t p_{j-1}$ by (6). Thus the union of these intervals is $[1, n_{j-1}]$, as required.

Suppose that we have already showed that for $s > 1$, $E_s + i$ assumes all the values in $[1, n_s]$. Now $E_{s-1} + i = i + g_{s-1} p_{s-1} + \ldots + g_{j-1} p_{j-1} = i + g_{s-1} p_{s-1} + E_s$. For fixed values of $g_{s-1} \in \{0, \ldots, k_{s-1}\}$, we let $i$ range again over $[1, n_j]$. This produces the intervals $[1 + E_s, n_j + E_s], [1 + p_{s-1} + E_s, n_j + p_{s-1} + E_s], \ldots, [1 + k_{s-1} p_{s-1} + E_s, n_j + k_{s-1} p_{s-1} + E_s]$. Let $I$ denote the union of these intervals. By the induction hypothesis,

$$I = [1, n_s] \cup [1 + p_{s-1}, n_s + p_{s-1}] \cup \ldots \cup [1 + k_{s-1} p_{s-1}, n_s + k_{s-1} p_{s-1}].$$

Since $n_s + k_{s-1} p_{s-1} = n_{s-1}$, the first of the intervals begins with 1 and the last ends with $n_{s-1}$. Moreover, every two consecutive intervals overlap: $1 + (t+1) p_{s-1} \leq n_s + t p_{s-1} = n_{s-1} - k_{s-1} p_{s-1} + t p_{s-1}$ by (6), proving the assertion. ∎

**Corollary 2.** (i) $n_j + k_1 p_1 + \ldots + k_{j-1} p_{j-1} = n_1$,
(ii) $n_j + k_2 p_2 + \ldots + k_{j-1} p_{j-1} = n_2$,
(iii) *For $i \in [1, n_1]$, $g_\ell \in \{0, \ldots, k_\ell\}$ ($1 \leq \ell < j$), $E_1 + i = i + g_1 p_1 + \ldots + g_{j-1} p_{j-1}$ assumes all the values in $[1, n_1]$.*

**Proof.** By (3), (4), $n_j + k_1 p_1 + \ldots + k_{j-1} p_{j-1} = n_{j-1} + k_1 p_1 + \ldots + k_{j-2} p_{j-2} = \ldots = n_2 + k_1 p_1 = n_1$, proving the first identity. The second identity is proved similarly. The third part is the case $s = 1$ of Lemma 5. ∎

The following is our main result.

**Theorem 1.** *Let $x_1$ be a word with long periods $p_1 < q_1$; $j$ as defined in Lemma 2(iii). Then the prefix $x_j = x_1[1 \ldots n_j]$ of $x_1$ is a word with Type I periods $p_j, q_j$ satisfying $n_j > q_j > p_j$. For $i = 1, \ldots, n_j$ we have,*

$$x_1[i] = x_1[i + g_1 p_1 + \ldots + g_{j-1} p_{j-1}], \tag{10}$$

*for all choices of $g_\ell \in \{0, \ldots, k_\ell\}$, $\ell = 1, \ldots, j - 1$.*

**Proof.** Note that $i + g_1p_1 + \ldots + g_{j-1}p_{j-1} \leq n_j + k_1p_1 + \ldots + k_{j-1}p_{j-1} = n_1$ by Corollary 2(i), so (10) is well-defined. If $x_1$ has Type II periods $q_1 > p_1$, then Corollary 1(iii) implies that $x_2$ has long periods $p_2, q_2$ satisfying $n_2 > q_2 > p_2 > 0$. We proceed by induction on $j$. For $j = 2$ (i.e., the periods $p_2, q_2$ of $x_2$ are Type I), (8) is (10) and we are done.

For $j > 2$ we may apply the induction hypothesis to $x_2$, to conclude that for $i = 1, \ldots, n_j$,

$$x_1[i] = x_1[i + g_2p_2 + \ldots + g_{j-1}p_{j-1}]. \tag{11}$$

We have $i + g_2p_2 + \ldots + g_{j-1}p_{j-1} \leq n_j + k_2p_2 + \ldots + k_{j-1}p_{j-1} = n_2$ by Corollary 2(ii), so (11) is well-defined.

Since $x_1$ has period $p_1$, we can add $g_1p_1$ to $i$ on both sides of (11), to get:

$$x[i + g_1p_1] = x[i + g_1p_1 + g_2p_2 + \ldots + g_{j-1}p_{j-1}].$$

Relation (10) now follows from this and from (8). ∎

*Example 7.* Let $x_1 = 1101110110111011$ of length $n_1 = 16$. It has periods $7, 11, 14, 15$. Note that $(7, 11)$ are Type II. Then $(n_1, p_1, q_1, k_1) = (16, 7, 11, 1)$. We have $(n_2, p_2, q_2, k_2) = (9, 4, 7, 1)$, with $(4, 7)$ being Type II periods. Then $(n_3, p_3, q_3, k_3) = (5, 3, 4, 1)$, where $(3, 4)$ are Type I. Thus $j = 3$, so by Theorem 1, $x[i] = x[i + 7g_1 + 4g_2]$, $g_1, g_2 \in \{0, 1\}$ for $i = 1, \ldots, 5$.

## 3 Maximum Subword Complexity

A word $x$ with long periods $p, q$ can exist over various alphabets $A$. In this section we determine, for any given $x$ with long periods, the maximum alphabet size $|A|$ such that $x$ exists over $A$ and every letter of $A$ appears in $x$. An alphabet $A$ which is maximum in this sense and every letter of $A$ does appear in $x$ will be called a *proper alphabet with respect to $p, q$*. We then compute the subword complexity of $x$ over a proper alphabet with respect to $p, q$.

Note that for different long periods of the same word $x$ we will, in general, have a proper alphabet of different size, as well as a different subword complexity.

First, given $p_j, q_j, n_j$ satisfying (1), (2) and (5), we construct a proper alphabet $A$ and $x_j$ of size $n_j$ with Type I periods $p_j < q_j$ over $A$. Let $A = \{a_1, \ldots, a_{p_j+q_j-n_j}\}$ be an alphabet, where the $a_i$ are its distinct letters. Put $x_1[1 \ldots p_{j+1}] = a_1 \ldots a_{p_{j+1}}$. By Lemma 2(v), $p_{j+1} < n_{j+1}$. Let $x_1[p_{j+1} + i] = x_1[i]$ for $1 \leq n_{j+1} - p_{j+1} (= n_j - q_j)$. Then $x_{j+1} = x_1[1 \ldots n_{j+1}]$ is periodic with period $p_{j+1}$, consistent with Corollary 1(i). Set $x_1[n_{j+1} + 1 \ldots p_j] = z = a_{p_{j+1}} + 1 \ldots a_{p_j+q_j-n_j}]$.

Let $x_j = (x_{j+1}z)^{k_j}x_{j+1}$. Since $|x_{j+1}z| = p_j$, we have $|x_j| = k_jp_j + n_{j+1} = n_j$. Moreover, $x_j$ is periodic with period $p_j$. To show that $x_j$ is a word with Type I periods $p_j < q_j$, it suffices to show that it has period $q_j$. By the $p_{j+1}$-periodicity of $x_{j+1}$, $x_1[i] = x_1[i + p_{j+1}]$ for $i \in [1, n_{j+1} - p_{j+1}] = [1, n_j - q_j]$.

Also $x_1[i+p_{j+1}] = x_1[i+q_j - k_jp_j] = x_1[i+q_j]$ by the $p_j$-periodicity of $x_j$. Thus $x_1[i] = x_1[i+q_j]$ for $i \in [1, n_j - q_j]$.

The alphabet $A$ has size $p_j + q_j - n_j = p_1 + q_1 - n_1$ by Lemma 2(ii). In view of the $p_j$-periodicity of $x_j$, $A$ cannot be any larger. It is also proper.

Secondly, given $j \geq 2$, $p_i, q_i, n_i$ for $1 \leq i \leq j$, where $p_i, q_i$ are Type II periods for $1 \leq i < j$ and $p_j, q_j$ are Type I periods, we construct a proper alphabet $B$ and word $x_1$ with Type II periods $p_i < q_i$ ($1 \leq i < j$) and Type I periods $p_j, q_j$ over $B$. Since $x_1$ contains a Type I factor as a prefix by Theorem 1, $|B| \leq |A|$. We will see that actually $B = A$. We apply the procedure (4) to get $p_i, q_i, k_i, n_i, x_i$ for $1 \leq i \leq j$, where $j$ is as in Lemma 2(iv), and where $p_i < q_i$ are Type II periods for $1 \leq i < j$. Construct the first prefix $x_j = x_1[1 \ldots n_j]$ of $x_1$ over $A$, as above. Longer prefixes are constructed iteratively by descent. Suppose that for some $\ell \in \{2, \ldots, j\}$ we have already constructed $x_\ell = x_1[1 \ldots n_\ell]$. To construct $x_{\ell-1}$, put $x_1[i] = x_1[i - p_{\ell-1}]$ for $i = n_\ell + 1, \ldots, n_{\ell-1}$. Since the prefix $x_\ell$ of $x_{\ell-1}$ has period $q_\ell = p_{\ell-1}$ by hypothesis, the entire factor $x_{\ell-1}$ has also period $p_{\ell-1}$. We now show that it has also period $q_{\ell-1}$.

Let $i \in [1, n_{\ell-1}]$. Since $x_{\ell-1}$ has period $p_{\ell-1}$, we have $x_1[i + q_{\ell-1}] = x_1[i + q_{\ell-1} - k_{j-1}p_{j-1}] = x_1[i + p_\ell]$. Now $i + p_\ell \leq n_{\ell-1} - q_{\ell-1} + p_\ell = n_\ell$, so $x_1[i + p_\ell]$ for $i \in [1, n_{\ell-1}]$ lies in $x_\ell = x_1[1 \ldots n_\ell]$. Since $x_\ell$ has period $p_\ell$ by hypothesis, we get $x_1[i + q_{\ell-1}] = x_1[i]$, so $x_{\ell-1}$ has periods $p_{\ell-1} < q_{\ell-1}$ as required.

We have proved, constructively,

**Theorem 2.** *Any word $x_1$ of length $n_1$ with long periods $p_1 < q_1$ can be realized over a proper alphabet $A$ of size $|A| = p_1 + q_1 - n_1$.*

The subword complexity of a word $x_1$ with long periods $p_1 < q_1$ over a proper alphabet $A$ will be called *maximum subword complexity* with respect to $p_1, q_1$. We shall now compute the maximum subword complexity.

**Theorem 3.** *Let $x_1$ of length $n_1$ be a word with Type I periods $p_1, q_1$ over a proper alphabet $A$ with respect to $p_1, q_1$, and let $C(m)$ be its maximum subword complexity function with respect to $p_1, q_1$. Then*

$$C(m) = p_1 + q_1 - n_1 + m - 1 \quad for \quad 1 \leq m \leq n_1 - q_1 + 1,$$
$$C(m) = p_1 \quad for \quad n_1 - q_1 + 1 < m \leq n_1 - p_1 + 1,$$
$$C(m) = n_1 - m + 1 \quad for \quad n_1 - p_1 + 1 < m \leq n_1,$$

*and $|A| = p_1 + q_1 - n_1$.*

**Proof.** Theorem 2 implies that $|A| = C(1) = p_1 + q_1 - n_1$. We consider three cases.

(I) $1 \leq m \leq n_1 - q_1$. Induction on $m$. We have just seen that the result holds for $m = 1$. Assume it holds for $m - 1$ ($2 \leq m \leq n_1 - q_1 + 1$), i.e., $C(m-1) = p_1 + q_1 - n_1 + m - 2$. Thus there are $p_1 + q_1 - n_1 + m - 2$ distinct factors of length $m - 1$ in $x_1$. Since $x_1$ has period $p_1$, each of these factors has a copy whose first letter is in $x_1[1 \ldots p_1]$. Each of these copies is a prefix of a factor of length $m$, so $C(m) \geq p_1 + q_1 - n_1 + m - 2$. If we had $C(m) = C(m - 1)$, then $x_1$ would be periodic with period $p_1 + q_1 - n_1 + m - 2 \leq p_1 - 1$, contradicting the assumption of computing the subword complexity for a proper alphabet with respect to $p_1 < q_1$. Therefore $C(m) \geq p_1 + q_1 - n_1 + m - 1$.

We now show that this is actually an equality. We do this by showing that every factor of length $m$ has a copy in one of the $p_1 + q_1 - n_1 + m - 1$ factors which have their last letter in the interval $I = [n_1 - p_1 + 1, q_1 + m - 1]$.

Let $x_1[\ell \ldots \ell + m - 1]$ be any factor of $x_1$ of length $m$. Let $i$ be the smallest integer $\geq n_1 - p_1 + 1$ such that $x_1[i \ldots i + m - 1] = x_1[\ell \ldots \ell + m - 1]$. Since $x_1$ has period $p_1$, such $i$ does exist. We wish to show that $i \in I$. Suppose not. Then $i \geq q_1 + m$, so $i - q_1 \geq m \geq 2$. By the $q_1$-periodicity of $x_1$ we then have $x_1[i \ldots i + m - 1] = x_1[i - q_1 \ldots i + m - 1 - q_1]$.

By (7), $k_1 p_1 < q_1$, hence $x_1[i \ldots i + m - 1] = x_1[i - q_1 + k_1 p_1 \ldots i + m - 1 - q_1 + k_1 p_1]$. Now $i - (q_1 - k_1 p_1) < i$, and $i - (q_1 - k_1 p_1) \geq q_1 + m - q_1 + k_1 p_1 \geq k_1 p_1 + 2 > n_1 - p_1 + 1$, since $p_1, q_1$ are Type I. Thus $i - (q_1 - k_1 p_1)$ is smaller than $i$, yet has the desired properties. This contradiction shows that $i \in I$, so $C(m) = p_1 + q_1 - n_1 + m - 1$.

(II) $n_1 - q_1 < m \leq n_1 - p_1 + 1$. From the case $m = n_1 - q_1 + 1$ we see that the $p_1$ factors $x_1[1 \ldots n_1 - q_1 + 1], \ldots, x_1[p_1 \ldots n_1 - q_1 + p_1]$ of length $n_1 - q_1 + 1$ are all distinct. Therefore the same holds for the $p_1$ factors $x_1[1 \ldots m], \ldots, x_1[p_1 \ldots p_1 + m - 1]$ for every $m$ satisfying $p_1 + m - 1 \leq n_1$. There can be no other factors of this length, since $x_1$ has period $p_1$.

(III) $n_1 - p_1 + 1 < m \leq n_1$. As in the previous case, the $n_1 - m + 1$ factors $x[1 \ldots m], \ldots, x_1[n_1 - m + 1 \ldots n_1]$ of length $m$ are all distinct. Thus $C(m) = n_1 - m + 1$. ∎

# A Very Elementary Presentation of the Hannenhalli–Pevzner Theory

Anne Bergeron

LACIM, Université du Québec à Montréal
C.P. 8888 Succ. Centre-Ville, Montréal, Québec, Canada, H3C 3P8
`bergeron.anne@uqam.ca`

**Abstract.** In 1995, Hannenhalli and Pevzner gave a first polynomial solution to the problem of finding the minimum number of reversals needed to sort a signed permutation. Their solution, as well as subsequent ones, relies on many intermediary constructions, such as simulations with permutations on $2n$ elements, and manipulation of various graphs. Here we give the first completely elementary treatment of this problem. We characterize *safe reversals* and *hurdles* working directly on the original signed permutation. Moreover, our presentation leads to polynomial algorithms that can be efficiently implemented using bit-wise operations.

## 1 Introduction

In the last ten years, beginning with [6], many papers have been devoted to the subject of computing the *reversal distance* between two permutations. A *reversal* $\rho(i, j)$ transforms a permutation

$$\pi = (\ \pi_1 \ldots \underline{\pi_i \ \pi_{i+1} \ \ldots \ \pi_j} \ldots \pi_n \ )$$

$$\text{to } \pi' = (\ \pi_1 \ldots \pi_j \ \ldots \ \pi_{i+1} \ \pi_i \ldots \pi_n \ ).$$

and the reversal distance between two permutations is the minimum number of reversals that transform one into the other.

From a problem of unknown complexity, it graduated to an NP-Hard problem [2], but an interesting variant was proven to be polynomial [3]. In the *signed* version of the problem, each element of the permutation has a plus or minus sign, and a reversal $\rho(i, j)$ transforms $\pi$ to:

$$\pi' = (\ \pi_1 \ldots -\pi_j \ldots -\pi_{i+1} \ -\pi_i \ldots \pi_n \ ).$$

Permutations, and their reversals, are useful tools in the comparative study of genomes. The genome of a species can be thought of as a set of ordered sequences of genes – the ordering devices being the chromosomes –, each gene having an orientation given by its location on the DNA double strand. Different species often share similar genes that were inherited from common ancestors. However, these genes have been shuffled by mutations that modified the content of chromosomes, the order of genes within a particular chromosome, and/or

A. Amir and G.M. Landau (Eds.): CPM 2001, LNCS 2089, pp. 106–117, 2001.
© Springer-Verlag Berlin Heidelberg 2001

the orientation of a gene. Comparing two sets of similar genes appearing along a chromosome in two different species yields two (signed) permutations. It is widely accepted that the reversal distance between these two permutations faithfully reflects the evolutionary distance between the two species.

Computing the reversal distance of signed permutations is a delicate task since some reversals unexpectedly affect deep structures in permutations. In 1995, Hannenhalli and Pevzner proposed the first polynomial algorithm to solve it [3], developing along the way a theory of how and why some permutations were particularly resistant to sorting by reversals. It is of no surprise that the label *fortress* was assigned to specially acute cases.

Hannenhalli and Pevzner relied on several intermediate constructions that have been simplified since [4], [1], but grasping all the details remains a challenge. All the criteria given for choosing a *safe* reversal involve the construction of an associate permutation on $2n$ points, and the analysis of cycles and/or connected component of graphs associated with this permutation.

In this paper, we present a very elementary treatment of the sorting of the *oriented components* of a permutation, together with an elementary definition of the concept of *hurdle* that further simplifies the definition given in [4]. Our first algorithm is so simple that, for example, sorting a permutation of length 20, *by hand*, should be easy and straightforward.

The next section presents the basic algorithms. Section 3 contains the necessary links to the Hannenhalli–Pevzner theory, and the proofs of the claims of the Section 2. Finally, in the last section, we discuss complexity issues, and we give a *bit-vector* implementation of the sorting algorithm that runs in $\mathcal{O}(n^2)$.

## 2   Basic Sorting

The problem of sorting by reversal a signed permutation $\pi$ is to find $d(\pi)$, its reversal distance from the identity permutation $(+1 \; +2 \ldots +n)$. As usual, we will *frame* a permutation $\pi = (\pi_1 \; \pi_2 \ldots \pi_n)$ with 0 and $n+1$, yielding the permutation: $(0 \; \pi_1 \; \pi_2 \ldots \pi_n \; n+1)$.

Given a signed permutation $\pi = (0 \; \pi_1 \; \pi_2 \ldots \pi_n \; n+1)$, an *oriented pair* $(\pi_i, \pi_j)$ is a pair of adjacent integers, that is $|\pi_i| - |\pi_j| = \pm 1$, with opposite signs. For example, the oriented pairs of the permutation:

$$( \; 0 \; +3 \; +1 \; +6 \; +5 \; -2 \; +4 \; +7 \; )$$

are $(+1, -2)$ and $(+3, -2)$.

Oriented pairs are useful in the sense that they indicate reversals that create consecutive elements. For example, the pair $(+1, -2)$ induces the reversal:

$$( \; 0 \; +3 \; +1 \; \underline{+6 \; +5 \; -2} \; +4 \; +7 \; )$$

$$( \; 0 \; +3 \; +1 \; +2 \; -5 \; -6 \; +4 \; +7 \; )$$

creating the consecutive sequence $+1 \; +2$.

In general, the reversal induced by an oriented pair $(\pi_i, \pi_j)$ will be,

$$\rho(i, j - 1), \text{ if } \pi_i + \pi_j = +1, \text{ and}$$
$$\rho(i + 1, j), \text{ if } \pi_i + \pi_j = -1.$$

Note that reversals that create consecutive pairs of integer are always induced by oriented pairs. Such a reversal is called an *oriented* reversal. We define the *score* of an (oriented) reversal as the number of oriented pairs in the resulting permutation. For example, the score of the reversal:

$$( 0 \underline{+3 +1 +6 +5} -2 +4 +7 )$$

$$( 0 -5 -6 -1 -3 -2 +4 +7 )$$

is 4, since the resulting permutation has 4 oriented pairs. Computing the score of a reversal is tedious but elementary, and we will discuss efficient algorithms to do so in Section 4. The fact that oriented reversals have a beneficial effect on the ordering of a permutation suggests a first sorting strategy:

**Algorithm 1.** As long as $\pi$ has an oriented pair, choose the oriented reversal that has maximal score.

For example, the two oriented pairs of the permutation:

$$( 0 \underline{+3 +1 +6 +5} -2 +4 +7 )$$

are $(+1, -2)$, $(+3, -2)$, and their score are respectively 2 and 4. So we choose the reversal induced by $(+3, -2)$, yielding the new permutation:

$$( 0 -5 -6 -1 \underline{-3 -2} +4 +7 ) .$$

This permutation has now four oriented pairs $(0, -1)$, $(-3, +4)$, $(-5, +4)$ and $(-6, +7)$, all of which have score 2, except $(-3, +4)$. Acting on this pair yields:

$$( 0 -5 -6 \underline{-1} +2 +3 +4 +7 ) .$$

which has four oriented pairs. Note here that the score of the pair $(0, -1)$ is 0. The corresponding oriented reversal would produce a permutation with no oriented pair, and the algorithm would stop, in this case with an unsorted permutation. Fortunately, the pair $(-1, +2)$ has a positive – and maximal – score, and we get, in a similar way, the last two necessary reversals to sort the permutation:

$$( 0 -5 \underline{-6 +1} +2 +3 +4 +7 )$$

$$( 0 -5 \underline{-4 -3 -2 -1} +6 +7 )$$

$$( 0 +1 +2 +3 +4 +5 +6 +7 )$$

Interestingly enough, this elementary strategy is sufficient to optimally sort most random permutations and almost all permutations that arise from biological data. The strategy is also optimal, and we will prove in the next section the following claim.

**Claim 1:** If the strategy of Algorithm 1 applies $k$ reversals to a permutation $\pi$, yielding a permutation $\pi'$, then $d(\pi) = d(\pi') + k$.

The output of Algorithm 1 will be a permutation of positive elements. Most reversal applied to such permutations will create oriented pairs, but the choice of an optimal reversal is delicate. We discuss this problem in the next paragraph.

## 2.1    Sorting Positive Permutations

Let $\pi$ be a signed permutation with only positive elements, and assume that $\pi$ is *reduced*, that is $\pi$ does not contain consecutive elements. Suppose also that $\pi$ is framed by 0 and $n + 1$ and consider, as in [4], the circular order induced by setting 0 to be the successor of $n + 1$.

Define a *framed interval* in $\pi$ as an interval of the form:

$$i \ \ \pi_{j+1} \ \pi_{j+2} \ \ldots \pi_{j+k-1} \ \ i + k$$

such that all integers between $i$ and $i + k$ belong to the interval $[i \ldots i + k]$. For example, consider the permutation:

$$(\ 0\ 2\ 5\ 4\ 3\ 6\ 1\ 7\ ).$$

The whole permutation is a framed interval by construction. But we have also the interval: 2 5 4 3 6, which can be reordered as 2 3 4 5 6, and, by circularity, the interval 6 1 7 0 2, which can be reordered as 6 7 0 1 2, since 0 is the successor of 7.

**Definition 1.** If $\pi$ is reduced, a *hurdle* in $\pi$ is a framed interval that properly contains no framed interval.

**Claim 2:** Hurdles as defined in Definition 1 are the same hurdles that are defined in [3] and [4].

When a permutation has only one or two hurdles, one reversal is sufficient to create enough oriented pairs in order to completely sort the permutation with Algorithm 1. Two operations are introduced in [3], the first one is *hurdle cutting* which consist in reversing one internal element, say $\pi_{j+1}$, of a hurdle:

$$i \ \underline{\pi_{j+1}} \ \pi_{j+2} \ \ldots \pi_{j+k-1} \ i + k.$$

This reversal is sufficient to sort all the interval using Algorithm 1. For example, the following permutation contains only one hurdle:

$$(0\ 2\ 4\ 3\ 1\ 5).$$

The reversal of element 2 cuts the hurdle, and the resulting permutation

$$(0\ -2\ 4\ 3\ 1\ 5)$$

can be sorted with 4 reversals by Algorithm 1.

The second operation is *hurdle merging*, which acts on the end points of two hurdles:

$$i\ldots\underline{i+k}\ldots\underline{i'}\ldots i'+k'$$

and does the reversal $\rho(i+k, i')$. If a permutation has only two hurdles, merging them will produce a permutation that can be completely sorted by Algorithm 1.

Thus, for example, merging the two hurdles in the permutation

$$(0\ 2\ 5\ 4\ 3\ 6\ 1\ 7).$$

yields the permutation:

$$(0\ 2\ 5\ 4\ 3\ -6\ 1\ 7).$$

which can be sorted in 5 reversal using Algorithm 1.

Merging and cutting hurdles in a permutation that contains more than 2 hurdles must be managed carefully. Indeed, cutting some hurdles can create new ones!

**Definition 2.** A *simple* hurdle is a hurdle whose cutting decreases the number of hurdles. Hurdles that are not simple are called *super hurdles*.

For example, the permutation $(0\ 2\ 5\ 4\ 3\ 6\ 1\ 7)$ has two hurdles. Cutting the hurdle 2 5 4 3 6 yields the permutation,

$$(0\ 2\ 3\ 4\ 5\ 6\ 1\ 7)$$

which, by collapsing the sequence 2 3 4 5 6 is reduced to:

$$(0\ 2\ 1\ 3),$$

which has only one hurdle. However, the permutation $(0\ 2\ 4\ 3\ 5\ 1\ 6\ 8\ 7\ 9)$ contains two hurdles, and if one cuts the hurdle 2 4 3 5, the resulting reduced permutation will be

$$(0\ 2\ 1\ 3\ 5\ 4\ 6)$$

which still has two hurdles.

The following algorithm is adapted from [4], and is discussed originally in [3].

**Algorithm 2.** If a permutation has $2k$ hurdles, $k \geq 2$, merge any two non-consecutive hurdles. If a permutation has $2k+1$, $k \geq 1$, then if it has one simple hurdle, cut it; If it has none, merge two non-consecutive hurdles, or consecutive ones if $k = 1$.

Together with Algorithm 1, Algorithm 2 can be used to optimally sort any signed permutation. This completes the first part of the paper, and, in the next section, we turn to the task of proving our various claims.

## 3   Selected Results from the Hannenhalli–Pevzner Theory

The exposition of the complete results of the Hannenhalli–Pevzner theory is beyond the scope of this paper, and the reader is referred to the original paper [3], or the book on computational molecular biology by Pevzner [5]. Instead, we will show the soundness of our algorithms by directly using the *overlap graph* introduced in [4].

The first step in the construction of the overlap graph is to simulate a signed permutation on $n$ elements with an unsigned permutation on $2n$ elements. Each positive element $x$ in the permutation is replaced by the sequence $2x - 1\ 2x$, and each negative element $-x$ by the sequence $2x\ 2x - 1$. For example, the permutation:

$$\pi = (\ 0\ -1\ +3\ +5\ +4\ +6\ -2\ +7\ )$$

becomes:

$$\pi' = (\ 0\ 2\ 1\ 5\ 6\ 9\ 10\ 7\ 8\ 11\ 12\ 4\ 3\ 13\ )$$

Reversals $\rho(i, j)$ of $\pi$ are simulated by unsigned reversals $\rho(2i - 1, 2j)$ in $\pi'$.

The *overlap graph* associated with a permutation $\pi$ has $n$ vertices labeled by $(0, 1), (2, 3), \ldots, (2n, 2n + 1)$, with an edge between two vertices $(a, b)$ and $(c, d)$ iff, in the unsigned permutation, the interval corresponding to the positions of $a$ and $b$ overlaps – without proper containment – the interval corresponding to the positions of $b$ and $d$. For example, if one draws arcs joining the end points of the pairs $(0, 1), (2, 3), \ldots, (2n, 2n + 1)$ in the permutation $\pi'$:

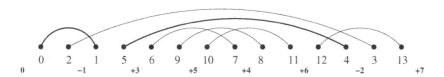

The overlap graph can then be easily drawn by tracing an edge for each intersecting arcs in the above diagram, yielding:

There is a natural bijection between the vertices of the overlap graph and pairs of adjacent integers $(\pi_i, \pi_j)$. in the original permutation. Indeed, a pair of adjacent integers will generate four consecutive integers in the unsigned permutation: $2x - 1$, $2x$, $2x + 1$, and $2x + 2$. The vertex $(2x, 2x + 1)$ is associated with

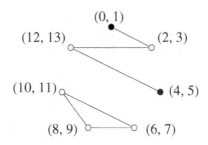

the pair $(\pi_i, \pi_j)$. For example, the oriented pair $(3, -2)$ in $\pi$ corresponds to the vertex $(4, 5)$ in the overlap graph. Vertices corresponding to oriented pairs are naturally called *oriented vertices*, and are denoted by solid dots in the overlap graph. Moreover, we will refer to the *reversal induced by a vertex* meaning the reversal induced by the oriented pair corresponding to the vertex. The following facts, mostly from [4], pinpoint the important relations between a signed permutation and its overlap graph.

**Fact 1**: *A vertex has an odd degree iff it is oriented.*

*Proof.* Let $2x - 1$, $2x$, $2x + 1$, and $2x + 2$, be the four integers associated with the oriented pair $(\pi_i, \pi_j)$. Since $\pi_i$ and $\pi_j$ have different signs, the positions of $2x$ and $2x + 1$ will not have the same parity in the unsigned permutations. Thus, the interval between $2x$ and $2x + 1$ has an odd length, implying that it overlaps an odd number of other intervals. On the other hand, any interval that overlaps an odd number of intervals must have an odd length. Therefore the positions of its end points must have different parities, implying that the corresponding pair of adjacent integers is oriented. ∎

**Fact 2**: *If one performs the reversal corresponding to an oriented vertex $v$, the effect on the overlap graph will be to complement the subgraph of $v$ and its adjacent vertices.*

*Proof.* The reversal corresponding to an oriented vertex $v$ has the effect of collapsing the associated interval, thus $v$ will become isolated. Let $u$ and $w$ be two intervals overlapping $v$, meaning that exactly one of their end points lies in the interval spanned by $v$. The reversal induced by $v$ will reverse these two points. Here, a picture is worth a thousand words: ∎

**Fact 3**: *If one performs the reversal corresponding to an oriented vertex $v$, each vertex adjacent to $v$ will change its orientation.*

*Proof.* Since $v$ is oriented, it has an odd number $2k + 1$ of adjacent vertices. Let $w$ be a vertex adjacent to $v$, with $j$ neighbors also adjacent to $v$. With the reversal, $w$ will loose $j + 1$ neighbors, and gain $2k - j$ new ones. Thus the degree of $w$ will change by $2k - 2j - 1$, changing its orientation. ∎

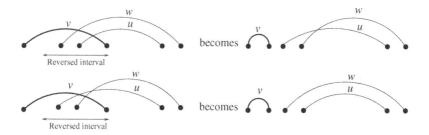

**Fact 4**: *The score of the oriented reversal corresponding to an oriented vertex $v$ is given by:*

$$T + U - O - 1$$

*where $T$ is the total number of oriented vertices in the graph, $U$ is the number of unoriented vertices adjacent to $v$, and $O$ is the number of oriented vertices adjacent to $v$.*

*Proof.* This follows trivially from the preceding facts.    ∎

We now state a basic result that is proven, in different ways, both in [3] and [4]. Define an *oriented component* of the overlap graph as a connected component that contains at least one oriented vertex. A *safe* reversal is a reversal that does not create new unoriented components, except for isolated vertices.

**Proposition 1 (Hannenhalli and Pevzner).** *Any sequence of oriented safe reversals is optimal.*

The difficulties in sorting oriented components lie in the detection of safe reversals. Hannenhalli and Pevzner deal with the problem by computing several statistics on cycles and breakpoints of various graphs. Kaplan et al. solve it by searching for particular cliques in the overlap graph. The next theorem argues that the elementary strategy of choosing the reversal with maximal score is optimal, thus proving Claim 1.

**Theorem 1.** *An oriented reversal of maximal score is safe.*

*Proof.* Suppose that vertex $v$ has maximal score, and that the reversal induced by $v$ creates a new unoriented component $C$ containing more than one vertex. At least one of the vertices in $C$ must have been adjacent to $v$, since the only edges affected by the reversal are those between vertices adjacent to $v$. So, let $w$ be a vertex formerly adjacent to $v$ and contained in $C$, and consider the scores of $v$ and $w$:

$$score(v) = T + U - O - 1$$
$$score(w) = T + U' - O' - 1$$

All unoriented vertices adjacent to $v$ must be adjacent to $w$. Indeed, an unoriented vertex adjacent to $v$ and not to $w$ will become oriented, and connected to $w$, contrary to the assumption that $C$ is unoriented. Thus, $U' \geq U$.

All oriented vertices adjacent to $w$ must be adjacent to $v$. If this was not the case, an oriented vertex adjacent to $w$ but not to $v$ would remain oriented, again contradicting the fact that $C$ is unoriented. Thus, $O' \leq O$.

Now, if both $O' = O$ and $U' = U$, vertices $v$ and $w$ have the same set of adjacent vertices, and complementing the subgraph of $v$ and its adjacent vertices will isolate both $v$ and $w$. Therefore, we must have that $score(w) > score(v)$, which is a contradiction. ∎

### 3.1   Hurdles

In this section, we assume that $\pi$ is a positive and reduced permutation. These assumptions are equivalent to say that the overlap graph has no oriented components – all of which can be cleared by Algorithm 1 –, and no isolated vertices.

Consider again the circular order, this time on the interval $[0..2n-1]$, induced by setting 0 to be the successor of $2n - 1$. The *span* of a set of vertices $X$ in the overlap graph is the minimum interval that contains, in the circular order, all the intervals of vertices in $X$. For example, the three connected components of the following overlap graph have spans $[4, 15] = [4\ 7\ 8\ 11\ 12\ 9\ 10\ 13\ 14\ 5\ 6\ 15]$, $[8, 13] = [8\ 11\ 12\ 9\ 10\ 13]$, and $[16, 3] = [16\ 1\ 2\ 17\ 0\ 3]$.

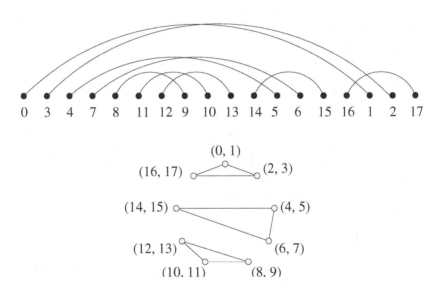

Hurdles are defined in [3] as unoriented components which are minimal with respect to span inclusion. Moreover, in [4], it is shown that the span of a connected component is always of the form $[2i, 2j - 1]$. The following Lemmas and Theorem detail the relationships between connected components and framed intervals, substantiating the second claim of Section 1.

**Lemma 1.** *Framed intervals of the form $[i, j]$ in a permutation on $n$ elements are in one-to-one correspondence with framed intervals of the form $[2i, 2j - 1]$ in the corresponding unsigned permutation on $2n$ elements.*

*Proof.* The end points $i$ and $j$ of a framed interval $[i, j]$ will be mapped, respectively, to the pairs $2i - 1, 2i$, and $2j - 1, 2j$. All integers between $i$ and $j$ appear in the interval $[i, j]$, if and only if all the integers between $2i$ and $2j - 1$ appear in the interval $[2i, 2j - 1]$.

**Lemma 2.** *Any framed interval* $[2i, 2j - 1]$ *is the span of a union of connected components.*

*Proof.* If $[2i, 2j - 1]$ is a framed interval, it contains exactly the integers between $2i$ and $2j - 1$, thus the only arcs in this interval are: $(2i, 2i + 1)$, $(2i + 2, 2i + 3), \ldots, (2j - 2, 2j - 1)$, and no other arc intersects this set. Therefore, the corresponding set of vertices is not connected to any other vertex. ∎

**Lemma 3.** *The span* $[2i, 2j - 1]$ *of a connected component is always a framed interval.*

*Proof.* If vertex $(2i, 2i + 1)$ is connected to $(2j - 2, 2j - 1)$, there must be a sequence of intersecting arcs linking $2i$ to $2j - 1$.

Any arc with only one end point between $2i$ and $2j - 1$ would therefore intersect one of the arcs in the sequence, so there are none. Thus, if integer $2k$ is in the interval, then $2k + 1$ is also in the interval, and if $i \le k < j$, then $2k + 2$ is also in the interval. ∎

**Theorem 2.** *If $\pi$ is reduced, an unoriented component is minimal iff its span is a framed interval that contains no other.*

*Proof.* By Lemma 3, the span of a connected component is always a framed interval. If the component is minimal with respect to span inclusion, by Lemma 2, its span cannot contain properly another framed interval.

On the other hand, a framed interval $[2i, 2j-1]$ that contains no other yields a single connected component $C$ whose vertices endpoints are exactly the integers between $2i$ and $2j - 1$. Thus the vertices of $C$ are consecutive on the circle, and component $C$ is minimal. ∎

The main consequence of Theorem 2 is to give an elementary characterization of the concept of hurdles, that does not need the construction of the overlap graph.

## 4   Settling Scores

We now turn to the problem of computing the score of a reversal. In the following, we assume that the overlap graph of a permutation is explicitly represented as a bit matrix, and consider, for a vertex $v$, the bit vector $\boldsymbol{v}$ whose $w^{th}$ coordinate is 1 iff vertex $v$ is adjacent to vertex $w$.

We also assume that the score and parity of each vertex are stored in the vectors $\boldsymbol{s}$ and $\boldsymbol{p}$, respectively. It is only necessary to keep track of the *variable* part of the score that is, for both oriented and unoriented vertices, the expression $U_v - O_v$, where $U_v$ is the number of unoriented vertices adjacent to $v$, and $O_v$ is the number of oriented vertices adjacent to $v$.

As an example, consider the permutation

$$( \ 0 \ +3 \ +1 \ +6 \ +5 \ -2 \ +4 \ +7 \ )$$

its overlap graph. and its associated data structure.

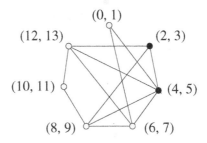

| | $(0,1)$ | $(2,3)$ | $(4,5)$ | $(6,7)$ | $(8,9)$ | $(10,11)$ | $(12,13)$ |
|---|---|---|---|---|---|---|---|
| $(0,1)$ | 0 | 0 | 1 | 1 | 0 | 0 | 0 |
| $(2,3)$ | 0 | 0 | 1 | 0 | 1 | 0 | 1 |
| $(4,5)$ | 1 | 1 | 0 | 1 | 1 | 0 | 1 |
| $(6,7)$ | 1 | 0 | 1 | 0 | 1 | 0 | 1 |
| $(8,9)$ | 0 | 1 | 1 | 1 | 0 | 1 | 0 |
| $(10,11)$ | 0 | 0 | 0 | 0 | 1 | 0 | 1 |
| $(12,13)$ | 0 | 1 | 1 | 1 | 0 | 1 | 0 |
| $\boldsymbol{p}$ | 0 | 1 | 1 | 0 | 0 | 0 | 0 |
| $\boldsymbol{s}$ | 0 | 1 | 3 | 2 | 0 | 2 | 0 |

Clearly, initializing such a structure requires $\mathcal{O}(n^2)$ steps. Given it, finding the reversal with maximal score is trivial. The interesting part is the effect of a reversal on the structure.

Suppose that we choose to perform the reversal corresponding to the oriented vertex $v$. Since $v$ will become unoriented and isolated, each vertex incident to $v$ will automatically gain a point of score. Thus we first set:

$$\boldsymbol{s} \leftarrow \boldsymbol{s} + \boldsymbol{v}$$

Next, if $w$ is a vertex incident to $v$, the vector $\boldsymbol{w}$ is changed according to:

$$\boldsymbol{w} \leftarrow \boldsymbol{w} \oplus \boldsymbol{v}$$

where $\oplus$ is the *exclusive or* bit wise operation. Indeed, vertices adjacent to $w$ after the reversal are either existing vertices that were not adjacent to $v$, or vertices that were adjacent to $v$ but not to $w$. The exceptions to this rule are $v$ and $w$ themselves, and this problem can easily be solved by setting the diagonal bit $\boldsymbol{v}_v$ to 1, and $\boldsymbol{w}_w$ to 1 before computing the direct sum.

Now, if $w$ is unoriented, each of its former adjacent vertices will *loose* one point of score, since $w$ will become oriented, and each of its new adjacent vertices will *loose* one point of score. Note that a vertex that *stays* connected to $w$ will loose a total of two points. We can thus write the effect of the change of orientation of $w$ on the vector $\boldsymbol{s}$ of scores as:

$$\boldsymbol{s} \leftarrow \boldsymbol{s} - \boldsymbol{w}$$
$$\boldsymbol{w}_w \leftarrow 1$$
$$\boldsymbol{w} \leftarrow \boldsymbol{w} \oplus \boldsymbol{v}$$
$$\boldsymbol{s} \leftarrow \boldsymbol{s} - \boldsymbol{w}$$

where the subtractions are performed component wise on the vector of scores. If $w$ is oriented, the losses are converted to gains, so the subtractions are converted to additions.

Finally, the parity vector $\boldsymbol{p}$ is updated by the equation:

$$\boldsymbol{p} \leftarrow \boldsymbol{p} \oplus \boldsymbol{v}.$$

The algorithm requires eventually $\mathcal{O}(n)$ *vector* operations for each reversal, and these operations can be implemented very efficiently using bit operations widely available in processors.

Kaplan et al., in [4], give an algorithm based on [1] that clears the hurdles from a permutation in less than $\mathcal{O}(n^2)$, and that can be used in conjunction with the above algorithm. But since we already have an extensive representation on the overlap graph, we can use it to keep track of the connected components, and to detect hurdles.

# References

1. Piotr Berman, Sridhar Hannenhalli, *Fast Sorting by Reversal.* CPM 1996, LNCS 1075: 168-185.
2. Alberto Caprara, *Sorting by reversals is difficult.* RECOMB 1997, ACM Press: 75-83.
3. Sridhar Hannenhalli, Pavel A. Pevzner, *Transforming Cabbage into Turnip: Polynomial Algorithm for Sorting Signed Permutations by Reversals.* JACM 46(1): 1-27 (1999).
4. Haim Kaplan, Ron Shamir, Robert Tarjan, *A Faster and Simpler Algorithm for Sorting Signed Permutations by Reversals.* SIAM J. Comput. 29(3): 880-892 (1999).
5. Pavel Pevzner, Computational Molecular Biology, MIT Press, Cambridge, Mass., 314 p., (2000).
6. David Sankoff, *Edit Distances for Genome Comparisons Based on Non-Local Operations.* CPM 1992, LNCS 644,: 121-135.

# Tandem Cyclic Alignment

Gary Benson*

Department of Biomathematical Sciences
The Mount Sinai School of Medicine
New York, NY 10029-6574
benson@ecology.biomath.mssm.edu

**Abstract.** We present a solution for the following problem. Given two sequences $X = x_1 x_2 \cdots x_n$ and $Y = y_1 y_2 \cdots y_m$, $n \leq m$, find the best scoring alignment of $X' = X^k[i]$ vs $Y$ over all possible pairs $(k, i)$, for $k = 1, 2, \ldots$ and $1 \leq i \leq n$, where $X[i]$ is the cyclic permutation of $X$, $X^k[i]$ is the concatenation of $k$ complete copies of $X[i]$ ($k$ tandem copies), and the alignment must include all of $Y$ and all of $X'$. Our algorithm allows any alignment scoring scheme with *additive* gap costs and runs in time $O(nm \log n)$. We have used it to identify related tandem repeats in the *C. elegans* genome as part of the development of a multi-genome database of tandem repeats.

## 1   Introduction

### 1.1   Problem Description

The problem we solve is the following:
**Tandem Cyclic Alignment**

**Given:** Two sequences $X = x_1 x_2 \cdots x_n$ and $Y = y_1 y_2 \cdots y_m$, $n \leq m$ and an alignment scoring scheme with *additive* gap costs.

**Find:** The best scoring alignment of $X' = X^k[i]$ vs $Y$ over all possible pairs $(k, i)$, for $k = 1, 2, \ldots$ and $1 \leq i \leq n$, where $X[i]$ is the cyclic permutation of $X$,

$$X[i] = x_i x_{i+1} \cdots x_n x_1 \cdots x_{i-1},$$

$X^k[i]$ is the concatenation of $k$ complete copies of $X[i]$ ($k$ tandem copies), and the alignment must include all of $Y$ and all of $X'$.

Let $X$ and $Y$ be two strings over an alphabet $\Sigma$. An alignment of $X$ and $Y$ (see section 3.1 for an example) is a pair of equal length sequences $\hat{X}, \hat{Y}$ over the alphabet $\Sigma \cup \{-\}$ where $-$ is a *gap* character and $X, Y$ are obtained from $\hat{X}, \hat{Y}$ by removing the gap characters. An alignment can be interpreted as a sequence $Q$ of *edit operations* [6] that transform $X$ into $Y$. The allowed operations are 1) insert a symbol into $X$, 2) delete a symbol in $X$ and 3) replace a symbol in $X$ with a (possibly identical) symbol from $\Sigma$. A scoring scheme defines a weight

---

* Partially supported by NSF grants CCR-9623532 and CCR-0073081.

A. Amir and G.M. Landau (Eds.): CPM 2001, LNCS 2089, pp. 118–130, 2001.

for each possible operation and the alignment score is the sum of the weights assigned to the operations in $Q$.

There are two widely used classes of scoring schemes, 1) *distance scoring*, in which identical replacement has weight $= 0$, all other operations have weight $\geq 0$ and the best alignment has minimum score, and 2) *similarity scoring*, in which "good" replacements have weight $> 0$, all other operations have weight $\leq 0$ and the best alignment has maximum score. Within these classes, scoring schemes are further characterized by the treatment of gap costs. A *gap* is the result of the deletion of one or more consecutive characters in one of the sequences (insertion into the other sequence). *Additive* gap costs assign a constant weight to each of the consecutive characters. Other gap functions have been found useful for biological sequences, including affine gap costs ($\alpha + \beta k$ for a gap of $k$ consecutive characters where $\alpha$ and $\beta$ are constants) and concave gap costs ($\alpha + \beta f(k)$ where $f()$ is a concave function such as square root). The solution in this paper assumes a scoring scheme with *additive* gap costs. For ease of discussion, we will, for the remainder of the paper, assume distance scoring although the results apply as well to similarity scoring.

Our motivation for this problem arises from an ongoing effort to construct a multi-genome database of tandem repeats (TRDB). A central task is the clustering of tandem repeats into families *i.e.* repeats that occur in different locations in a genome but have identical or very similar underlying patterns. Grouping these repeats will facilitate identification and study of their common properties. Tandem repeat families have been detected in both prokaryotes and eukaryotes, including the *E. coli*, *S. cerevisiae*, *C. elegans* and human genomes.

Clustering requires an effective and consistent means of measuring the similarity or distance between repeats. Standard comparison methods are not easily applied to tandem repeats because they contain *repetitive, approximate copies* of an underlying pattern. In addition, comparison of related repeats often reveals a scrambling of the left to right order of the slightly different internal copies. An accurate comparison method should be insensitive to copy number and copy order and we have therefore chosen to abstract the repeats as either 1) consensus patterns or 2) profiles and then compare them using alignment.

Because repeat copies are adjacent, the *designation of first position* in a consensus or profile *is arbitrary*. This is not just a theoretical abstraction, the number of copies in a repeat is often not a whole number and distinct repeats which are obviously similar often do not start and end at the same relative positions. Therefore, comparison must allow cyclic permutation of one pattern so that its first position can be arbitrarily aligned with any position in the other.

Once families are constructed, we can determine interfamily evolutionary relationships by comparing patterns from different families. In particular, we can determine if one pattern consists of multiple approximate copies of the other, again with the property of cyclic permutation. It is this comparison that Tandem Cyclic Alignment addresses.

## 1.2   Background

The Tandem Cyclic Alignment problem is a merger of two classes of pairwise alignment problems, 1) *tandem alignment*, in which one of the sequences consists of an indeterminate number of tandem copies of a pattern and 2) *cyclic alignment*, in which cyclic permutation of one of the patterns is allowed. Three related problems from these classes are:

**Pattern Local, Text Global Tandem Alignment.** Given a pattern $X$, a text $Y$ and a scoring scheme for alignment, find the best scoring alignment of $X' = X^k[1]$ vs $Y$ over all $k = 1, 2, \ldots$, where all of $Y$ must occur in the alignment, but where the part of $X'$ aligned with $Y$ need not contain a whole number of copies of $X[1]$. The alignment, rather, may start and end on any index of $X$.

**Pattern and Text Global Tandem Alignment.** Given a pattern $X$, a text $Y$, an index $i$, $1 \leq i \leq |X|$, and a scoring scheme for alignment, find the best scoring alignment of $X' = X^k[i]$ vs $Y$, over all $k = 1, 2, \ldots$, where all of $Y$ and all of $X'$ must occur in the alignment.

**Cyclic Global Alignment.** Given sequences $X$ and $Y$ and a scoring scheme for alignment, find the best scoring alignment of $X[i]$ vs $Y$ over all possible $i$, $1 \leq i \leq |X|$ where all of $Y$ and exactly one whole (cyclically permuted) copy of $X$ must occur in the alignment.

The tandem alignment problems are both solved by wraparound dynamic programming (WDP) [8,3] in $O(mn)$ time when the scoring function has additive or affine gap costs. The cyclic alignment problem can be solved naively in $O(n^2m)$ time by separately computing the alignment of $X[i]$ vs $Y$ for every value of $i$. Maes [7] presented a $O(nm \log n)$ time solution for scoring schemes with additive gap costs by observing that there exists a set of best scoring alignments, one for each $1 \leq i \leq n$ such that the alignments are pairwise *non-crossing* (below). Landau, Myers and Schmidt [5] gave a $O(n + km)$ algorithm for unit cost differences (edit distance) when the score of the best alignment is bounded by $k$. Their algorithm, although theoretically efficient, has a large constant factor and is difficult to implement because it requires constructing a suffix tree preprocessed for least common ancestor queries. Schmidt [9] gave a rather complicated $O(nm)$ algorithm for similarity scoring where each insertion/deletion character costs $-s$ and match/mismatch weights are in the interval $[-s, m]$ for fixed positive integer values $m$ and $s$. This method can not be used to compute general distance scores more efficiently than the Maes algorithm.

It seems natural to adapt the Maes solution to our problem, except for one difficulty: in tandem cyclic alignment, there may be no set of best scoring alignments which are all pairwise non-crossing. What this means is that the number of copies of $X$ used in an alignment can vary depending on the starting position $i$. (For an example see Section 3.1). We show, though, that no alignment can

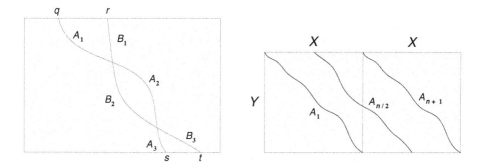

**Fig. 1.** (left) Alignments do not cross; (right) the Maes algorithm.

cross the "same" alignment more than once. This leads to a $O(mn \log n)$ time solution using adaptations of the Maes algorithm and WDP.

The remainder of the paper is organized as follows. In section 2 we give brief descriptions of the non-crossing alignments property, the Maes algorithm and wraparound dynamic programming. In section 3 we give the main theorem about crossing tandem cyclic alignments. In section 4 we then apply this property to obtain our algorithm. Finally, in section 5 we show an example from our analysis applied to tandem repeats from the *C. elegans* genome.

## 2    Preliminaries

### 2.1    Non-crossing Alignments

When gap costs are additive, a simple non-crossing property of optimal paths in the two dimensional alignment matrix applies [4,1,7]. We present one variation appropriate for this paper.

**Definition.** Two paths $A$ and $B$ in an alignment matrix *cross* if there exist two rows $e$ and $f$ such that in row $e$ all matrix cells in path $A$ are left of all cells in path $B$ and in row $f$ all cells in path $B$ are left of all cells in path $A$. The paths share one or more common cells where they cross.

Note that *sharing cells* is not the same as crossing.

**Property:** Given an alignment matrix (see figure 1, left) and four cells $q, r, s$ and $t$ with $q$ left of $r$ in the top row and $s$ left of $t$ in the bottom row, for any optimal scoring path $A$ from $q$ to $s$, there exists an optimal scoring path $B$ from $r$ to $t$ such that the two paths do not cross.

*Proof.* By contradiction, suppose all optimal scoring paths from $r$ to $t$ cross $A$. Let $B$ be one such path. $A$ and $B$ must cross an even number of times. Consider the separate subpaths (labeled $A_2$ and $B_2$ in the figure) in which $A$ is first right of $B$ proceeding from the top row.

Claim 1: Cost of $B_2$ is equal or worse than cost of $A_2$. Otherwise $A$ is not optimal because joining subpaths $A_1$, $B_2$ and $A_3$ is better.

Claim 2: Cost of $B_2$ is better than cost of $A_2$. Otherwise, joining subpaths $B_1$, $A_2$ and $B_3$ gives a path with score no worse than $B$, but which does not cross $A$. Such a path was assumed not to exist.

Clearly Claims 1 and 2 lead to a contradiction.

## 2.2    The Maes Algorithm for Cyclic Global Alignment

The Maes algorithm [7] capitalizes on the non-crossing property to *bound* the area of the alignment matrix that must be computed for each index $i$ in the alignment of $X[i]$ vs $Y$.

First the alignment of $X[1]$ vs $Y$ (call it $A_1$) is computed in $O(nm)$ time. A new matrix is then constructed which uses two concatenated copies of $X$ vs $Y$ (figure 1, right). The alignment $A_1$ shifted right (call it $A_{n+1}$) optimally aligns the second copy of $X$ with $Y$.

$A_1$ and $A_{n+1}$ bound any alignments which start and end between them. Specifically, they bound the alignment of $X[n/2]$ vs $Y$ (call it $A_{n/2}$). It is easy to see that this procedure can be followed recursively, for a logarithmic number of steps, subdividing $X$ into halves, then fourths, etc. always at the midpoints between bounding alignments. In each step, the alignment score calculations in a matrix cell are computed once, except for matrix cells on a bounding path, where they are computed twice (once for the computation in the interval to the left and once to the right), yielding $O(nm \log n)$ as the overall time of the algorithm.

## 2.3    Wraparound Dynamic Programming (WDP)

WDP [8,3] models the similarity computation of $Y$ with an unrestricted number of copies of $X$ while using an alignment matrix of size $nm$ rather than of size $m^2$, *i.e.* using only one copy of $X$. WDP computes in matrix $S[i, j]$ the optimal score that would be obtained by aligning $Y_1 \cdots Y_i$ with $X^* X_1 \cdots X_j$, where $X^*$ indicates zero or more tandem copies of $X$. The correctness proof hinges on the observation that any optimal scoring alignment will *not* contain a single deletion of $h \geq n$ characters of $X$ ( $n = |X|$). This is so because otherwise, another alignment exists, identical except for having a deletion of only $h - n$ characters, and possesing a better score. Since WDP examines all alignments with deletions in $X$ of size $< n$, it produces the optimal scoring alignment.

The technique involves computing *two* passes through each row. In both passes, all cells but the first are treated normally. In the first pass, cell $S[i, 1]$ (corresponding to $Y_i$ and $X_1$) is given the better of 1) a value derived from the cell $S[i-1, 1]$, the first cell in the row above (corresponding to a deletion of $Y_i$) and 2) a value derived from cell $S[i-1, n]$, the last cell in the row above (corresponding of a pairing of $Y_i$ and $X_1$). This later is a wraparound value. In the second pass, $S[i, 1]$ receives the maximum of 1) its current value, and 2) a value derived from $S[i, n]$, the last cell in its row (corresponding to a deletion of $X_1$). This is also a wraparound value.

# 3   Crossing Tandem Cyclic Alignments

Here we show that although tandem cyclic alignments may cross, no alignment can cross the "same" alignment more than once. "Same" in this case means an alignment that has been shifted one or more full copies of the pattern left or right, similar to the shifting of the alignment $A_1$ to become $A_{n+1}$ in the Maes algorithm.

## 3.1   An Example

Let the pattern $X$ and text $Y$ be

$$X = gaccga \quad Y = accgatacgagacccgagaacgagaccg.$$

Then, using an edit distance scoring scheme, (match=0, mismatch, indel=1), the *only* best scoring alignment of $X^k[1]$ vs $Y$ (with a score of 6) uses 5 copies of $X[1]$:

```
         *                        *
     gaccga gaccga ga-ccga gaccga gaccga
     -accga ta-cga gacccga gaacga gaccg-
```

while the *only* best scoring alignment of $X^h[4]$ vs $Y$ (with a score of 8) uses 4 copies of $X[4]$:

```
         *                    *
     --cgagac cgaga-c cgagac cgaga-c-
     accgata- cgagacc cgagaa cgagaccg
```

Since the alignments use a different number of copies of $X$, they cross and there is no set of best scoring pairwise non-crossing alignments.

## 3.2   No Alignment Crosses the "Same" Alignment more than Once

**Theorem 1.** *Given two sequences, $X$ and $Y$ and an index $i$, $1 \leq i \leq n$, let $c_i$ be the number of copies of $X[i]$ in a best scoring alignment of $X' = X^k[i]$ vs $Y$ over all $k = 1, 2, \ldots$, where all of $Y$ and all of $X'$ must be included in the alignment (i.e. $c_i = k$ in that best scoring alignment). Then, for any $j$, $1 \leq j \leq n$ there exists a best scoring alignment of $X' = X^h[j]$ vs $Y$ over all $h = 1, 2, \ldots$ such that $c_j = h$ in that alignment and $|c_i - c_j| \leq 1$.*

In other words, if a best scoring alignment of $X^k[i]$ vs $Y$ uses $c$ copies of $X[i]$, then for any $j$, there is a best scoring alignment of $X^h[j]$ vs $Y$ which uses one of $\{c - 1, c, c + 1\}$ copies of $X[j]$.

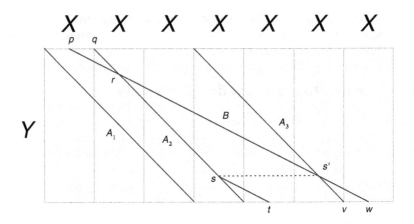

**Fig. 2.** An illustration of Theorem 1.

*Proof.* Assume that $|c_i - c_j| > 1$. We show a contradiction if $c_j > c_i + 1$. A similar argument holds for $c_j < c_i - 1$. Refer to figure 2. Let $A_1$ be a best scoring alignment of $X^{c_i}[i]$ vs $Y$ and let $B$ be a best scoring global alignment of $X^{c_j}[j]$ vs $Y$ with smallest $c_j$ and let $c_j > c_i + 1$. Let $A_2$ be a duplication of $A_1$ shifted to the right by one copy of $X[i]$ and let $A_3$ be the rightmost shifted copy crossed by $B$. (By assumption, $A_2$ and $A_3$ are distinct.)

Let $r$ and $s'$ be the points, respectively, where $B$ crosses $A_2$ and $A_3$ and let $s$ correspond to the point on $A_2$ matching $s'$. Call $x \succ y$ the part of an alignment from point $x$ to point $y$. Let $s \succ t$ be a duplication of $s' \succ w$ in $B$ shifted to the left. Finally call $cost(x \succ y)$ the alignment score for $x \succ y$ (and recall that we are assuming distance scoring so that smaller cost is better).
Claim 1: $cost(r \succ s') \geq cost(r \succ s)$. Otherwise, piece together $q \succ r$, $r \succ s'$ and $s' \succ v$ to get a better scoring alignment than $A_2$. But, $A_2$ is optimal.
Claim 2: $cost(r \succ s') < cost(r \succ s)$. Otherwise, piece together $p \succ r$, $r \succ s$ and $s \succ t$ to get an alignment with score no worse than $B$, and using less than $c_j$ copies of $X[j]$. But $B$ uses minimal copies.
Claims 1 and 2 produce a contradiction.

## 4    The Tandem Cyclic Alignment Algorithm

The Tandem Cyclic Alignment problem is solved in three steps. Each step requires first finding a *guide* alignment and then implementing the Maes algorithm using the guide as alignment $A_1$. Since we are using tandem copies of the pattern, the Maes algorithm will be implemented as *Bounded Wraparound Dynamic Programming* (BWDP) which is described following the outline of the main algorithm:

**Step 1:** Use pattern and text global WDP (section 1.2) to find the best scoring alignment of $X^k[1]$ vs $Y$ for $k = 1, 2, \ldots$. Call this alignment $A$. Let the number of copies of $X$ used in $A$ be $c$. Use $A$ as $A_1$ in the BWDP version of the Maes algorithm to find the remaining best scoring non-crossing alignments for $X^c[i]$ vs $Y$ for $i = 2, \ldots, n$. Call the best scoring alignment from this step $B_c$. Save $B_c$.

Call $c^*$ the number of copies in the solution to the tandem cyclic alignment problem. At this point, we have saved the best from the set of cyclic alignments each of which uses $c$ copies of $X$, but we do not know if $c = c^*$. However, by Theorem 1, we know that $c^* \in \{c - 1, c, c + 1\}$.

**Step 2:** Using $A$ (from step 1) and a copy of $A$ shifted to the *right* one pattern length, find the best scoring alignment of $X^{c+1}[1]$ vs $Y$ using BWDP. Call this alignment $A^+$ (figure 3). Use $A^+$ as $A_1$ in the BWDP version of the Maes algorithm to find the remaining best scoring, non-crossing alignments for $X^{c+1}[i]$ vs $Y$ for $i = 2, \ldots, n$. Call the best scoring alignment from this step $B_{c+1}$. Save $B_{c+1}$.

**Step 3:** Using $A$ (again from step 1) and a copy of $A$ shifted to the *left*, find the best scoring alignment of $X^{c-1}[1]$ vs $Y$ using BWDP. Call this alignment $A^-$. Use $A^-$ as $A_1$ in the BWDP version of the Maes algorithm to find the remaining best scoring, non-crossing alignments for $X^{c-1}[i]$ vs $Y$ for $i = 2, \ldots, n$. Call the best scoring alignment from this step $B_{c-1}$. Save $B_{c-1}$.

**Step 4:** Choose the best scoring alignment from $B_c$, $B_{c+1}$ and $B_{c-1}$.

**Time Complexity.** Each of the three main steps starts with finding a guide alignment using WDP or BWDP in time $O(nm)$. Then each step finds the remaining alignments using the BWDP version of the Maes algorithm in $O(nm \log n)$ time. The total time is therefore $O(nm \log n)$.

## 4.1   Bounded Wraparound Dynamic Programming (BWDP)

BWDP is computed in an alignment matrix $W[i, j]$ of size $(m+1)(2n+1)$, *i.e.* it uses *two* copies of $X$. We are given two alignments $L$ and $R$ as boundaries. We assume that $L$ and $R$ are both alignments of $X^c[j]$ vs $Y$ for a fixed $c$ and different $j$ and that neither crosses outside the pair of "master" bounding alignments $X^c[1]$ vs $Y$ and its duplicate shifted right one copy of $X$ (or alternately $X^c[1]$ vs $Y$ and its duplicate shifted left).

We use $L$ and $R$ to obtain, for each row $i = 0, \ldots, m$ in the matrix, the left-most, $L[i]$, and the rightmost, $R[i]$, boundary columns between which alignment scores will be computed. Finally, we are given an index $k$, $L[0] \leq k \leq R[0]$ as the starting column for the alignment. Figure 4, left side, shows the bounded computation as it would appear if we use an unrestricted number of copies of $X$. Note that for some $i$, $L[i]$ may be left of the starting position $k$ or $R[i]$ may be right of the ending position $k + cn$. In this case, we contract the boundaries to the appropriate values.

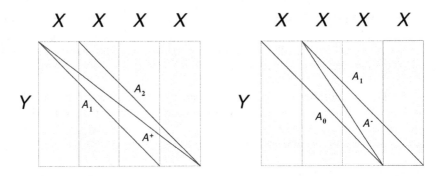

**Fig. 3.** Finding the guide alignments: $A^+$ (left) and $A^-$ (right). BWDP achieves the same result using only one copy of $X$.

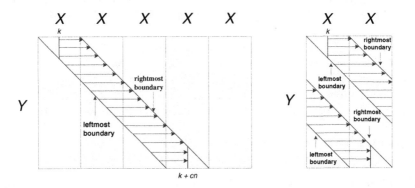

**Fig. 4.** Bounded wraparound dynamic programming simulates computation with an unrestricted number of copies of $X$.

Figure 4, right side, shows the same bounded computation, but this time, in an array which contains only two copies of $X$. When the boundaries exceed the first two copies of $X$, the computations wrap around. There is only one question which must be addressed to guarantee that the BWDP result is the same as the unrestricted-copies-of-$X$ result. Can the boundaries collide or cross in the space of only two copies of $X$ (*i.e.* can $R$ catch up with $L$ as they wrap around)?

**Definition:** The *width* of the computation space is the maximum difference $R[i] - L[i] + 1$, $i = 0, \ldots, m$ in the unrestricted-copies-of-$X$ computation.

**Lemma 2.** The maximum width of the computation space is $2n$.

*Proof.* Note that all the boundaries must lie within the "master" boundaries so it suffices to show the maximum width for the masters. Since the master alignments are duplicates separated by one copy of $X$, corresponding positions in the alignments are $n$ columns apart, *i.e.* they occur in columns $c$ and $c + n$ (figure 5). Consider a row $i$ in which the alignment moves horizontally in the matrix (a deletion of characters in $X$). If $L[i]$ is in column $c$, then $R[i]$ is in column $c+n+h$ where $h$ is the length of the horizontal move. As stated previously (section 2.3), $h \leq n - 1$, so the maximum possible width of the computation is $2n$.

**Corollary 3.** The bounding alignments can not collide in the BWDP array which has 2n columns (excluding column zero which is not used after the boundaries wrap around).

## 5  Application to Tandem Repeats from the *C. Elegans* Genome

We implemented the tandem cyclic alignment algorithm and used it to analyze the consensus patterns of tandem repeats found in the *C. elegans* genome. Our goal was to identify pairs of patterns, one of which is a multiple approximate copy of the other. The individual repeats were obtained with the Tandem Repeats Finder (TRF) program [2] which identifies approximately 25,000 tandem repeats in *C. elegans*. From these, we selected nearly 5300 repeats in four groups with nominal pattern sizes of 70 base pairs (bp), 51bp, 35bp, and 17bp. (Repeats within a group had pattern sizes within 3bp of the nominal size.) Each repeat was paired with every repeat from all groups of smaller nominal pattern size (except 70 bp which was not paired with 51 bp and 51 bp which was not paired with 35 bp) and tandem cyclic alignment was run on all pairs.

DNA consists of two strands, one of which is the *reverse complement* of the other. In a reverse complement, the direction of the sequence is reversed, the

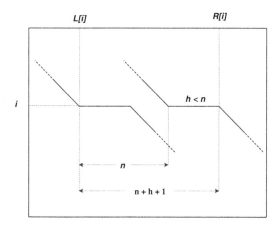

**Fig. 5.** The maximum width of the computation space is $2n$.

As and Ts are swapped and the Cs and Gs are swapped. Since similar repeats may have been found as reverse complements, for every pair, we first align the patterns as they appear and then reverse complement one pattern and align them again.

The total number of alignments (including reverse complements) was 16.3 million. On a 500 Mhz PC, the alignments took 17 and 3/4 hours or just over an hour per million alignments.

Figure 6 illustrates an example of the relationships found in this search. It consists of the alignment of the consensus patterns from 3 repeats, from widely scattered genomic locations, that were found to be related. Pattern 1176 is 19 bp long. Four copies are shown (indicated by alternate shading). Pattern 197 is 34 bp long. Two copies are shown. Pattern 8989 is 68 bp long. One copy is shown. For the latter two patterns, only the differences with the top pattern are indicated. A dash $(-)$ means that there is no character that corresponds to the character in the top line. This is an insertion (into the top line) or a deletion (from the second or third line).

Notice first that pattern 8989 is almost identical to two copies of pattern 197, differing only in the substitution of A and C. TRF is able to find such closely related patterns (of different sizes) for the same repeat and in fact reported that repeat 8989 also had a pattern of size 34 that was identical to 197.

Next compare patterns 1176 and 197. Pattern 197 consists of two copies of 1176 with 6 differences. Because the two halves of 197 were quite different, TRF did not report a pattern of size 19 (or any other similar size) for repeat 197. The following scenario (highly speculative!) may have occurred. Two identical 19 bp copies existed in an ancestral repeat and one of those copies was extensively mutated, including the deletion of 4 nucloetides. The resulting pair of repeats, now 34 bp long was subsequently transposed to another location in the genome where it duplicated, forming a tandem repeat. Some evidence for this scenario exists in one of the copies of repeat 1176 which contains the adjacent two nucleotide deletion seen in pattern 197.

Without the tandem cyclic alignment algorithm, the relationship between patterns 197 and 1176 would be less clear. Our goal in comparing these patterns is to obtain an *accurate* measure of the distance between them. If we used the pattern local, text global tandem alignment algorithm (section 1.2), the results would depend on the presentation of the text *i.e.* the cyclic permutation in which it appears in the input. If pattern 197 (the text) were presented starting just after a deletion (at the AAT following the two nucleotide deletion for example), then the algorithm would fail to align any characters from pattern 1176 with the positions of the two deleted characters (which do not actually occur in pattern 197). On the other hand, if pattern 197 were presented starting as it does in figure 6 at GCAA, then all the deleted characters will appear in the alignment. The alignment score will be different in these two cases.

Adjustment of alignment score obtained by the pattern local, text global tandem alignment algorithm is possible, but not necessarily straightforward. As an illustration, consider the pattern local, text global alignment (left below) and

**Fig. 6.** An example of aligned consensus patterns of different sizes.

the tandem cyclic alignment (right below) of text $X$ and pattern $Y$:

$$X = gggtgg \quad Y = gggt.$$

```
gggt gg          gggt gg--
gggt gg          gggt gggt
```

The former can be transformed into the later by merely adding the deleted characters and the cost for the gap. But, a different situation occurs if the last character of the text is changed to $t$:

$$\hat{X} = gggtgt.$$

```
gggt gt          gggt g--t
gggt gg          gggt gggt
```

Now the score changes not only by the cost of a gap, but also by the loss of a mismatched pair and the gain of a matched pair. More complicated situations are not difficult to construct. The tandem cyclic alignment algorithm however always gives the correct score without manipulation regardless of the presentation of the pattern or text.

## 6    Conclusion

We have defined a new alignment problem, tandem cyclic alignment and provided an algorithm which solves this problem in $O(nm \log n)$ time for two sequences of length $n$ and $m$, $n \leq m$ when using any alignment scoring scheme with additive gap costs. The algorithm was used to compare tandem repeats from the *C. elegans* genome in order to identify pairs of repeats with an evolutionary relationship where the consensus pattern of one is a multiple of the consensus pattern of the other. We showed an example of one such relationship which would not be reliably recognized with other alignment algorithms.

# References

1. A. Apostolico, M.J. Atallah, L.L. Larmore, and S. Mcfaddin. Efficient parallel algorithms for string editing and related problems. *SIAM J. Comput.*, 19:968–988, 1990.
2. G. Benson. Tandem repeats finder: a program to analyze DNA sequences. *Nucleic Acids Research*, 27:573–580, 1999.
3. V. Fischetti, G. Landau, J. Schmidt, and P. Sellers. Identifying periodic occurrences of a template with applications to a protein structure. In A. Apostolico, M. Crochemore, Z. Galil, and U. Manber, editors, *Proc. 3rd annual Symp. on Combinatorial Pattern Matching, Lecture Notes in Computer Science*, volume 644, pages 111–120. Springer-Verlag, 1992.
4. H. Fuchs, Z. Kedem, and S. Uselton. Optimal surface reconstruction from planar contours. *CACM*, 20:693–702, 1977.
5. G.M. Landau, E.W. Myers, and J.P. Schmidt. Incremental string comparison. *SIAM J. Comput.*, 27:7–82, 1998.
6. V.I. Levenshtein. Binary codes capable of correcting deletions, insertions and reversals. *Soviet Phys. Dokl.*, 10:707–710, 1966.
7. M. Maes. On a cyclic string-to-string correction problem. *Information Processing Letters*, 35:73–78, 1990.
8. W. Miller and E. Myers. Approximate matching of regular expressions. *Bulletin of Mathematical Biology*, 51:5–37, 1989.
9. J. Schmidt. All shortest paths in weighted grid graphs and its application to finding all aproximate repeats in strings. *SIAM J. Comput.*, 27:972–992, 1998.

# An Output-Sensitive Flexible Pattern Discovery Algorithm

Laxmi Parida, Isidore Rigoutsos, and Dan Platt

IBM Thomas J. Watson Research Center
Yorktown Heights, NY10598, USA
parida@us.ibm.com

**Abstract.** Given an input sequence of data, a motif is a repeating pattern, possibly interspersed with "dont care" characters and a flexible motif could have a variable (as opposed to fixed) number of "dont care" characters. Given a sequence of records with $F$ fields each, an association rule is a common set of $f$ fields, $f \leq F$, with identical (or similar) repeating values. The data in either case could be a sequence of characters or sets of characters or even real values. It is well known that the *number* of motifs or association rules, say $N$, could potentially be exponential in the size of the input sequence or number of records, say $n$. In this paper we present a new algorithm to discover all flexible motifs or association rules in the input. A novel feature of this algorithm is that its running time is linear in the size of the output (ignoring polylog factors). More precisely, the complexity of the algorithm is $O((n^5 + N)\log n)$. This is the first algorithm for motif discovery with a proven output sensitive complexity bound. The discovery algorithm works in two phases: in the first phase it detects a linear number of core motifs in time polynomial in the input size $n$ and in the second phase it detects all the remaining motifs $N'$ in $O(N' \log n)$ time. The core motifs of the first phase are also characterized as being those of "highest specificity": loosely speaking, a pattern with higher specificity has less "dont care" characters. Some applications (for instance the ones that require the study of those portions of the input sequence that contribute to the non-gapped regions of motifs ) require only the core motifs. Hence for such applications, the first phase of the algorithm suffices. However, the general problem is of use in motif discovery tasks in gene or protein sequences, or discovery of association rules from gene expression data or in data mining.

## 1 Introduction

Given a sequence of data, a "rigid" motif is a repeating pattern, possibly interspersed with dont-care characters, that has the same length in every occurrence in the input sequence. Pattern or motif discovery in data is widely used as a means of "understanding" large volumes of data such as DNA or protein sequences [Roy92,JCH95,NG94,SV96,WCM+96,SNJ95,RF98,Cal00]. We refer the reader to [BJEG98] for an extensive survey of existing motif discovery algorithms and implementations.

A. Amir and G.M. Landau (Eds.): CPM 2001, LNCS 2089, pp. 131–142, 2001.
© Springer-Verlag Berlin Heidelberg 2001

Allowing the motifs to have a variable number of gaps (or "dont care" characters), termed flexible motifs, further increases the expressibility of the motifs [Roy92,JCH95,NG94,SNJ95]. For example given a string $s = abcdaXcdabbcd$, $m = a.cd$ is a rigid pattern that occurs twice in the data at positions 1 and 5 in $s$. In the above example, the flexible motif would occur three times at positions 1, 5 and 9. At position 9 the dot character represents two gaps instead of one. The definition of the flexible pattern used in this paper is an extension of the *generalized regular pattern* described in [BJEG98] in the sense that the discovery algorithm can also handle a sequence of real numbers.

The task of discovering patterns must be clearly distinguished from that of matching a given pattern in a database. In the latter situation we know what we are looking for, while in the former we do not know what is being sought. Typically, the higher the self similarity in the sequence, greater is the number of patterns or motifs in the data. Motif discovery on such data, such as repeating DNA or protein sequences, is indeed a source of concern since these exhibit a very high degree of self-similarity (repeating patterns). The *number* of rigid motifs could potentially be exponential in the size of the input sequence and in the case where the input is a sequence of real numbers, there could be uncountably infinite number of motifs (assuming two real numbers are equal if they are within some $\delta > 0$ of each other).

Usually, this problem of a large number of motifs is tackled by pre-processing the input, using heuristics, to remove the repeating or self-similar portions of the input or using a "statistical significance" measure [JCH95,BG98]. However, due to the absence of a good understanding of the domain, there is no consensus over the right model to use. Thus there is a trend towards model-less motif discovery in different fields [WCM+96,RF98,SV96]: we use the same approach to the pattern discovery problem in this paper. There have been empirical evidence showing that the run time is linear in the output size for the rigid motifs [RF98,Cal00] and experimental comparisons between available implementations [Cal00,BJEG98]. However, none of the currently known algorithms have proven output sensitive complexity bounds and the only known complexity bounds are all exponential in the input size $n$.

In order to apply motif discovery techniques to real life situations one has to deal with the fact that in many applications the input is known with a margin of error. Many amino acids in protein sequences, for instance, are easily interchanged by evolution without loss of function [DSO78]. Also the use of distance matrices in the context of DNA sequences such as PAM [SD78] and BLOSUM [HH92] is common. For example, a character $a$ can be viewed as $a$ or $b$ for pattern detection purposes (but a $b$ cannot be viewed as an $a$). In all these situations it is possible to view the input as a string of sets of characters instead of just characters. For instance a sequence of the form baccta can be viewed as b{a,b}cct{a,b}. In some other applications, the input is an array of real numbers as in the case of micro-array chips [De96,Le96,BB99,We97,BDY99], and two distinct real numbers are deemed identical for pattern detection purposes, if they are within some given $\delta > 0$ of each other. Conventional motif discovery algorithms deal with these situations in an ad hoc manner and there is a need for a uniform framework such that the same algorithm can tackle all the scenarios described above.

The main contributions of the paper are as follows. We present an $O((n^5 + N) \log n)$ algorithm for the flexible motif discovery problem, where $N$ is the size of the output. This is the first known algorithm with a proven output-sensitive complexity bound even in the restricted case of rigid motifs. Moreover, we provide a uniform framework to handle situations where the input is known with a margin of error. As a result our flexible-pattern discovery algorithm effortlessly generalizes to sequences of sets (homologous characters) or real numbers. This significantly increases the domain of applicability of our motif discovery algorithm.

The pattern discovery algorithm is based on identifying a unique core set of maximal motifs. This idea was introduced in [PRF+00] for the case of rigid motifs. It was shown that the size of the core is $O(n)$, although the total number of maximal motifs could be exponential in the input size. In [Par00] a uniform framework is presented that encompasses both rigid and flexible motifs. It is further shown that even in the case of flexible motifs there exists a unique core set of motifs which is of size $O(n)$. This result holds even under added constraints on the set of motifs which is central to our algorithm here.

The algorithm works in two phases: We first detect the core motifs that takes no more than $O(n^5 \log(n))$ time for flexible motifs in the worst case. In the second phase we detect all the other motifs in time linear, ignoring polylog factors, in the number of the new motifs, using the core motifs.

## 2   Some Preliminary Definitions

Let $s$ be a sequence of sets of characters from an alphabet $\Sigma$, '.' $\notin \Sigma$. The '.' is called a "dont care" or a dot character and any other element is called *solid*. Also, $\sigma$ will refer to a singleton character or a set of characters from $\Sigma$. For brevity of notation, a singleton set is not enclosed in curly braces. For example let $\Sigma = \{A, C, G, T\}$, then $s_1 = ACTGAT$ and $s_2 = \{A, T\}CG\{T, G\}$ are two possible sequences. The $j^{\text{th}}$ $(1 \leq j \leq |s|)$ element of the sequence is given by $s[j]$. For instance in the above example $s_2[1] = \{A, T\}$, $s_2[2] = \{C\}$, $s_2[3] = \{G\}$ and $s_2[4] = \{T, G\}$. Also, if $x$ is a sequence, then $|x|$ denotes the length of the sequence and if $x$ is a set of elements then $|x|$ denotes the cardinality of the set. Hence $|s_1| = 6$, $|s_2| = 4$, $|s_1[1]| = 1$ and $|s_2[4]| = 2$.

**Definition 1.** *($e_1 \preceq e_2$) $e_1 \preceq e_2$ if and only if $e_1$ is a "dont care" character or $e_1 \subseteq e_2$.*

The flexibility of a motif is due to the variability in the number of dot characters and this is done by annotating the dot characters.

**Definition 2.** *(Annotated dot Character, $.^{\alpha}$) An annotated "." character is written as $.^{\alpha}$ where $\alpha$ is a set of non-negative integers $\{\alpha_1, \alpha_2, \ldots, \alpha_k\}$ or an* *interval* $\alpha = [\alpha_l, \alpha_u]$, *representing all integers between* $\alpha_l$ *and* $\alpha_u$ *including* $\alpha_l$ *and* $\alpha_u$.

To avoid clutter, the annotation superscript $\alpha$, will be an integer interval.

**Definition 3.** *(Rigid, Flexible String) Given a string* $m$, *if at least one dot element, is annotated,* $m$ *is called a* flexible *string, otherwise* $m$ *is called* rigid.

**Definition 4.** *(Realization) Let* $p$ *be a flexible string. A rigid string* $p'$ *is a realization of* $p$ *if each annotated dot element* $.^{\alpha}$ *is replaced by* $l$ *dot elements where* $l \in \alpha$.

For example, if $p = a.^{[3,6]}b.^{[2,5]}cde$, then $p' = a...b...cde$ is a realization of $p$ and so is $p'' = a...b.....cde$.

**Definition 5.** *(p Occurs at l) A rigid string* $p$ *occurs at position* $l$ *on* $s$ *if* $p[j] \preceq$ $s[l + j - 1]$ *holds for* $1 \le j \le |p|$. *A flexible string* $p$ *occurs at position* $l$ *in* $s$ *if there exists a realization* $p'$ *of* $p$ *that occurs at* $l$.

If $p$ is flexible then $p$ could possibly occur *multiple times* at a location on a string $s$. For example, if $s = axbcbc$, then $p = a.^{[1,3]}b$ occurs twice at position 1 as a**x**bcbc and a**x**bc**b**c. This multiplicity of occurrence increases the complexity of the algorithm over that of rigid motifs in the discovery process as discussed in the next section.

**Definition 6.** *(Motif* $m$, *Location List* $\mathcal{L}_m$*) Given a string* $s$ *on alphabet* $\Sigma$ *and a positive integer* $k$, $k \le |s|$, *a string (flexible or rigid)* $m$ *is a motif with location list* $\mathcal{L}_m = (l_1, l_2, \ldots, l_p)$, *if* $m[1] \ne$ '*.*', $m[|m|] \ne$ '*.*' [1] *and* $m$ *occurs at each* $l \in \mathcal{L}_m$ *and there exists no* $l'$, $l' \notin \mathcal{L}_m$ *and* $m$ *occurs at* $l'$ *with* $p \ge k$.

**Definition 7.** *(Realization of a Motif* $m$*) Given a motif* $m$ *on an input string* $s$ *with a location list* $\mathcal{L}_m$ *and* $m'$ *a realization of the string* $m$, *then* $m'$ *is a realization of the motif* $m$ *if and only if there exists some* $k \in \mathcal{L}_m$ *such that* $m'$ *occurs at* $k$ *in* $s$.

Notice that because of our notation of annotating a dot character with an integer interval (instead of a set of integers), not *every* realization of the flexible motif occurs in the input string. In the remaining discussion we will use this stricter definition of motif realization (Definition 7) unless othewise specified.

**Definition 8.** *($m_1 \preceq m_2$) Given two motifs* $m_1$ *and* $m_2$ *with* $|m_1| \le |m_2|$, $m_1 \preceq m_2$ *holds if for every realization* $m_1'$ *of motif* $m_1$ *there exists a realization* $m_2'$ *of motif* $m_2$ *such that* $m_1'[j] \preceq m_2'[j]$, $1 \le j \le |m_1|$.

---

[1] The first and last characters of the motif are solid characters; if "dont care" characters are allowed at the ends, the motifs can be made arbitrarily long in size without conveying any extra information.

For example, let $m_1 = AB..E$, $m_2 = AK..E$ and $m_3 = ABC.E.G$. Then $m_1 \preceq m_3$, and $m_2 \npreceq m_3$. The following lemma is straightforward to verify.

**Definition 9.** $(m_1 = m_2)$ *Given two motifs $m_1$ and $m_2$ with $|m_1| = |m_2|$, $m_1 = m_2$ holds if for every realization $m_1'$ of motif $m_1$ there exists a realization $m_2'$ of motif $m_2$ such that $m_1'[j] = m_2'[j]$, $1 \leq j \leq |m_1|$.*

**Lemma 1.** *If $m_1 \preceq m_2$, then $\mathcal{L}_{m_1} \supseteq \mathcal{L}_{m_2}$. If $m_1 \preceq m_2$ and $m_2 \preceq m_3$, then $m_1 \preceq m_3$.*

**Definition 10.** *(Sub-motifs of Motif $m$) Given a motif $m$ let $m[j_1]$, $m[j_2]$, ... $m[j_l]$ be the $l$ solid elements in the motif $m$. Then the sub-motifs of $m$ are given as follows: for every $j_i, j_k$, the sub-motif is obtained by dropping all the elements before (to the left of) $j_i$ and all elements after (to the right of) $j_k$ in $m$.*

A very natural notion of maximality defined earlier [BJEG98,Par00] is given below for completeness.

**Definition 11.** *(Maximal Motif) Let $p_1, p_2, \ldots, p_k$ be the motifs in a sequence $s$. Define $p_i[j]$ to be '.', if $j > |p_i|$. A motif $p_i$ is maximal in composition if and only if there exists no $p_l$, $l \neq i$ with $\mathcal{L}_{p_i} = \mathcal{L}_{p_l}$ and $p_i \preceq p_l$. A motif $p_i$, maximal in composition, is also maximal in length if and only if there exists no motif $p_j$, $j \neq i$, such that $p_i$ is is a sub-motif of $p_j$ and $|\mathcal{L}_{p_i}| = |\mathcal{L}_{p_j}|$. A maximal motif is maximal both in composition and in length.*

## 3   Algorithm Prerequisites

It is quite clear that the number of maximal flexible motifs could be exponential in the size of the input $s$. It has been shown in [Par00] that there is a small basis set of motifs of size $O(n)$ for every input of size $n$. The remaining motifs can be computed from this set of motifs. We recall the definition and the statement of the theorem for the sake of completeness here.

We now define the notion of redundancy and the basis set. Informally speaking, we call a motif $m$ redundant if $m$ *and* its location list $\mathcal{L}_m$ can be deduced from the other motifs *without* studying the input string $s$. We introduce such a notion below and the section on "generating operation" describes how the redundant motifs and the location lists can be computed from the irredundant motifs.

**Definition 12.** *(Redundant, Irredundant Motif) A maximal motif $m$, with location list $\mathcal{L}_m$, is redundant if there exist maximal motifs $m_i$, $1 \leq i \leq p$, $p \geq 1$, such that $\mathcal{L}_m = \mathcal{L}_{m_1} \cup \mathcal{L}_{m_2} \ldots \cup \mathcal{L}_{m_p}$ and $m \preceq m_i$ for all $i$. A maximal motif that is not redundant is called an* irredundant *motif.*

Notice that for a rigid motif $p > 1$ ($p$ in Definition 12) since each location list corresponds to *exactly one* motif whereas for a flexible motif $p$ could have a value 1. For example, let $s = axfygsbapgrftb$. Then $m_1 = a.^{[1,3]}f.^{[1,3]}b$, $m_2 = a.^{[1,3]}g.^{[1,3]}b$, $m_3 = a....b$ with $\mathcal{L}_{m_1} = \mathcal{L}_{m_2} = \mathcal{L}_{m_3} = \{1,8\}$. But $m_3$ is redundant since $m_3 \preceq m_1, m_2$. Also $m_1 \not\preceq m_2$ and $m_2 \not\preceq m_1$, hence both $m_1$ and $m_2$ are irredundant although $\mathcal{L}_{m_1} = \mathcal{L}_{m_2}$. This also illustrates the case where one location list corresponds to two distinct flexible motifs (motifs $m_1$ and $m_2$ are distinct if $m_1 = m_2$ does not hold).

*Generating Operations.* The redundant motifs need to be generated from the irredundant ones, if required. We define the following generating operations. The binary OR operator $\otimes$ is used in the algorithm in the process of motif detection and the AND operator $\oplus$ in the generation of redundant motifs from the basis.

Given an input sequence $s$, let $m$, $m_1$ and $m_2$ be motifs.

Binary AND operator, $m_1 \oplus m_2$: $m = m_1 \oplus m_2$, where $m$ is such that $m \preceq m_1, m_2$ and there exists no motif $m'$ with $m \preceq m'$. For example if $m_1 = A.D.^{[2,4]}G$ and $m_2 = AB.^{[1,5]}FG$. Then, $m = m_1 \oplus m_2 = A...^{[2,4]}G$.

Binary OR operator, $m_1 \otimes m_2$: $m = m_1 \otimes m_2$, where $m$ is such that $m_1, m_2 \preceq m$ and there exists no motif $m'$ with $m' \preceq m$. For example if $m_1 = A..D..G$ and $m_2 = AB...FG$. Then, $m = m_1 \otimes m_2 = AB.D.FG$.

**Definition 13.** *(Basis) Given an input sequence $s$, let $\mathcal{M}$ be the set of all maximal motifs on $s$. A set of maximal motifs $\mathcal{B}$ is called a basis of $\mathcal{M}$ iff the following hold:*

1. *for each $m \in \mathcal{B}$, $m$ is irredundant with respect to $\mathcal{B} - \{m\}$, and,*
2. *let $\mathbf{G}(\mathcal{X})$ be the set of all the redundant maximal motifs generated by the set of motifs $\mathcal{X}$, then $\mathcal{M} = \mathbf{G}(\mathcal{B})$.*

The following theorem has been proved in [Par00] and we give only the statement here.

**Theorem 1.** *Let $s$ be a string with $n = |s|$ and let $\mathcal{B}$ be a basis or a set of irredundant flexible motifs. Then $\mathcal{B}$ is unique and $|\mathcal{B}| = O(n)$.*

We give a useful corollary to this theorem below.

**Corollary 1.** *Given an input sequence of length $n$, let $M$ be a set of motifs[2], not necessarily maximal, with the following properties:,*

1. *For each $p, q \in M$, $p \neq q$, let $p'$ be a suffix string of $p$ and $p' \not\preceq q$, unless $|\mathcal{L}_p| \neq |\mathcal{L}_q|$*
2. *there does not exist $p \in M$ such that $\mathcal{L}_p = \cup \mathcal{L}_{q_i}$ and $p \preceq q_i$ for all $i$.*

*Then $|M| = O(n)$.*

---

[2] For example if $m_1 = ABC$ with $\mathcal{L}_{m_1} = \{1,5,7\}$ with $m_1 \in M$ and $m_2 = AB$ with $\mathcal{L}_{m_2} = \{1,5,7\}$ then $m_2 \notin M$. However, if $\mathcal{L}_{m_2} = \{1,5,7,12\}$ ($\mathcal{L}_{m_1} \subset \mathcal{L}_{m_2}$), then $m_2$ could belong to $M$.

This result is used in the algorithm to bound the number of non-maximal motifs at each iteration of the algorithm. We next describe two problems on sets: the Set Intersection Problem (SIP) and the Set Union Problem (SUP) which are used in the pattern discovery algorithm in the next section.

*The Set Intersection Problem, SIP(n, m, l).* Given $n$ sets $S_1, S_2, \ldots, S_n$, on $m$ elements, find all the $N$ distinct sets of the form $S_{i_1} \cap S_{i_2} \cap \ldots \cap S_{i_p}$ with $p \geq l$. Notice that it is possible that $N = O(2^n)$. We give an $O(N \log n + mn)$ algorithm, described in Appendix A, to obtain all the intersection sets.

*The Set Union Problem, SUP(n, m).* Given $n$ sets $S_1, S_2 \ldots, S_n$ on $m$ elements each, find all the sets $S_i$ such that $S_i = S_{i_1} \cup S_{i_2} \cup \ldots \cup S_{i_p}$ $i \neq i_j$, $1 \leq j \leq p$. We present an algorithm in Appendix B to solve this problem in time $O(n^2 m)$.

## 4    The Pattern Discovery Algorithm

The algorithm can be described as follows. It begins by computing one-character patterns and then successively grows them by concatenating with other patterns until it cannot be grown any further. However, the drawback of this simplistic approach is that the number of patterns at each step grows very rapidly. We solve the problem by first computing only the basis set. This is done by trimming the number of growing patterns at each step and using theorem 1 to bound their number by $O(n)$. Thus in time $O(n^5 \log n)$, the basis can be detected. In the next step the remaining motifs from the basis is computed in time "proportional" to their number.

*Input Parameters.* The input parameters are: (1) the string $s$ [3], (2) the minimum number of times a pattern must appear $k$, (3) the flexibility of the dot characters $\Delta$. The flexibility property has the following interpretation: given a flexibility of $\Delta$, we accept dot character annotations of the form $[\alpha_1, \alpha_2]$ where $(\alpha_2 - \alpha_1) \leq \Delta$. For the rest of the algorithm we will assume that the alphabet size $|\Sigma| = O(1)$.

We use the following notation, given a motif $m$ (not necessarily maximal): $F(m)$ denotes the first element of $m$ and $E(m)$ denotes the last element of $m$. Note that $F(m) \neq$ '.' and $E(m) \neq$ '.'. $\mathcal{L}'_m = \{(i, j) | m'$ is the realization of $m$ that occurs at $i$ and ends at $j\}$. Note that $\mathcal{L}_m = \{i | (i, -) \in \mathcal{L}_m\}$.

### 4.1    Computing the Basis

This proceeds in two major phases as follows:
    Step 1: Pattern initialization phase

---

[3]  Recall that each element of $s$ is a character or a set of characters from the alphabet $\Sigma$ or even real numbers. In the case of the input is a sequence of real numbers we have shown in [PRF$^+$00] that this problem can be mapped onto an instance of a pattern discovery problem on strings of sets of characters. Thus the treatment discussed in this paper also extends to flexible patterns on real number sequences.

Repeat Until Done:
Step 2: Pattern concatenation phase

*Pattern Initialization Phase.* This proceeds in the following steps.

Step 1.1. For every $\sigma \in \Sigma$, construct $m = \sigma$ and $\mathcal{L}'_m = \{(i,i)|s[i] = \sigma\}$. $F(m) = E(m) = \sigma$.
This takes $O(n)$ time.

Step 1.2. This step is required only while dealing with strings on sets of characters. For example if $m_1 = \{b,c,d\}$ and $m_2 = \{b,c,e\}$, we need to check if there exists $m = \{b,c\}$. Note that $\mathcal{L}_m = \mathcal{L}_{m_1} \cup \mathcal{L}_{m_2}$ while the characters in $m$ are the intersection of the sets of characters in $m_1$ and $m_2$. We solve this using the Set Intersection Problem SIP($|\Sigma|, k, 2$).
Assuming $|\Sigma| = O(1)$, this takes $O(n^2)$ time.

Step 1.3. Let $m = m_{1d2}$ denote the string obtained by concatenating the elements $m_1$ followed by $d$ '.' characters followed by the element $m_2$.
For $d = 0 \ldots n$, construct the motif $m = m_{idj}$ and $\mathcal{L}'_{m_{idj}} = \{(x, x+d)|(x,x) \in \mathcal{L}'_{m_i}, (x+d, x+d) \in \mathcal{L}'_{m_j}\}$ $F(m) = F(m_i)$, $E(m) = E(m_j)$.
This takes $O(n^2)$ time and the number of motifs at this step is $O(n)$.

Step 1.4. In the case of flexible motifs, construct the following flexible motifs. Construct sets of motifs $P$ such that for all $m_i, m_j \in P$, $F(m_i) = F(m_j)$ and $E(m_i) = E(m_j)$. For each such set $P$, for $l = 0 \ldots n - \Delta$, $m = m_i.^{[l,l+\Delta]} m_j$ and $\mathcal{L}'_m = \cup_{k=l}^{k=l+\Delta} \mathcal{L}'_{m_{ilj}}$ and $F(m) = F(m_i)$, $E(m) = E(m_j)$.
This takes $O(n^2)$ time and the number of motifs at this step is $O(n)$.

Step 1.5. This is the pruning step. We do two kinds of pruning: (1) where all suffix motifs are removed and (2) where all the "redundant" motifs are removed. For the former, we offset every location list to zero and check for identity of location lists. The latter is described below. Let $\mathcal{L}$ denote all the location lists of the motifs constructed in Step 1.3. Using the Set Union Problem SUP($|\mathcal{L}|, n$) remove all the motifs whose location list is exactly the union of some other location lists. If $\mathcal{L}_m = \cup \mathcal{L}_{m_i}$, remove $m$ and update each $m_i$ as $m_i = m_i \bigotimes m$ and if $|m| > |m_i|$, $E(m_i) = E(m)$.
For example, if $m_1 = a.b$, $m_2 = a..c$ and $m = a...d$ with $\mathcal{L}_m = \mathcal{L}_{m_1} \cup \mathcal{L}_{m_2}$ then $m_1$ is updated as $m_1 = a.b.d$ and $m_2$ is updated as $m_2 = a..cd$.
This step takes $O(n^3)$ time and the number of motifs at this step is $O(n)$.

*Pattern Concatenation Phase.* This proceeds in the following steps.

Step 2.1. Consider every pair of motifs $m_1$ and $m_2$ with $E(m_1) \preceq F(m_2)$ or $F(m_1) \preceq E(m_2)$. Let $l = |m_1|$. Define $m = m_1 + m_2$ as follows:
If $E(m_1) \preceq F(m_2)$ then

$$m[i] = \begin{cases} m_1[i] & i \le l \\ m_2[i - l + 1] & i > l \end{cases}$$

If $F(m_1) \preceq E(m_2)$ then

$$m[i] = \begin{cases} m_1[i] & i < l \\ m_2[i - l + 1] & i \geq l \end{cases}$$

Further, $\mathcal{L}'_m = \{(i, j)|(i, k) \in \mathcal{L}_{m_i} \text{ and } (k, j) \in \mathcal{L}_{m_j}\}$, $F(m) = F(m_i)$ and $E(m) = E(m_j)$.

For rigid motif $m$, $|\mathcal{L}'_m| < n$ and for flexible motif $m$, $|\mathcal{L}'_m| < n^2$, hence this step takes $O(n^3)$ time for rigid motifs and takes $O(n^4)$ time for flexible motifs.

Step 2.2. This is the pruning phase and is the same as Step 1.4. This takes $O(n^3)$ time. Using Corollary 1 (prune (1) takes care of the first and prune (2) takes care of the second condition in the Corollary), at the end of this step there are $O(n)$ patterns.

The number of iterations is $\log J$ where $J$ is the length of the longest motif in $s$. Since $J$ is bounded by $n$, the algorithm takes $O(n^4 \log n)$ to detect the basis for rigid motifs and $O(n^5 \log n)$ in the case of flexible motifs.

## 4.2    Computing the Redundant Maximal Patterns

A redundant maximal motif $m$ is of the form $m_1 \oplus m_2 \oplus \ldots \oplus m_p$ for some $p$ and $\mathcal{L}_m = \mathcal{L}_{m_1} \cup \mathcal{L}_{m_2} \cup \ldots \cup \mathcal{L}_{m_p}$. We give an example below to show that a straightforward approach of combining (using the operator $\oplus$) compatible [4] motifs does not give the desired time complexity.

**Example 1.** Let $m_1 = ab...d$, $m_2 = a...cd$, $m_3 = a.e..d$, $m_4 = a..f.d$ with $\mathcal{L}_{m_1} = \{10, 20\}$, $\mathcal{L}_{m_2} = \{30, 40\}$, $\mathcal{L}_{m_3} = \{20, 40\}$, $\mathcal{L}_{m_4} = \{10, 30\}$. Then $\mathcal{L}_{m_5} = \{\mathcal{L}_{m_1} \cup \mathcal{L}_{m_2} \cup \mathcal{L}_{m_3} \cup \mathcal{L}_{m_4}\}$, $\mathcal{L}_{m_6} = \{\mathcal{L}_{m_2} \cup \mathcal{L}_{m_3} \cup \mathcal{L}_{m_4}\}$, $\mathcal{L}_{m_7} = \{\mathcal{L}_{m_1} \cup \mathcal{L}_{m_3} \cup \mathcal{L}_{m_4}\}$, $\mathcal{L}_{m_8} = \{\mathcal{L}_{m_1} \cup \mathcal{L}_{m_2} \cup \mathcal{L}_{m_4}\}$, $\mathcal{L}_{m_9} = \{\mathcal{L}_{m_1} \cup \mathcal{L}_{m_2} \cup \mathcal{L}_{m_3}\}$ are such that $m_5 = m_6 = m_7 = m_8 = m_9 = a....d$. In other words, the motif $m_5$ is constructed at least four more times than required.

We give below an output-sensitive algorithm to compute all the redundant motifs.

Given $B$ the set of all the irredundant motifs, construct $\mathcal{P}$ a set of subsets of $B$ as follows: $P \in \mathcal{P}$, if for each motif $m_i, m_j \in P$, without loss of generality, $F(m_i) \preceq F(m_j)$ and $m_i \not\preceq m_j$, and $P$ is the largest such set. For each $P \in \mathcal{P}$, we construct an instance of the Set Intersection Problem SIP as follows. We claim that the union of the solutions to each of the SIP gives all the maximal redundant motifs in time $O(N \log n)$. We illustrate this through two examples and omit the formal arguments due to space constraints. Recall that $N$ is the number of maximal motifs and $n$ is the length of the input sequence.

For each $P \in \mathcal{P}$ do the following. Let $l = \max_{m \in P} |m|$. Construct $\bar{m}[i]$, $2 \leq i \leq l$ as follows. $\bar{m}[i] = \{\sigma \neq `.`|\sigma \preceq p[i], p \in P\}$. Note that it is possible that

---

[4]    Two motifs $m_1$ and $m_2$ are compatible, without loss of generality, if $m_1[1] \preceq m_2[1]$ and there is $i$ s.t $m_1[i] \neq `.`$, $m_2[i] \neq `.`$ and $m_1[i] \preceq m_2[i]$, $1 < i \leq \min(|m_1|, |m_2|)$.

$\bar{m}[i] = \{\}$ for some $i$. Now construct an instance of $SIP(N', M, 2)$ as follows. The $M$ elements on which the sets are built is a subset of the basis set and $M = |P|$. The $N'$ sets are constructed as follows. $S_e^j = \{m_i | \bar{m}[j] = e\}$ for all possible values of $j$ and $e$ and $|S_e^j| \geq 2$. Assuming that $\Sigma = O(1)$, the number of such sets $N' = O(n)$. Recall that $n$ is length of the input string $s$ whose motifs are being discovered. Each $S_e^j$ with $|S_e^j| \geq 2$ corresponds to a maximal redundant motif. Although, same location lists may give distinct flexible motifs, this does not cause any problems since we use the solid characters of the motifs in $P$.

As an illustration we first show an example involving rigid motifs.

**Example 2.** Let $m_1 = abc.d$, $m_2 = abe$, $m_3 = add.d$, $m_4 = ad..e$, $m_5 = ab..d$. Here $l = 5$ and $S_b^2 = \{m_1, m_2, m_5\}$, $S_d^2 = \{m_3, m_4\}$, $S_d^5 = \{m_1, m_3, m_5\}$.

|       | 1 | 2 | 3 | 4 | 5 |
|-------|---|---|---|---|---|
| $m_1$ | a | b | c | . | d |
| $m_2$ | a | b | e |   |   |
| $m_3$ | a | d | d | . | d |
| $m_4$ | a | d | . | . | e |
| $m_5$ | a | b | a | . | d |

Each of the set corresponds to a maximal redundant motif. For example $S_b^2$ gives the maximal redundant motif $m_1 \oplus m_2 \oplus m_5 = ab$ with location list $\mathcal{L}_{m_1} \cup \mathcal{L}_{m_2} \cup \mathcal{L}_{m_5}$, $S_d^2$ gives $m_3 \oplus m_4 = ad$ with location list $\mathcal{L}_{m_3} \cup \mathcal{L}_{m_4}$, $m_1 \oplus m_3 \oplus m_5 = a...d$ with $S_d^5$ gives location list $\mathcal{L}_{m_1} \cup \mathcal{L}_{m_3} \cup \mathcal{L}_{m_5}$, The results from SIP give the unique intersection set $\{m_1, m_5\}$ and this corresponds to the motif $m = m_1 \oplus m_5 = ab..d$ with $\mathcal{L}_m = \mathcal{L}_{m_1} \cup \mathcal{L}_{m_5}$.

Consider an example using flexible motifs.

**Example 3.** Let $m_1 = ab.^{[2,3]}ec$, $m_2 = ab.^{[1,2]}bc$, $m_3 = ab.^{[1,3]}be$ and $m_4 = a.^{[1,3]}cb$.

|       | 1 | 2 | 3 | 4 | 5 |
|-------|---|-------------|-------------|---|---|
| $m_1$ | a | b | $.^{[2,3]}$ | e | c |
| $m_2$ | a | b | $.^{[1,2]}$ | b | c |
| $m_3$ | a | b | $.^{[1,3]}$ | b | e |
| $m_4$ | a | $.^{[1,3]}$ | c | b |   |

Here $l = 5$. The different sets are $S_b^2 = \{m_1, m_2, m_3\}$, $S_b^4 = \{m_2, m_3, m_4\}$, $S_c^5 = \{m_1, m_2\}$.

Each of the sets corresponds to a maximal redundant motif. $S_b^2$ gives $m_1 \oplus m_2 \oplus m_3 = ab$, with location list $\mathcal{L}_{m_1} \cup \mathcal{L}_{m_2} \cup \mathcal{L}_{m_3}$, $S_b^4$ gives $m_2 \oplus m_3 \oplus m_4 = a.^{[1,3]}.^{[1,2]}b = a.^{[2,5]}b$ [5] with location list $\mathcal{L}_{m_2} \cup \mathcal{L}_{m_3} \cup \mathcal{L}_{m_4}$, $S_c^5$ gives $m_1 \oplus m_2 = ab.^{[1,3]}.^{[1,2]}c = ab.^{[2,5]}c$ with location list $\mathcal{L}_{m_1} \cup \mathcal{L}_{m_2}$. The intersection results from SIP gives $\{m_2, m_3\}$ with $m = m_2 \oplus m_3 = ab.^{[1,3]}b$ and location list $\mathcal{L}_{m_2} \cup \mathcal{L}_{m_3}$.

---

[5] Since $x.^{[\alpha_1, \alpha_2]}.^{[\alpha_3, \alpha_4]}y = x.^{[\alpha_1+\alpha_3, \alpha_2+\alpha_4]}y$.

# References

BB99.       P.O. Brown and D. Botstein. Exploring the new world of the genome with
            DNA microarrays. *Nature Genetics*, 21:33–37, 1999.
BDY99.      A. Ben-Dor and Z. Yakhini. Clustering gene expression patterns. *Pro-
            ceedings of the Annual Conference on Computational Molecular Biology
            (RECOMB'99)*, pages 33–42, 1999.
BG98.       T.L. Bailey and M. Gribskov. Methods and statistics for combining motif
            match scores. *Journal of Computational Biology*, 5:211–221, 1998.
BJEG98.     Alvis Brazma, Inge Jonassen, Ingvar Eidhammer, and David Gilbert. Ap-
            proaches to the automatic discovery of patterns in biosequences. *Journal
            of Computational Biology*, 5(2):279–305, 1998.
Cal00.      Andrea Califano. SPLASH: structural pattern localization algorithm by
            sequential histogramming. *Bioinformatics (under publication)*, 2000.
De96.       J. DeRisi and L. Penland et al. Use of a cDNA microarray to analyse gen
            expression patterns in human cancer. *Nat. genetics*, 14(4):457–460, 1996.
DSO78.      M.O. Dayhoff, R.M. Schwartz, and B.C. Orcutt. A model of evolutionary
            change in proteins. *Atlas of Protein Sequence and Structure*, pages 345–352,
            1978.
HH92.       S. Henikoff and J.G. Henikoff. Amino cid substitution matrices from protein
            blocks. *Proc. Natl. Acad. Sci.*, 89:10915–10919, 1992.
JCH95.      I.J.F. Jonassen, J.F. Collins, and D.G. Higgins. Finding flexible patterns
            in unaligned protein sequences. *Protein Science*, pages 1587–1595, 1995.
Le96.       D.J. Lockhart and H. Dong et al. Expression monitoring by hybridization
            to high density oligonucleotide arrays. *Nat. biotechnol.*, 14(13):1675–1680,
            1996.
NG94.       A.F. Neuwald and P. Green. Detecting patterns in protein sequences. *Jour-
            nal of Molecular Biology*, pages 698–712, 1994.
Par00.      Laxmi Parida. Some results on flexible-pattern matching. In *Proc. of the
            Eleventh Symp. on Comp. Pattern Matching*, volume 1848 of *Lecture Notes
            in Computer Science*, pages 33–45. Springer-Verlag, 2000.
PRF+00.     Laxmi Parida, Isidore Rigoutsos, Aris Floratos, Dan Platt, and Yuan Gao.
            Pattern discovery on character sets and real-valued data: linear bound on
            irredundant motifs and an efficient polynomial time algorithm. In *Pro-
            ceedings of the eleventh ACM-SIAM Symposium on Discrete Algorithms
            (SODA)*, pages 297 – 308. ACM Press, 2000.
RF98.       I. Rigoutsos and A. Floratos. Motif discovery in biological sequences with-
            out alignment or enumeration. In *Proceedings of the Annual Conference
            on Computational Molecular Biology (RECOMB'98)*, pages 221–227. ACM
            Press, 1998.
Roy92.      M.A. Roytberg. A search for common patterns in many sequences.
            *CABIOS*, pages 57–64, 1992.
SD78.       R.M. Schwartz and M.O. Dayhoff. Matrices for detecting distance relation-
            ships. *Atlas of Protein Sequence and Structure*, pages 353–358, 1978.
SNJ95.      M. Suyama, T. Nishioka, and O. Juníchi. Searching for common sequence
            patterns among distantly related proteins. *Protein Engineering*, pages 366–
            385, 1995.
SV96.       M.F. Sagot and A. Viari. A double combinatorial approach to discover-
            ing patterns in biological sequences. *Proceedings of the 7th symposium on
            combinatorial pattern matching*, pages 186–208, 1996.

WCM+96.  J. Wang, G. Chirn, T.G. Marr, B.A. Shapiro, D. Shasha, and K. Jhang.
          Combinatorial pattern discovery for scientific data: some preleminary re-
          sults. *Proceedings of the ACM SIGMOD conference on management of
          data*, pages 115–124, 1996.
We97.     L. Wodicka and H. Dong et al. Genome-wide expression monitoring in
          saccharomyces cerevisiae. *Nat. biotechnol.*, 15(13):1359–1367, 1997.

# A  The Set Intersection Problem, SIP$(n, m, l)$

Given $n$ sets $S_1, S_2, \ldots, S_n$, on $m$ elements, find all the $N$ distinct sets of the
form $S_{i_1} \cap S_{i_2} \cap \ldots \cap S_{i_p}$ with $p \geq l$. We give an $O(N \log n + mn)$ algorithm
below to obtain all the intersection sets.

Let the elements be numbered $1 \ldots m$. Construct a binary tree $\mathcal{T}$ using the
subroutine CREATE-NODE shown below.

Assume a function CREATE-SET$(\mathcal{S})$ which creates $\mathcal{S}$, a subset of $S_1, S_2, \ldots,$
$S_n$ in an appropriate data structure $\mathcal{D}$ (say a tree). A query of the form if a subset
$\mathcal{S} \in \mathcal{D}$ (DOES-EXIST$(\mathcal{S})$) returns a True/False in time $O(\log n)$.

Node CREATE-NODE $(\mathcal{S}, h, l)$
{
    (1) New(this-node)
    (2) CREATE-SET$(\mathcal{S})$
    (3) Let $\mathcal{S}' = \{S_i \in \mathcal{S} | h \in S_i\}$
    (4) if $((|\mathcal{S}'| \geq l)$ and not DOES-EXIST$(\mathcal{S}')$ and $(h \geq 2))$
        (5) Left-child = CREATE-NODE$(\mathcal{S}', h - 1, l)$
    (6) Right-child = CREATE-NODE$(\mathcal{S}, h - 1, l)$
    (7) return (this-node)
}

For $l = 2$, there is exactly one node the tree $\mathcal{T}$. For $l > 2$, the initial call is
CREATE-NODE$(\{S_1, S_2, \ldots, S_n\}, m, l)$. Clearly, all the unique intersection sets,
which are $N$ in number are at the leaf node of this tree $\mathcal{T}$. Also, the number
of internal nodes can not exceed the number of leaf nodes, $N$. Thus the total
number of nodes of $\mathcal{T}$ is $O(N)$. The cost of query at each node is $O(\log n)$ (line
(4) of CREATE-NODE). The size of the input data is $O(nm)$ and each data
item is read exactly once in the algorithm (line (3) of CREATE-NODE) Hence
the algorithm takes $O(N \log n + nm)$ time.

# B  The Set Union Problem, SUP$(n, m)$

Given $n$ sets $S_1, S_2 \ldots, S_n$ on $m$ elements each, find all the sets $S_i$ such that
$S_i = S_{i_1} \cup S_{i_2} \cup \ldots \cup S_{i_p}$, $i \neq i_j, 1 \leq j \leq p$.

This is a very straightforward algorithm (this contributes an additive term
to the overall complexity of the pattern detection algorithm): For each set $S_i$,
we first obtain the sets $S_j$ $j \neq i, j = 1 \ldots n$ such that $S_j \subset S_i$. This can be done
in $O(nm)$ time (for each $i$). Next, we check if $\cup_j S_j = S_i$. Again this can be done
in $O(nm)$ time. Hence the total time taken is $O(n^2 m)$.

# Episode Matching*

Zdeněk Troníček

Dept. of Comp. Science and Eng., FEE CTU Prague
Karlovo nám. 13, 121 35 Prague 2, Czech Republic
tronicek@fel.cvut.cz

**Abstract.** The episode matching problem is considered and the method for preprocessing the text is presented. Once the text is preprocessed, an episode substring can be found in time linear to the length of pattern (episode).

A subsequence of a string $T$ is any string obtainable by removing zero or more symbols from $T$. Given two strings, pattern $S$ and text $T$, an *episode substring* is a minimal substring $\alpha$ of $T$ that contains $S$ as a subsequence. Minimal means that no proper substring of $\alpha$ contains $S$ as a subsequence. The *episode matching problem* is to find all episode substrings. All strings in this paper are considered on alphabet $\Sigma$ of size $\sigma$.

The problem arises in analyzing sequences of events, *e.g.* alarms from a telecommunication network, actions from a user, or records from a WWW-server log file. Knowledge of frequent episode substrings can then be used to describe or predict the sequence. The first notion about the problem comes probably from Mannila, Toivonen and Verkamo [5]. Their solution requires $O(nmk)$ time, where $m$ is the length of $S$, $n$ is the length of $T$, and $k$ is the number of episodes (in our case is $k = 1$). Das *et al.* [3] proposed several algorithms with the following time complexities: $O(nm)$, $O(\frac{nm}{\log m})$, and $O(n + s + \frac{nm \log \log s}{\log \frac{s}{m}})$ when additional space is limited to $O(s)$. Boasson *et al.* [2] showed that the problem is linear to $n$ and designed an algorithm which is exponential to $m$ and linear to $n$. All algorithms mentioned so far either do no preprocessing or preprocess the pattern.

The presented approach is based on preprocessing the text. We build the *Episode Directed Acyclic Subsequence Graph (EDASG)* which allows to find an episode substring in $O(m)$ time. Building the EDASG requires $O(n\sigma)$ time. Let $T = t_1 t_2 \ldots t_n$ and $S = s_1 s_2 \ldots s_m$.

First, we shortly recall the Directed Acyclic Subsequence Graph (DASG) which is used in the subsequence matching problem. Given a text $T$, the DASG is a finite automaton that accepts all subsequences of $T$. A finite automaton is, in this paper, a 5-tuple $(Q, \Sigma, \delta, q_0, F)$, where $Q$ is a finite set of states, $\Sigma$ is an input alphabet, $\delta$ is a transition function, $q_0$ is the initial state, and $F \subseteq Q$ is the set of final states.

---

* This research has been supported by GAČR grant No.201/01/1433.

A. Amir and G.M. Landau (Eds.): CPM 2001, LNCS 2089, pp. 143–146, 2001.

Let $Q = \{q_0, q_1, \ldots, q_n\}$ and $F = Q$. For each $a \in \Sigma$ and $i \in \langle 0, n \rangle$ we define the transition function $\delta$ as follows:
$\delta(q_i, a) = q_j$ if there exists $k > i$ such that $a = t_k$ and $j$ is minimal such $k$,
$\delta(q_i, a) = \emptyset$ otherwise.

Then the automaton $A = (Q, \Sigma, \delta, q_0, F)$ accepts a pattern $S$ if and only if $S$ is a subsequence of $T$. Two algorithms, right-to-left [1] and left-to-right [6], are known for building the DASG in $O(n\sigma)$ time. The DASG for $T$ allows to check whether $S$ is a subsequence of $T$ in $O(m)$ time. We note that each state $q \in Q$ corresponds to a prefix of $T$ such that this prefix is the longest path from $q_0$ to $q$.

**Lemma 1.** *Let $u, v, x, y \in \Sigma^*$ such that $u$ is a subsequence of $v$ and $x$ is a subsequence of $y$. Then $ux$ is a subsequence of $vy$.*

*Proof.* The set of all subsequences of a string $u_1 u_2 \ldots u_l$ can be described by regular expression $(\varepsilon + u_1)(\varepsilon + u_2) \ldots (\varepsilon + u_l)$. The lemma directly follows.    □

**Lemma 2.** *Given $i \in \langle 0, n \rangle$, the automaton $A_i = (Q, \Sigma, \delta, q_i, F)$ accepts $S$ iff $S$ is a subsequence of $t_{i+1} \ldots t_n$.*

*Proof.* Let $T_i = t_{i+1} \ldots t_n$. We prove two implications:
1. If $A_i$ accepts $S$ then $S$ is a subsequence of $T_i$. $A_i$ accepts $S$, thus $A$ accepts $t_1 \ldots t_i s_1 \ldots s_m$. Since $t_1 \ldots t_i s_1 \ldots s_m$ is a subsequence of $t_1 \ldots t_n$, we get that $s_1 \ldots s_m$ is a subsequence of $t_{i+1} \ldots t_n$.
2. If $S$ is a subsequence of $T_i$ then $A_i$ accepts $S$. $S$ is a subsequence of $T_i$, hence $A$ accepts $t_1 \ldots t_i s_1 \ldots s_m$. When $t_1 \ldots t_i$ has been read, the automaton $A$ is in state $q_i$ and consequently $A_i$ accepts $s_1 \ldots s_m$.    □

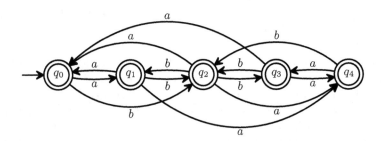

**Fig. 1.** The EDASG for text $T = abba$.

**Lemma 3.** *Let $i \in \langle 0, n \rangle$ and let $q_j$ be the active state of the automaton $A_i = (Q, \Sigma, \delta, q_i, F)$ after reading through $S$, i.e. $\delta^*(q_i, s_1 \ldots s_m) = q_j$. Then $j$ is minimal $k$ such that $t_{i+1} \ldots t_k$ contains $S$ as a subsequence, i.e. $j = min\{k : s_1 \ldots s_m$ is a subsequence of $t_{i+1} \ldots t_k\}$.*

*Proof.* (by contradiction) Suppose that there exists $k < j$ such that $s_1 \ldots s_m$ is a subsequence of $t_{i+1} \ldots t_k$. Then $s_1 \ldots s_m t_{k+1} \ldots t_n$ is a subsequence of $t_{i+1} \ldots t_n$ and therefore it should be accepted by $A_i$. We note that $A_i$ is in $q_j$ after reading through $S$. But $A_j$ accepts strings of length at most $n - j$, which contradicts the hypothesis that $k < j$. □

Let $A^R = (Q^R, \Sigma, \delta^R, q_n, F^R)$ be the DASG for the reversed text $T^R = t_n \ldots t_1$, i.e. $Q^R = \{q_n, q_{n-1}, \ldots, q_0\}, F^R = Q^R$ and $\delta^R(q_i, a) = q_j$ if there exists $k < i$ such that $a = t_k$ and $j$ is maximal such $k$, $\delta^R(q_i, a) = \emptyset$ otherwise.

The lemma 3 says that the active state $q_j$ of $A$ after reading through $S$ determines the end of episode substring. Its begin can be found using the DASG $A_j^R = (Q^R, \Sigma, \delta^R, q_j, F^R)$. Hence, to find an episode substring we need two DASGs: for text $T$ and for reversed text $T^R$. Since $Q = Q^R$, we can let both automata $A$ and $A^R$ share the set of states and combine them to the Episode Directed Acyclic Subsequence Graph. Formally we can say that the EDASG is a 6-tuple $(Q, \Sigma, \delta, \delta^R, q_0, F)$. One can also note that we extend the automaton $A$ with transitions of the automaton $A^R$. An example of the EDASG is in Fig. 1.

**procedure** BUILD_EDASG $(T)$
**input:** text $T = t_1 t_2 \ldots t_n \in \Sigma^*$
**output:** the EDASG for text $T$
1: **for all** $a \in \Sigma$ **do**
2:    $f[a] \leftarrow 0$
3:    $l[a] \leftarrow -1$
4: **end for**
5: create state $q_0$ and mark it as final
6: **for** $i = 1$ to $n$ **do**
7:    add state $q_i$ and mark it as final
8:    **for** $j = f[t_i]$ to $(i - 1)$ **do**
9:       add $\delta$ transition labeled $t_i$ between states $q_j$ and $q_i$
10:   **end for**
11:   $f[t_i] \leftarrow i$
12:   $l[t_i] \leftarrow i - 1$
13:   **for all** $a \in \Sigma, l[a] \neq -1$ **do**
14:      add $\delta^R$ transition labeled $a$ between states $q_i$ and $q_{l[a]}$
15:   **end for**
16: **end for**

Algorithm 1: Building the EDASG.

**Lemma 4.** *Let $q_j$ be the active state of the automaton $A = (Q, \Sigma, \delta, q, F)$ after reading through $S$, and $q_i$ be the active state of the automaton $A_j^R = (Q^R, \Sigma, \delta^R, q_j, F^R)$ after reading through $S^R = s_m \ldots s_1$. Then $t_{i+1} \ldots t_j$ is a minimal substring of $T$ that contains $S$ as a subsequence.*

*Proof.* We use lemma 3 for $A$ and $A_j^R$.    □

The algorithm for building the EDASG is a combination of right-to-left and left-to-right algorithms for building the DASG.

Once the EDASG has been built, we can find an episode substring in two steps: (1) we find the end of the substring (the EDASG simulates the DASG for $T$: it reads $S$ and uses transitions $\delta$), (2) we find the begin of the substring (the EDASG simulates the DASG for $T^R$: it reads $S^R$ and uses transitions $\delta^R$). The substring is determined by the active states after the first and second step. If $q_i$ is the active state after the second step, we can start to look for a next episode substring at state $q_{i+1}$.

For analysis of the algorithm we assume that alphabet $\Sigma$ has no useless symbols, *i.e.* all symbols of $\Sigma$ occur in $T$. Since each EDASG consists of two ordinary DASGs, the numbers of its $\delta$ and $\delta^R$ transitions are both $O(n\sigma)$. Thus, building the EDASG requires $O(n\sigma)$ time and $O(\sigma)$ extra space.

Finding an episode substring needs $O(m)$ time. Since there can be $O(n)$ episode substrings, we need $O(nm)$ time to find all episode substrings in the worst case. But once we preprocess the text, finding an episode substring is extremely fast, which is the major advantage of the presented approach.

If the alphabet is not known in advance, we use a balanced tree for values $f[a]$ and $l[a]$ ($a \in \Sigma$). The time complexities mentioned above are then multiplied by factor $\log \sigma$.

# References

1. R.A. Baeza-Yates. Searching subsequences. *Theor. Comput. Sci.*, 78(2):363–376, 1991.
2. L. Boasson, P. Cegielski, I. Guessarian, and Y. Matiyasevich. Window-accumulated subsequence matching problem is linear. In *Proceedings of the 18th ACM SIGACT-SIGMOD-SIGART Symposium on Principles of Database Systems*, pages 327–336, Philadelphia, Pennsylvania, 1999. ACM Press.
3. G. Das, R. Fleischer, L. Gąsieniec, D. Gunopulos, and J. Kärkkäinen. Episode matching. In A. Apostolico and J. Hein, editors, *Proceedings of the 8th Annual Symposium on Combinatorial Pattern Matching*, number 1264 in Lecture Notes in Computer Science, pages 12–27, Aarhus, Denmark, 1997. Springer-Verlag, Berlin.
4. H. Hoshino, A. Shinohara, M. Takeda, and S. Arikawa. Online construction of subsequence automata for multiple texts. In *Proceedings of the Symposium on String Processing and Information Retrieval 2000*, La Coruña, Spain, 2000. IEEE Computer Society Press.
5. H. Mannila, H. Toivonen, and A.I. Verkamo. Discovering frequent episodes in sequences. In *Proceedings of the 1st International Conference on Knowledge Discovery and Data Mining*, pages 210–215, Montreal, Canada, 1995. AAAI Press.
6. Z. Troníček and B. Melichar. Directed acyclic subsequence graph. In J. Holub and M. Šimánek, editors, *Proceedings of the Prague Stringology Club Workshop '98*, pages 107–118, Czech Technical University, Prague, Czech Republic, 1998. Collaborative Report DC–98–06.

# String Resemblance Systems:
# A Unifying Framework for String Similarity
# with Applications to Literature and Music

Masayuki Takeda

Department of Informatics, Kyushu University 33, Fukuoka 812-8581, Japan
and Japan Science and Technology Corporation
takeda@i.kyushu-u.ac.jp

## 1 Introduction

Identification of similar objects from a large collection of objects is one fundamental technique in several different areas in computer science, e.g., the case-based reasoning and the machine discovery. *Strings* are the most basic representations of objects inside computers, and thus string similarity is one of the most important topics in computer science.

Similarity measure must be sensitive to the kind of differences we wish to quantify. The *weighted edit distance* is one such framework in which the measure can be varied by altering weight assignment to each edit operation depending on symbols involved. However, it does not suffice to solve 'real problems' (see e.g., [2]). It is considered that two objects have necessarily a common structure if they seem similar, and the degree of similarity depends upon how valuable the common structure is. Based on this intuition, we present a unifying framework, named *string resemblance system* (SRS, for short). In this framework, similarity of two strings can be viewed as the maximum score of pattern that matches both of them. The differences among the measures are therefore the choices of (1) *pattern set* to which common patterns belong, and (2) *pattern score function* which assigns a score to each pattern.

For example, if we choose the set of *patterns with variable length don't cares* and define the score of a pattern to be the number of symbols in it, then the obtained measure is the length of the longest common subsequence (LCS) of two strings. In fact, the strings acdeba and abdac have a common pattern a⋆d⋆a⋆ which contains three symbols. With this framework one can easily design and modify his/her measures. In this paper we briefly describe SRSs and then report successful results of applications to literature and music.

## 2 Unifying Framework for String Similarity

In practical applications such as biological sequence comparisons, it is often preferred to measure *similarity* rather than distance between two given strings. We shall regard a distance measure as a similarity measure by multiplying the distance values by $-1$. Gusfield [2] pointed out that in dealing with string similarity

A. Amir and G.M. Landau (Eds.): CPM 2001, LNCS 2089, pp. 147–151, 2001.
© Springer-Verlag Berlin Heidelberg 2001

the language of alignments is often more convenient than the language of edit operations. Here, we generalize the alignment based scheme and propose a new scheme which is based on the notion of *common patterns*. Before describing our scheme, we need to introduce some notation. The set of strings over an alphabet $\Sigma$ is denoted by $\Sigma^*$. The length of a string $u$ is denoted by $|u|$. The string of length 0 is called the *empty string*, and denoted by $\varepsilon$. Let $\Sigma^+ = \Sigma^* - \{\varepsilon\}$. Let us denote by $\boldsymbol{R}$ the set of real numbers.

**Definition 1.** *A* string resemblance system *(SRS) is a 4-tuple* $\langle \Sigma, \Pi, L, \Phi \rangle$, *where:*

1. *$\Sigma$ is an alphabet;*
2. *$\Pi$ is a set of descriptions called* patterns*;*
3. *$L$ is a function called* interpretation *that maps a pattern in $\Pi$ to a language over $\Sigma$, i.e., a subset of $\Sigma^*$;*
4. *$\Phi$ is a function that maps a pattern in $\Pi$ to a real number called* score*.*

*The* similarity *between strings $x$ and $y$ with respect to $\langle \Sigma, \Pi, L, \Phi \rangle$ is defined by*

$$\boldsymbol{SIM}(x, y) = \sup\{\Phi(\pi) \mid \pi \in \Pi \text{ and } x, y \in L(\pi) \}.$$

We would assume that, for any $x, y \in \Sigma^*$, the set $\{\Phi(\pi) \mid \pi \in \Pi \text{ and } x, y \in L(\pi)\}$ is non-empty, bounded upwards, and contains the least upper bound as a member. This assumption guarantees that for any $x, y \in \Sigma^*$ there always exists a pattern $\pi \in \Pi$ common to $x$ and $y$ that maximizes the score $\Phi(\pi)$. Thus, computation of similarity is regarded as *optimal pattern discovery* in our framework. In this sense our framework bridges a gap between similarity computation and pattern discovery.

**Definition 2.** *An SRS $\langle \Sigma, \Pi, L, \Phi \rangle$ is said to be homomorphic if*

1. *$\Pi = (\Sigma \cup \Delta)^*$, where $\Delta$ is a set of* wildcards*.*
2. *$L : \Pi \to 2^{\Sigma^*}$ is a homomorphism such that $L(c) = \{c\}$ for any $c \in \Sigma$ and $L(\pi_1 \pi_2) = L(\pi_1)L(\pi_2)$ for any $\pi_1, \pi_2 \in \Pi$.*
3. *$\Phi : \Pi \to \boldsymbol{R}$ is a homomorphism such that $\Phi(\pi_1 \pi_2) = \Phi(\pi_1) + \Phi(\pi_2)$ for any $\pi_1, \pi_2 \in \Pi$.*

Note that when $\Sigma$ is fixed, a homomorphic SRS is determined by specifying (1) the set $\Delta$ of wildcards, (2) the values $L(\gamma)$ for all $\gamma \in \Delta$, and (3) the values $\Phi(\gamma)$ for all $\gamma \in \Sigma \cup \Delta$.

   The class of homomorphic SRSs covers most of the known similarity (dissimilarity) measures. For example, the edit distance falls into this class. Let $\Delta = \{\psi\}$ where $\psi$ is the wildcard that matches the empty string and any symbol in $\Sigma$, namely, $L(\psi) = \Sigma \cup \{\varepsilon\}$. Let $\Phi(\psi) = -1$ and $\Phi(c) = 0$ for all $c \in \Sigma$. Then, the similarity measure defined by this homomorphic SRS is the same as the edit distance except that the values are non-positive. Similarly, the Hamming distance can be defined by using the wildcard $\phi$ that matches any symbol in $\Sigma$.

   We can define the LCS measure by using the wildcard $\star$ that matches any string in $\Sigma^*$. Namely, the homomorphic SRS such that (1) $\Delta = \{\star\}$, (2) $L(\star) =$

$\Sigma^*$, and (3) $\Phi(\star) = 0$ and $\Phi(c) = 1$ for any $c \in \Sigma$ gives the LCS measure. Although another definition is possible for this measure which uses the wildcard $\psi$ with $L(\psi) = \Sigma \cup \{\varepsilon\}$, but the common patterns obtained are much simpler.

The weighted edit distance can also be defined as a homomorphic SRS in which the wildcards $\phi(a|b)$ $(a, b \in \Sigma \cup \{\varepsilon\}$ and $a \neq b)$ such that $L(\phi(a|b)) = \{a, b\}$ are introduced, and $\Phi(\phi(a|b))$ is the weight assigned to each pair of $a$ and $b$.

Next, we extend the class of homomorphic SRSs by easing the restriction on the pattern score functions as follows. A pattern score function $\Phi$ defined on $\Pi = (\Sigma \cup \Delta)^*$ is said to be *semi-homomorphic* if there exists a subset $\mathcal{D}$ of $\Pi$ with $\varepsilon \notin \mathcal{D}$ and $\Pi = \mathcal{D}^*$, and a function $g : \mathcal{D} \to \mathbf{R}$ such that, for any $\pi \in \Pi$,

$$\Phi(\pi) = \max\left\{ \sum_{i=1}^{\ell} g(\pi_i) \,\middle|\, \ell \geq 0, \pi_i \in \mathcal{D} \ (i = 1, \ldots, \ell), \text{ and } \pi = \pi_1 \cdots \pi_\ell \right\}.$$

**Definition 3.** *An SRS $\langle \Sigma, \Pi, L, \Phi \rangle$ is said to be* semi-homomorphic *if*

1. *$\Pi = (\Sigma \cup \Delta)^*$, where $\Delta$ is a set of wildcards.*
2. *$L : \Pi \to 2^{\Sigma^*}$ is a homomorphism such that $L(c) = \{c\}$ for any $c \in \Sigma$ and $L(\pi_1 \pi_2) = L(\pi_1)L(\pi_2)$ for any $\pi_1, \pi_2 \in \Pi$.*
3. *$\Phi : \Pi \to \mathbf{R}$ is semi-homomorphic.*

Computation of the weighted edit distance between two given strings $x$ and $y$ can be viewed as computation of the lowest scoring paths from node $(0, 0)$ to node $(|x|, |y|)$ in the *weighted edit graph* (see, e.g., [2]), a directed (acyclic) weighted graph where the vertices are the $(|x| + 1) \times (|y| + 1)$ points of the grid with rows $0, \ldots, |x|$ and columns $0, \ldots, |y|$. The computation can be done by standard dynamic programming in $O(|x||y|)$ time.

A similar discussion is possible for (semi-)homomorphic SRSs, with appropriate modifications in the definition of weighted edit graph. The construction time of such graph depends upon the response time for membership query "$w \in L(\gamma)$" for a wildcard $\gamma$ in $\Delta$ and upon that of $\Phi$. However, once such graph is constructed, the best score can be computed in linear time with respect to the number of edges in the graph, which varies depending upon $\Delta$ (and upon $\mathcal{D}$ in the case of semi-homomorphic SRSs). It would be interesting to reveal the hierarchy of subclasses of SRSs from the viewpoint of computational complexity, but this is beyond the scope of the present paper.

As demonstrated so far, we can handle a variety of string (dis)similarity by changing the pattern set $\Pi$ and the pattern score function $\Phi$. The pattern sets discussed above are, however, restricted to the form $\Pi = (\Sigma \cup \Delta)^*$, where $\Delta$ is a set of wildcards. Here we shall mention pattern sets of other types. An *order-free pattern* is a multiset $\{u_1, \ldots, u_k\}$ such that $k > 0$ and $u_1, \ldots, u_k \in \Sigma^+$, and is denoted by $\pi[u_1, \ldots, u_k]$. The language of pattern $\pi[u_1, \ldots, u_k]$ is defined to be the union of the languages $\Sigma^* u_{\sigma(1)} \Sigma^* \cdots \Sigma^* u_{\sigma(k)} \Sigma^*$ over all permutations $\sigma$ of $\{1, \ldots, k\}$. For example, the language of the pattern $\pi[abc, de]$ is $\Sigma^* abc \Sigma^* de \Sigma^* \cup \Sigma^* de \Sigma^* abc \Sigma^*$. The membership problem for order-free patterns is NP-complete, and therefore the similarity computation is impractical generally. However the problem is polynomial-time solvable when $k$ is fixed.

The pattern languages, introduced by Angluin [1], is also interesting for our framework. A *pattern* is a string in $\Pi = (\Sigma \cup V)^+$, where $V$ is an infinite set $\{x_1, x_2, \dots\}$ of variables and $\Sigma \cap V = \emptyset$. The language of a pattern $\pi$ is the set of strings obtained by replacing variables in $\pi$ by non-empty strings. The membership problem is NP-complete for the class of patterns as shown in [1], but it is polynomial-time solvable when the number of variables occurring more than once within $\pi$ is bounded by a fixed number $k$.

# 3    Discovery from Literary Works

Waka is a form of traditional Japanese poetry with a 1300-year history. A Waka poem has five lines and thirty-one syllables, arranged thus: 5-7-5-7-7. In [5] we attempted to semi-automatically discover similar poems from an accumulation of about 450,000 Waka poems in a machine-readable form. One reasonable approach is to arrange all possible pairs of poems in decreasing order of their similarity, and to scholarly scrutinize a first part. One of the aims here is to discover unheeded instances of Honkadori (poetic allusion), one important rhetorical device in Waka poems based on specific allusion to earlier famous poems.

We tested three similarity measures for dealing with similarity between Waka poems, which were newly designed along with our framework. The first measure is based on line-order alternation and on the *modified LCS measure* for quantifying affinity between lines, which is defined as a semi-homomorphic SRS such that $\Delta = \{\star\}$, $L(\star) = \Sigma^*$, and the pattern score function $\Phi$ defined by $\mathcal{D} = \Sigma^+ \cup \{\star\}$, and $g(\pi) = |\pi| - s$, if $\pi \in \Sigma^+$; otherwise, $g(\pi) = 0$, where $s$ $(0 < s \leq 1)$ is a penalty for break in continuity of symbols. This measure was proved suitable for finding instances of poetic allusion.

The second and the third measures are based on the order-free patterns defined in the previous section, in order to cope with word-order alternation. These two measures differ in the respect that the pattern score function of the third measure depends on the *rarity* of common pattern within a given large collection of poems, whereas that of the second one is defined syntactically. The idea of rarity is proved to be effective in identifying only close affinities which are hardly seen elsewhere, possibly excluding known stereotype expressions.

The first measure is especially favored by Waka researchers and used in discovering affinities of some unheeded poems with some earlier ones. The discovered affinities raise an interesting issue for Waka studies: (1) We have proved that one of the most important poems by Fujiwara-no-Kanesuke, one of the renowned thirty-six poets, was in fact based on a model poem found in Kokin-Shū. The same poem had been interpreted just to show "frank utterance of parents' care for their child." Our study revealed the poet's techniques in composition half hidden by the heart-warming feature of the poem by extracting the same structure between the two poems. (2) We have compared Tametada-Shū, the mysterious anthology unidentified in Japanese literary history, with a number of private anthologies edited after the middle of the Kamakura period (the 13th-century) using the same method, and found that there are about 10 pairs of similar poems between Tametada-Shū and Sōkon-Shū, an anthology by Shōtetsu. The result suggests that the mysterious anthology was edited by a poet in the early Muromachi period (the 15th-century). There have been surmised dispute about the editing date since one scholar suggested the middle of Kamakura period as a probable one. We have had a strong evidence about this problem.

# 4   Finding Affinities from Musical Scores

Any monophonic score can be regarded as a string of ordered pairs consisting of the pitch of the note and its length. Mongeau and Sankoff [4] proposed a dissimilarity measure for monophonic scores, which is a variant of the weighted edit distance where additional two edit operations, *fragmentation* and *consolidation*, are allowed to associate multiple notes with a single note or vice versa. It is reported in [4] that the measure arranges the variations on a theme by Mozart in a reasonable order which coincides with subjective impressions. However, it turned out from our experimental results that a problem arises when dealing with the mixtures of variations on several themes.

In [3] we tested three similarity measures and showed that the third one could cope with this problem. As a preprocessing, each note in a musical score is replaced with a sequence of notes of a unit length (16th note) to obtain simply a string of pitches. The measures are respectively based on three measures to quantify the affinities between two phrases of uniform length, each falls into the class of the semi-homomorphic SRSs. The set $\Pi = (\Sigma \cup \{\phi\})^*$ is commonly used in the three. While the pattern score function of the first measure is the one which simply counts up matches (i.e., the number of symbols in a pattern), those of the second and third measures are sensitive to the continuity of matches. More precisely, in the second measure it is defined by $\mathcal{D} = \Sigma^+ \cup \{\phi\}$, and $g(\pi) = |\pi|$, if $\pi \in \Sigma^+$ and $|\pi| \geq s$; otherwise, $g(\pi) = 0$, where $s$ is a threshold. In the third measure, $\mathcal{D} = \{\pi \in (\Sigma \cup \{\phi\})^+ \mid \pi \text{ does not contain } \phi^{t+1}\}$, and $g(\pi)$ is the number of symbols within $\pi$, if $|\pi| \geq s$; otherwise, $g(\pi) = 0$, where $s, t$ are thresholds. Despite its simplicity, the third measure is better than Mongeau and Sankoff's one in the sense that it is able to exclude variations on other themes.

# References

1. D. Angluin. Finding patterns common to a set of strings. *J. Comput. Sys. Sci.*, 21:46–62, 1980.
2. D. Gusfield. *Algorithms on Strings, Trees, and Sequences: Computer Science and Computational Biology.* Cambridge University Press, New York, 1997.
3. T. Kadota, A. Ishino, M. Takeda, and F. Matsuo. On melodic similarity. *IPSJ SIG Notes*, 2000(49):15–24, 2000. (in Japanese).
4. M. Mongeau and D. Sankoff. Comparison of musical sequences. *Computers and the Humanities*, 24(3):161–175, 1990.
5. K. Tamari, M. Yamasaki, T. Kida, M. Takeda, T. Fukuda, and I. Nanri. Discovering poetic allusion in anthologies of classical Japanese poems. In *Proc. 2nd International Conference on Discovery Science (DS'99)*, pages 128–138, 1999.

# Efficient Discovery of Proximity Patterns
# with Suffix Arrays
## (Extended Abstract)

Hiroki Arimura[1,2], Hiroki Asaka[1], Hiroshi Sakamoto[1], and Setsuo Arikawa[1]

[1] Department of Informatics, Kyushu University
Fukuoka 812-8581, Japan
{arim,arikawa}@i.kyushu-u.ac.jp
[2] PRESTO, Japan Science and Technology Corporation, Japan

**Abstract.** We describe an efficient implementation of a text mining algorithm for discovering a class of simple string patterns. With an index structure, called the *virtual suffix tree*, for pattern discovery built on the top of the suffix array, the resulting algorithm is simple and fast in practice compared with the previous implementation with the suffix tree.

## 1 Introduction

In this extended abstract, we investigated a text mining problem based on the framework of *optimal pattern discovery*[13]. We give an efficient implementation of an existing text mining algorithm [2] for finding $k$-*proximity d-phrase patterns* from a large collection of texts using a data structure, called the virtual suffix tree. The *virtual suffix tree* is a space-efficient alternative to the suffix-tree built on the top of the suffix array [5,12] and can be used for bottom-up traversal of the suffix tree allowing reconstruction of the index [10].

Our algorithm runs in almost linear time with poly-log factor in the total length of a text collection in average case for fixed $k, d \geq 0$. From the practical point of view, the algorithm is conceptually simple, easy to implement, and scalable on large text data due to the use of the virtual suffix tree than the previous implementation with the suffix tree.

## 2 The Class of Patterns

We introduce the class of patterns to discover [2]. Let $\Sigma$ be a constant alphabet of letters. A $k$-*proximity d-phrase association patterns* ($(k, d)$-proximity pattern) is a sequence of $d$ phrases $(\langle p_1, \cdots, p_d \rangle, k)$ with the bounded gap length $k$, called *proximity*. Each $p_i$ is a substring of a text, called a *phrase*. The pattern $\pi$ *matches* a text if the phrases $p_1, \ldots, p_d$ occur in the text in this order and the distance between the occurrences of the consecutive pair of phrases does not exceed $k$. For instance, $(\langle \mathtt{tata} \rangle, \langle \mathtt{cacag} \rangle, \langle \mathtt{caatcag} \rangle; 20)$ is an example of $(20, 3)$-patterns. $\mathcal{P}_k^d$ denotes the class of $(k, d)$-proximity patterns. The well-known *followed-by* patterns [5,11] are special case with $d = 2$. Another definition as *gap patterns* with with bounded gap length is possible, but slight change is needed.

A. Amir and G.M. Landau (Eds.): CPM 2001, LNCS 2089, pp. 152–156, 2001.

# 3   Framework of Text Mining

As the framework of data mining, we adopted *optimal pattern discovery* [13]. A *sample* is a pair $(S, \xi)$ of a collection of texts $S = \{s_1, \ldots, s_m\} \subseteq \Sigma^*$ and a binary labeling $\xi : S \to \{0, 1\}$, called *objective function*, which may be a pre-defined or human-specified category.

Let $\mathcal{P}$ be a class of *patterns* and $\psi : [0, 1] \to \mathbf{R}$ be a symmetric, concave, real-valued function called *impurity function* [4] to measure the skewness of the class distribution after the classification by a pattern. The *classification error* $\psi(x) = \min(p, 1 - p)$ and the *information entropy* $\psi(x) = -p \log p - (1 - p) \log(1 - p)$ are examples of $\psi$ [4,13]. We identify each $\pi \in \mathcal{P}$ to a classification function (matching function) $\pi : \Sigma^* \to \{0, 1\}$ as usual.

Given a sample $(S, \xi)$, a pattern discovery algorithm tries to find a pattern $\pi$ in the hypothesis space $\mathcal{P}$ that optimizes the *evaluation function* $G^{\psi}_{S,\xi}(\pi) = \psi(M_1/N_1)N_1 + \psi(M_0/N_0)N_0$ where tuple $(M_1, M_0, N_1, N_0)$ is the *contingency table* determined by $\pi$, $\xi$ over $S$, namely, $M_\alpha = \sum_{x \in S} [\pi(x) = 1 \wedge \xi(x) = \alpha]$ and $N_\alpha = \sum_{x \in S} [\pi(x) = \alpha]$, where $\alpha \in \{0, 1\}$ and $[Pred] \in \{0, 1\}$ is the indicator function. Now, we state our data mining problem.

**Optimal Pattern Discovery with $\psi$**
**Given:** a set $S$ of texts and an objective function $\xi : S \to \{0, 1\}$.
**Problem:** Find a pattern $\pi \in \mathcal{P}$ that minimizes the cost $G^{\psi}_{S,\xi}(\pi)$ within $\mathcal{P}$.

What is good to minimize the cost $G^{\psi}_{S,\xi}(\pi)$? For the case of the classification error $\psi$ used in machine learning, it is known that any algorithm that efficiently solves the above optimization problem can approximate an arbitrary unknown probability distribution and thus can work with noisy environments [8].

# 4   Previous Algorithms with Suffix Tree

In this poster, we consider only the case of fixed $d$, the maximum number of phrases. Otherwise, the problem becomes hard to approximate (MAXSNP-hard) for the class $\cup_{k,d} \mathcal{P}^d_k$ [2]. For fixed $d$, the optimal pattern discovery problem for $\mathcal{P}^d_k$ is solvable by a naive generate-and-test algorithm in $O(n^{2d+1})$ time. Although this can be reduced to $O(n^{d+1})$ time with the suffix tree [15], it is still too slow to apply real world problems.

In [2], we have developed an efficient algorithm, called *Split-Merge-with-Tree* (SMT) for $\mathcal{P}^d_k$. With the suffix tree [10], SMT uses the known correspondence between patterns in $\mathcal{P}^d_k$ and the axis-parallel rectangles in $d$-dim rank space $\{1, \ldots, n\}^d$ on a suffix array [11]. *SMT* searches a best rectangle by combining the $d$-dimensional orthogonal range tree and the suffix tree [2]. *SMT* is fast in theory, but slow in practice due to the huge space requirement and complicated implementation [3].

# 5    Reconstructible Suffix Dictionary

For overcoming the problem, we implemented the idea used in the *SMT* algorithm using a new indexing data structure called the virtual suffix tree described below.

Let $T$ be a text of length $n$ and $P = \{1, \ldots, n\}$ be the set of all positions (or index points) where are possible occurrences of patterns. We identify a suffix with the index point it starts, and also identify a text $T$ as the set of all of its suffixes. Let $Q \subseteq T$ be any subset of the suffixes and $p$ be any branching substring of $T$. Define $p$ belongs to $Q$ if $p$ is a prefix of some suffix in $Q$, and define the *info* of $p$ is any information from which we can retrieve $|p|$ in $O(1)$ time and can enumerate all $K$ occurrence of $p$ in $T$ in $O(K)$ time.

The essence of the previous algorithm *SMT* is summarized by an abstract data type called a *reconstructible suffix dictionary* $D(T)$ defined with the following operations.

*Reconstructible suffix dictionary*
- Given a text $T$ of length $n$, create the dictionary $D(T)$ for all suffixes of $T$ (Create) in $(n \log n)$ time (Note that we have identified $T$ and the set of its positions).
- Given a dictionary $D(Q)$ and a subset $Q' \subseteq Q$ of index points, reconstruct the dictionary $D(Q')$ for all suffixes in $Q'$ in $O(m \log m)$ time, where $m = |Q'|$ is the cardinality of $Q'$ (Reconstruct).
- Given $D(Q)$, enumerate the info's of all branching substring in $Q$ in $O(m)$ time, where $m = |Q|$ (Traverse).
- Given $D(Q)$ and a string $p$, detect all the occurrences (or the info) of $p$ in $Q$ in $O(|p| \log m)$ time if exists, where $m = |Q|$. (Search)

We implemented the above abstract data type by a data structure called a *virtual suffix tree*, which is just the *suffix array SA* coupled with the *height array* $Hgt$ [12], the inverse array $Rank$ of $SA$ [11], and the *range minima query* for constant-time lcp information [6,14]. Here, $Hgt$ is the array that stores the the length of the longest common prefixes of adjacent suffixes in $SA$ [12]. Although a reconstructible suffix dictionary can be implemented by the suffix tree [10] combined with marking technique, the virtual suffix tree has advantages over the suffix tree when simplicity, space-efficiency and quick reconstruction (restriction) are important.

For *Traverse* operation, we developed a linear time algorithm [7] that simulates bottom-up traversal of the suffix tree when $SA$ and $Hgt$ are given, while well-known simulation of the suffix tree by binary search takes $O(n^2)$ time and $O(n \log n)$ time without and with lcp info.[14]), resp. We also developed a linear time algorithm for building the array $Hgt$ from $SA$ and $T$ [7]. For *Reconstruct* operation, we use the inverse array $Rank$ and the sorting of ranks for update of $Pos$ array, and also use constant-time lcp information for update of $Hgt$ array in claimed worst-case time complexity.

---

*Algorithm Split-Merge-with-Array* (SMA):
- *Given*: Integers $k, d \geq 0$ and text $T$.
- Initialize $R = \{1, \ldots, n\}$ and create the virtual suffix tree $D(T)$ for text $T$. Invoke $Find(R, d, k, \varepsilon)$ below and report the patterns with the minimal cost.

*Procedure Find$(Q, d, k, \pi)$*:
- *Given*: A set $Q$ of index points, integers $d, k \geq 0$, and a sequence $\pi$ of phrases.
- If $d = 0$, then compute the cost $G^{\psi}_{S,\xi}(\pi)$ from $Q$, which corresponding to the set of the occurrences of $\pi$ in $T$. Record the pattern $(\pi; k)$ with its cost.
- Otherwise, $d > 0$. Reconstruct $D(Q)$ from $D(T)$. Then, traverse the info's of all branching substrings $p$. For each $p$ in $D(Q)$, do the followings:
  (i) Let $O$ be the set of all occurrences of $p$ in $Q$. Shift each position in $O$ at most $k$ positions to the right by adding the skip $|p| + g$ for every $0 \leq g \leq k$. Let $P$ be the resulting set of ranks.
  (ii) Invoke $Find(P, d - 1, \pi \cdot p)$ recursively.

---

**Fig. 1.** The algorithm for finding proximity phrase patterns with suffix arrays.

# 6   Fast Text Mining Algorithm with Suffix Array

Let $T = s_1 \$_1 \cdots s_m \$_m$ be a text obtained by concatenating all texts in $S = \{s_1, \ldots, s_m\}$ delimited with unique markers $\$'_i s$. Here, the document id $i$ and the label $\xi(s_i)$ are attached with each position. Fig. 1 shows our mining algorithm *Split-Merge-with-Array* (SMA). By an analysis similar to [2], we have the following theorem. Details of the proof will appear in elsewhere.

**Theorem 1.** *For every integers $d, k$, the algorithm SMA, given a sample $(S, \xi)$, computes all $K$ $(k, d)$-proximity patterns that minimize the cost $G^{\psi}_{S,\xi}$ in expected time $O(k^{d-1} n (\log n)^d + K)$ and space $O(dn)$ under the assumption that texts in $S$ are randomly generated from a memoryless source.*

By experiments on English texts (15.2MB) [9] and Web pages [1], we observed that an implementation of SMA runs in several seconds $(d = 1)$ to several minutes $(d = 4$ and $k = 8$ words) with WS (Ultra SPARC II, 300MHz, 256MB, Solaris 2.6, gcc). This is about $10^2$ to $10^3$ times speed-up to the previous implementation [3] of the algorithm SMT with the suffix tree and the range-tree.

## Acknowledgments

This work is partially supported by a Grant-in-Aid for Scientific Research on Priority Areas "Discovery Science" from the Ministry of Education, Science, Sports, and Culture in Japan. We would like to thank to Junichiro Abe, Ryouichi Fujino, Shimozono Shinich for fruitful discussions and comments on this issue.

# References

1. H. Arimura, J. Abe, R. Fujino, H. Sakamoto, S. Shimozono, S. Arikawa, Text Data Mining: Discovery of Important Keywords in the Cyberspace, In *Proc. IEEE Kyoto Int'l Conf. Digital Library*, 2001. (to appear)
2. H. Arimura, S. Arikawa, S. Shimozono, Efficient discovery of optimal word-association patterns in large text databases, *New Generation Computing*, Special issue on Discovery Science, 18, 49–60, 2000.
3. Arimura, H., Wataki, A., Fujino, R., Arikawa, S., A fast algorithm for discovering optimal string patterns in large text databases, In *Proc. the 9th Int. Workshop on Algorithmic Learning Theory* (ALT'98), LNAI 1501, 247–261, 1998.
4. L. Devroye, L. Gyorfi, G. Lugosi, *A Probablistic Theory of Pattern Recognition*, Springer-Verlag,1996.
5. G. Gonnet, R. Baeza-Yates and T. Snider, New indices for text: Pat trees and pat arrays, In William Frakes and Ricardo Baeza-Yates (eds.), *Information Retrieval: Data Structures and Algorithms*, 66–82, 1992.
6. D. Gusfield, Algorithms on Strings, Trees, and Sequences: Computer Science and Computational Biology, Cambridge University Press, New York, 1997.
7. T. Kasai, G. Lee, H. Arimura, S. Arikawa, K. Park, Linear-time longest-common-prefix computation in suffix arrays and its applications, In Proc. CPM'01, LNCS, Springer-Verlag, 2000 (this volumn). (A part of this work is also available as: T. Kasai, H. Arimura, S. Arikawa, Efficient substring traversal with suffix arrays, DOI-TR 185, 2001, `ftp://ftp.i.kyushu-u.ac.jp/pub/tr/trcs185.ps.gz`.)
8. M.J. Kearns, R.E. Shapire, L.M. Sellie, Toward efficient agnostic learning. *Machine Learning*, 17(2–3), 115–141, 1994.
9. D. Lewis, Reuters-21578 text categorization test collection, Distribution 1.0, AT&T Labs-Research, `http://www.research.att.com/~lewis/`, 1997.
10. E.M. McCreight, A space-economical suffix tree construction algorithm, *JACM*, 23(2):262-272, 1976.
11. U. Manber and R. Baeza-Yates, An algorithm for string matching with a sequence of don't cares. IPL 37, 1991.
12. U. Manber and G. Myers, Suffix arrays: A new method for on-line string searches, *SIAM J. Computing*, 22(5), 935–948 (1993).
13. S. Morishita, On classification and regression, In *Proc. Discovery Science '98*, LNAI 1532, 49–59, 1998.
14. B. Schieber and U. Vishkin, On finding lowest common ancestors: simplifications an parallelization, *SIAM J. Computing*, 17, 1253–1262, 1988.
15. J.T.L. Wang, G.W. Chirn, T.G. Marr, B. Shapiro, D. Shasha and K. Zhang, In *Proc. SIGMOD'94*, 115–125, 1994.

# Computing the Equation Automaton of a Regular Expression in $O(s^2)$ Space and Time

Jean-Marc Champarnaud and Djelloul Ziadi

LIFAR, Université de Rouen
76821 Mont-Saint-Aignan Cedex, France
{champarnaud,ziadi}@dir.univ-rouen.fr

**Abstract.** Let $E$ be a regular expression the size of which is $s$. Mirkin's prebases and Antimirov's partial derivatives lead to the construction of the same automaton, called the equation automaton of $E$. The number of states in this automaton is less than or equal to the number of states in the position automaton. On the other hand, it can be computed by Antimirov's algorithm with an $O(s^5)$ time complexity, whereas there exist $O(s^2)$ implementations for the position automaton. We present an $O(s^2)$ space and time algorithm to compute the equation automaton. It is based on the notion of canonical derivative which is related both to word and partial derivatives. This work is tightly connected to pattern matching area since the aim is, given a regular expression, to produce an as small as possible recognizer with the best space and time complexity.

## 1  Introduction

The conversion of a regular expression into an equivalent finite automaton [21] has many applications, especially in pattern matching [10]. The notion of word derivative of a regular expression [7] and the related notions of continuation [6] and of partial derivative [3] are suitable tools to study this problem. Three fundamental results lead to the construction of three well-known automata. First, the set of the aci-dissimilar word derivatives of a regular expression is finite [7], which leads to the definition of the deterministic derivatives automaton. Secondly, the continuations w.r.t. a given symbol $a$ in a linear expression (i.e. the non-null derivatives w.r.t. $ua$, for all words $u$) are aci-similar [6], which yields a constructive interpretation of the position automaton (classically computed by Glushkov [11] and McNaughton-Yamada [17] algorithms). And third, the set of the partial derivatives of a regular expression is finite [3], which leads to the construction of the equation automaton [18,3].

The notion of canonical derivative (or c-derivative) developed by the authors [9] enlightens the tight connection which exists between the set of partial derivatives of a regular expression and the set of word derivatives of its linearized version. This notion leads to the definition of the c-continuation automaton, whose main interest is that it yields the position automaton by an isomorphism and the equation automaton by a quotient. Let us point out that these theoretical results are not related to the work of Hromkovič *et al.* [14]: firstly, the

A. Amir and G.M. Landau (Eds.): CPM 2001, LNCS 2089, pp. 157–168, 2001.
© Springer-Verlag Berlin Heidelberg 2001

common follow sets automaton they define has more many states than the position automaton, and secondly they do not use any derivative-like tool; on the opposite, our results are based on a new algebraic tool: the c-derivatives.

Given a regular expression the size of which is $s$ the c-continuation automaton and its quotient can be computed with an $O(s^2)$ space and time complexity, which significantly improves the $O(s^5)$ complexity of Antimirov's algorithm [3], and takes up the challenge of computing the equation automaton, which is smaller than the position one, with the same complexity as the most efficient implementations of the position automaton. The set of states is deduced from a preprocessing of the starred subexpressions and from an implicit computing of the c-continuations. The computation of the set of transitions is based on the specific structure of the c-derivatives and on their connection to position sets. Let us notice that the techniques we use to handle c-continuations are necessarily different from the procedures used by Hagenah and Muscholl [12,13] to implement the common follow sets automaton. On the other hand, some refinements used in this paper to compute the set of transitions lay on the properties of the implicit structure [23,20] we designed in the past to represent the position automaton; a very closely related structure is used in [12,13].

Section 2 recalls the classical constructions of the position automaton and of the equation automaton. Section 3 summarizes theoretical results concerning c-derivatives and c-continuations of a regular expression, and their relations with word derivatives and partial derivatives. The definition of the c-continuation automaton is recalled, as well as the way it is connected to the position automaton and to the equation automaton. The new algorithm to build the equation automaton is developed in Section 4 which deals with the construction of the set of states and in Section 5 which presents algorithmic refinements to compute the set of transitions. The new algorithm has been implemented in language C; a full example of its output is provided in the Annex A.

## 2    Preliminaries

We assume terminology and basic results concerning regular languages and finite automata are known and refer to classical books [4,22] about these topics. We recall the classical constructions of the position automaton and of the equation automaton.

The *size* of an expression is the length of its suffixed form, and its *(alphabetic) width* is the number of symbol occurrences. Let $E$ be a regular expression of size $s$ and width $w$. Notice that $s$ and $w$ are linearly dependent as far as sequences of star operators and occurrences of the empty word and of the empty set are carefully handled. However, some computation steps depend on the size of the expression and other ones on the alphabetic width. This additional information may be helpful, for implementation purpose for instance, or for a deeper analysis of the number of operations. This is why we express the complexities w.r.t. both $s$ and $w$.

## 2.1   The Position Automaton of a Regular Expression

Let $E$ be a linear expression over $\Sigma$ and $\lambda(E)$ be defined by: $\lambda(E) = 1$ if $\varepsilon \in L(E)$ and 0 otherwise. We consider the following sets of symbols:
– $First(E)$, the set of symbols that match the first symbol of some word in $L(E)$.
– $Last(E)$, the set of symbols that match the last symbol of some word in $L(E)$.
– $Follow(E, x)$, for all $x$ in $\Sigma$: the set of symbols that follow the symbol $x$ in some word of $L(E)$.
If $E$ is a regular expression over $\Sigma$, and $Pos_E$ the set of its positions, we consider its linearized version $\overline{E}$, whose symbols are the positions of $E$, and we set: $First(E) = First(\overline{E})$, $Last(E) = Last(\overline{E})$, $Follow(E, x) = Follow(\overline{E}, x)$, for all $x$ in $Pos_E$. Let $h$ be the mapping from $Pos_E$ to $\Sigma$ induced by the linearization of $E$ over $Pos_E$. The position automaton $\mathcal{P}_E$ of $E$, whose states are the positions of $E$, and which recognizes $L(E)$, is defined as follows.

**Definition 1 (Position Automaton).** *The position automaton of $E$, $\mathcal{P}_E = (Q, \Sigma, i, T, \delta)$, is defined by:* $Q = Pos_E \cup \{0\}$, $i = \{0\}$, $T = [if\ \lambda(E) = 0\ then\ Last(E)\ else\ Last(E) \cup \{0\}]$, $\delta(0, a) = \{x \in First(E) \mid h(x) = a\}$, $\forall a \in \Sigma$, and $\delta(x, a) = \{y \mid y \in Follow(E, x)\ and\ h(y) = a\}$, $\forall x \in Pos_E$ and $\forall a \in \Sigma$.

The automaton $\mathcal{P}_E$ is classically built by an $O(s^3)$ algorithm due to Glushkov [11] and to McNaughton and Yamada [17]. There exist $O(s^2)$ implementations [5,8], [23,20].

*Example 1.* Let $E = x^*(xx + y)^*$ and $\overline{E} = x_1^*(x_2 x_3 + y_4)^*$. We have: $First(E) = \{x_1, x_2, y_4\}$, $Last(E) = \{x_1, x_3, y_4\}$, $\lambda(E) = 1$, $Follow(E, x_1) = \{x_1, x_2, y_4\}$, $Follow(E, x_2) = \{x_3\}$, $Follow(E, x_3) = Follow(E, y_4) = \{x_2, y_4\}$.

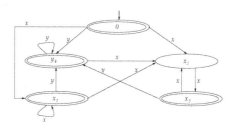

**Fig. 1.** The position automaton for $E = x^*(xx + y)^*$.

## 2.2   The Equation Automaton of a Regular Expression

The notion of partial derivative of a regular expression is due to Antimirov [3].

**Definition 2 (Set of Partial Derivatives w.r.t. a Word).** *Given a regular expression $E$ and a word $u$, the set of partial derivatives of $E$ w.r.t. $u$, written*

$\partial_u(E)$, is recursively defined on the structure of $E$ as follows:

$$\partial_a(0) = \emptyset = \partial_a(1)$$
$$\partial_a(x) = \{1\} \quad if\ a = x \quad \emptyset\ otherwise$$
$$\partial_a(F + G) = \partial_a(F) \cup \partial_a(G)$$
$$\partial_a(F \cdot G) = \begin{cases} \emptyset & if\ G = 0 \\ \partial_a(F) \cdot G & if\ G \neq 0\ and\ \lambda(F) = 0 \\ \partial_a(F) \cdot G \cup \partial_a(G) & otherwise \end{cases}$$
$$\partial_a(F^*) = \partial_a(F) \cdot F^*$$
$$\partial_\varepsilon(E) = \{E\} \quad and \quad \partial_{ua}(E) = \partial_a(\partial_u(E))$$

*Example 2.* Let $E = x^*(xx+y)^*$. We have: $\partial_x(E) = \{x^*(xx+y)^*,\ x(xx+y)^*\}$, $\partial_y(E) = \{(xx+y)^*\}$, $\partial_x(x(xx+y)^*) = \{(xx+y)^*\}$, $\partial_x((xx+y)^*) = \{x(xx+y)^*\}$, $\partial_y((xx+y)^*) = \{(xx+y)^*\}$.

Antimirov [3] has proved that the cardinality of the set $\mathcal{PD}(E)$ of all the partial derivatives of a regular expression $E$ is less than or equal to $w + 1$. Hence the definition of $\mathcal{E}_E$, the equation automaton of $E$, whose states are the partial derivatives of $E$, and which recognizes $L(E)$.

**Definition 3 (Equation Automaton).** *The equation automaton of a regular expression* $E$, $\mathcal{E}_E = (Q, \Sigma, i, T, \delta)$, *is defined by:* $Q = \mathcal{PD}(E)$, $i = E$, $T = \{p \mid \lambda(p) = 1\}$ *and* $\delta(p, a) = \partial_a(p)$, $\forall p \in Q$ *and* $\forall a \in \Sigma$.

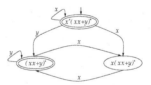

**Fig. 2.** The equation automaton for $E = x^*(xx + y)^*$.

# 3 The C-Continuation Automaton of a Regular Expression

The new algorithm we present to compute the equation automaton is based on the notion of c-derivative of a regular expression [9]. In this section, we review the main properties of c-derivatives, and we recall the definition of the c-continuation automaton.

## 3.1   C-Derivatives

**Definition 4 (C-Derivative).** *The c-derivative $d_u(E)$ of a regular expression $E$ w.r.t. a word $u$ is recursively defined as follows:*

$$d_a(0) = 0 = d_a(1)$$
$$d_a(x) = 1 \ \text{if} \ a = x, \ 0 \ otherwise$$
$$d_a(F + G) = d_a(F) \ \text{if} \ d_a(F) \neq 0, \ d_a(G) \ otherwise$$
$$d_a(F \cdot G) = d_a(F) \cdot G \ \text{if} \ d_a(F) \neq 0, \ \lambda(F) \cdot d_a(G) \ otherwise$$
$$d_a(F^*) = d_a(F) \cdot F^*$$
$$d_\varepsilon(E) = E \quad and \quad d_{u_1 \ldots u_n}(E) = d_{u_2 \ldots u_n}(d_{u_1}(E))$$

The two following propositions connect c-derivatives respectively to word derivatives and to partial derivatives.

**Proposition 1.** *Let $E$ be a linear expression, $a$ be any symbol in $\Sigma$, $u$ and $v$ be any words in $\Sigma^*$. The following properties hold:*
*(a) A non-null c-derivative of $E$ is either 1 or a subexpression of $E$ or a product of subexpressions of $E$.*
*(b) $u^{-1}E \sim_{aci} v^{-1}E \Leftrightarrow d_u(E) \equiv d_v(E)$*
*(c) The set of non-null c-derivatives $d_{ua}(E)$ of a linear expression $E$ reduces to a unique expression, the c-continuation of $a$ in $E$, denoted by $c_a$.*

**Proposition 2.** *Let $E$ be a regular expression over $\Sigma$, $\overline{E}$ be its linearized version and $h$ be the projection of $Pos_E$ on $\Sigma$. The set of partial derivatives of $E$ w.r.t. $u$ is equal to the set of the projections on $\Sigma$ of the non-null c-derivatives of $\overline{E}$ w.r.t. words $v$ over $Pos_E$ such that $h(v) = u$.*

As a corollary of Proposition 2, the set $\mathcal{PD}(E)$ of all the partial derivatives of $E$ is equal to the set of the projections on $\Sigma$ of the c-continuations in $\overline{E}$.

*Example 3.* Let $E = x^*(xx + y)^*$. We have:
$\mathcal{PD}(E) = \{x^*(xx + y)^*, \ x(xx + y)^*, \ (xx + y)^*\}$.
    The c-continuations in $\overline{E}$ are: $c_0 = \overline{E}$, $c_{x_1} = x_1^*(x_2x_3 + y_4)^*$, $c_{x_2} = x_3(x_2x_3 + y_4)^*$, $c_{x_3} = (x_2x_3 + y_4)^*$ and $c_{y_4} = (x_2x_3 + y_4)^*$.
    We verify that: $\mathcal{PD}(E) = \{h(c_0), \ h(c_{x_1}), \ h(c_{x_2}), \ h(c_{x_3}), \ h(c_{y_4})\}$.

## 3.2   The C-Continuation Automaton

Let $E$ a regular expression. Proposition 1.c leads to the definition of the non-deterministic automaton $\mathcal{C}_E$, called the c-continuation automaton of $E$. States are pairs $(x, c_x)$, where $x$ is in $Pos_E \cup \{0\}$ and $c_x$ is the c-continuation of $x$ in $\overline{E}$. Transitions are deduced from the c-derivation of c-continuations $c_x$.

**Definition 5 (C-Continuation Automaton).** *The c-continuation automaton of $E$, $\mathcal{C}_E = (Q, \Sigma, i, T, \delta)$, is defined by: $Q = \{(x, c_x) | x \in Pos_E \cup \{0\}\}$, $i = (0, c_0)$, $T = \{(x, c_x) \mid \lambda(c_x) = 1\}$, $\delta((x, c_x), a) = \{(y, c_y) \mid h(y) = a \ and \ d_y(c_x) \equiv c_y\}$, $\forall x \in Pos_E \cup \{0\}$ and $\forall a \in \Sigma$.*

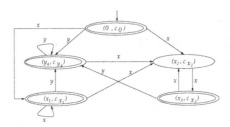

**Fig. 3.** The c-continuation automaton for $E = x^*(xx + y)^*$.

The c-continuation automaton $\mathcal{C}_E$ and the position automaton $\mathcal{P}_E$ are isomorphic. This property is a corollary of Berry and Sethi results [6], since a c-continuation is a particular continuation. The proof can also be directly deduced from the following proposition:

**Proposition 3.** *Let $E$ be a regular expression. The following equalities hold:*

1. $First(E) = \{y \in Pos_E \mid d_y(\overline{E}) \neq 0\}$;
2. $Last(E) = \{y \in Pos_E \mid \lambda(c_y(\overline{E})) = 1\}$;
3. $Follow(E, x) = \{y \in Pos_E \mid d_y(c_x(\overline{E})) \neq 0\}$.

### 3.3    The Quotient of the C-Continuation Automaton

Let $\sim$ be the equivalence relation on the set of states of $\mathcal{C}_E$, defined by: $(x, c_x) \sim (y, c_y) \Leftrightarrow h(c_x) \equiv h(c_y)$. This relation is proved to be right-invariant, thus the quotient automaton $\mathcal{C}_E/_\sim = (Q_\sim, \Sigma, i, T, \delta)$ is defined as follows:

**Definition 6 (Quotient Automaton).** $Q_\sim = \{[c_x] \mid x \in Pos_E \cup \{0\}\}$, $i = [c_0]$, $T = \{[c_x] \mid \lambda(c_x) = 1\}$, *and* $[c_y] \in \delta([c_x], a) \Leftrightarrow \exists c_z \mid c_z \in [c_y]$, $h(z) = a$ *and* $d_z(c_x) \equiv c_z\}$, $\forall[c_x], [c_z] \in Q_\sim$ *and* $\forall a \in \Sigma$.

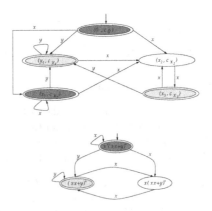

**Fig. 4.** $\mathcal{C}_E$ and $\mathcal{C}_E/_\sim$ for $E = x^*(xx + y)^*$.

From Proposition 2 it comes: the quotient automaton $\mathcal{C}_E/_\sim$ and the equation automaton $\mathcal{E}_E$ are identical. This result leads to a new construction of the equation automaton.

In the two following sections, we first show how to implement this new construction with a cubic time and space complexity. Then we present algorithmic refinements which lead to a quadratic time and space complexity, thus considerably improving Antimirov's algorithm.

## 4   Computation of the Set of States of $\mathcal{C}_E/_\sim$

We first describe an $O(s^3)$ explicit computation of the list of the c-continuations of $\overline{E}$. Then we introduce the notion of pseudo c-continuation and show that it leads to an $O(s^2)$ computation of the set of states of $\mathcal{C}_E/_\sim$.

### 4.1   Computing the List of the C-Continuations

As a consequence of Definition 4, the following property holds:

**Proposition 4.** *The c-derivative $d_u(E)$ of a linear expression $E$ w.r.t. a word $u$ of $\Sigma^+$ is either $0$ or such that:*

$$d_u(u) = 1$$
$$d_u(F + G) = d_u(F) \ if \ d_u(F) \neq 0, \ d_u(G) \ otherwise$$
$$d_u(F \cdot G) = \begin{cases} d_u(F) \cdot G & if \ d_u(F) \neq 0 \\ d_s(G) & otherwise \ (s \neq \varepsilon \ is \ some \ suffix \ of \ u) \end{cases}$$
$$d_u(F^*) = d_s(F) \cdot F^* \ \ (s \neq \varepsilon \ is \ some \ suffix \ of \ u)$$

This property implies that a c-continuation $c_a(E)$ in a linear expression $E$ is a product of distinct subexpressions $H_i$ of $E$, possibly reduced to a single subexpression or to 1. We now show how this product can be computed over the syntax tree $T(E)$ of $E$.

**Proposition 5.** *Let $F$ and $G$ be subexpressions of the linear expression $E$. Let $f$ be the mapping such that:*

$$f(F) = \begin{cases} F^* & if \ F \ is \ a \ son \ of \ F^* \ in \ T(E) \\ G & if \ F \ is \ a \ son \ of \ F \cdot G \ in \ T(E) \\ 1 & otherwise \end{cases}$$

*Let $\odot$ denote the concatenation of a list of expressions. We write $F \prec G$ if and only if $G$ is an ancestor of $F$. Then for all symbol $a$ in $\Sigma$, we have:*

$$c_a(E) = \bigodot_{a \preceq H \prec E} f(H)$$

The proof is by induction on the number of operators in $E$. The complexity of the computation of the c-continuations over $T(E)$ is given by the following proposition.

**Proposition 6.** *Let $E$ be a linear expression of size $s$ and alphabetic width $w$. The list of the c-continuations in $E$ can be computed with an $O(ws^2)$ space and time complexity. If $E$ is starfree, the complexity is $O(ws)$.*

The proof lays on the fact that the size of a c-continuation is bounded by $s^2$. Moreover, Aho *et al.* algorithm [1] or Paige and Tarjan refinement [19] allow to sort a list of strings in lexicographic order with a time complexity $O(\sigma + k)$, where $\sigma$ is the total sum of the sizes of the strings and $k$ is the size of the alphabet. We deduce an $O(ws^2)$ space and time procedure to identify equivalent c-continuations. Hence the sets of states of $\mathcal{C}_E$ and of $\mathcal{C}_E/\sim$ can be produced in $O(ws^2)$ space and time.

## 4.2   Preprocessing of Starred Subexpressions

According to Proposition 6, the computation of the list of the c-continuations is only $O(ws)$ when $E$ is starfree. We therefore preprocess starred subexpressions of $E$, and we compute the list of the c-continuations in the resulting starfree expression $E'$. Notice that the expression $E'$ is not explicitly computed: the aim is to substitute their star-names to starred subexpressions involved in a c-continuation, and it can be achieved by making use of the labels of star-links. Hence the procedure:

*Procedure* Stars($E$: regular expression)
   *Step 1:*   Process a topdown left to right traversal of $T(E)$,
               and store the starred subexpressions in a list $L$.
   *Step 2:*   Sort $L$ in lexicographic order
               and associate the same star-name to identical strings.
   *Step 3:*   Mark star-links (s. t. $f(F) = F^*$) by the star-name of $F^*$.

The complexity of this preprocessing is as follows:

**Proposition 7.** *Let $E$ a linear expression of size $s$ and alphabetic width $w$. The procedure Stars($E$) preprocesses the starred subexpressions of $T(E)$ with an $O(s^2)$ space and time complexity.*

## 4.3   Computing the List of the Pseudo-Continuations

We call pseudo-continuation w.r.t. a position $x$ the string $l_x$ which deduces from the c-continuation $c_x$ by substituting each starred subexpression by its star-name. More formally, let us denote by $S(H)$ the star-name of the star expression $H$, and by $s(H)$ the string associated to the expression $H$. We have the following definition:

**Definition 7 (Pseudo-Continuation).** *Let $E$ be a linear expression and $c_x$ be the c-continuation w.r.t. $x$ in $E$. The pseudo-continuation w.r.t. $x$ in $E$ is the string $l_x$ such that:*

$$c_x = H_1 \cdot H_2 \cdot \ldots H_l$$

*where $H_i$ is a subexpression of $E$, $1 \le i \le l$*

$$l_x = A_1 \cdot A_2 \cdot \ldots A_l$$

*where $A_i = \begin{cases} S(H_i) & \text{if } H_i \text{ is a star expression} \\ s(H_i) & \text{otherwise} \end{cases}$*

We first show that pseudo-continuations can be substituted to c-continuations inside the identification process. Let $S_*$ the alphabet of the star-names. A pseudo-continuation $l_x$ is a string over the alphabet $Y = Pos_E \cup S_* \cup \{0, 1, +, \cdot\}$. Let us extend $h$ as a mapping from $Pos_E \cup S_*$ to $\Sigma$ by setting $h(s) = s$ for all $s$ in $S_*$.

**Proposition 8.** *Let $x$ and $y$ in $Pos_E \cup \{0\}$. Then we have:*

$$h(l_x) \equiv h(l_y) \iff h(c_x) \equiv h(c_y)$$

**Proof.** ($\Rightarrow$) Obvious: identical star-names inside strings $h(l_x)$ and $h(l_y)$ necessarily have the same expansion in $h(c_x)$ and $h(c_y)$.

($\Leftarrow$) This is due to the fact that it is syntactically impossible for subexpressions $F$, $G$ and $H$ to verify $H^* \equiv F^* \cdot G^*$.

∎

Finally, the list of the pseudo-continuations $l_x$ in $\overline{E}$ and the set of their projections on $\Sigma$ can be computed by the following procedure:

*Procedure* Pseudocontinuations($E$: regular expression)
   *Step 1:*   Compute the set of links $f(F)$ in $T(E)$.
   *Step 2:*   Perform the procedure *Stars(E)*.
   *Step 3:*   For each position $x$ of $\overline{E}$, construct $l_x$.
   *Step 4:*   Sort the projections of the pseudo-continuations.

The complexity of this procedure deduces from Proposition 7 and Lemma 1.

**Lemma 1.** *Let $E$ be a regular expression of size $s$ and alphabetic width $w$. Then the alphabet of the pseudo-continuations in $\overline{E}$ has an $O(w + s)$ size. Pseudo-continuations have an $O(s)$ size.*

**Proposition 9.** *Let $E$ be a regular expression of size $s$ and alphabetic width $w$. The procedure Pseudocontinuations($E$) computes the list of the pseudo-continuations in $\overline{E}$ and the set of their projections on $\Sigma$ with an $O(s^2)$ space and time complexity.*

As a consequence the set of states of the quotient automaton $\mathcal{C}_E/_\sim$ can be computed with an $O(s^2)$ space and time complexity.

# 5  Computation of the Set of Transitions of $\mathcal{C}_E/_\sim$

According to Proposition 3, the set of transitions in $\mathcal{C}_E$ and consequently the set of transitions in $\mathcal{C}_E/_\sim$ can be deduced from the position sets of $E$. Let us point out that the structure and the properties of c-continuations provide a proof for some computational refinements such as the set disjoint unions used to build the ZPC structure [23,20].

## 5.1  Transitions in $\mathcal{C}_E$

Let $(x, c_x)$ be a state of $\mathcal{C}_E$. We consider the set $T_{(x,c_x)}$ of the positions associated to the targets of the out-going transitions of $(x, c_x)$: $T_{(x,c_x)} = \{y \mid y \in Pos_E$ and $d_y(c_x) \neq 0\}$.

Let $c_x = H_1 \cdot H_2 \cdot \ldots \cdot H_l$ be the decomposition of $c_x$ as a product of subexpressions of $E$. The set $T_{(x,c_x)}$ can be deduced from the First sets of the subexpressions $H_i$ as follows:

**Proposition 10.** *Consider the integer $s$, $1 \leq s \leq l$, such that $\lambda(H_i) = 1$ for all $1 \leq i < s$ and $\lambda(H_s) = 0$. Assume $s = l$ if $\lambda(H_i) = 1$ for all $1 \leq i \leq l$. Then we have:*

$$T_{(x,c_x)} = \bigcup_{1 \leq i \leq s} First(H_i)$$

The proof is based on Proposition 3.

We now show that the set $T_{(x,c_x)}$ can be computed as a disjoint union of First sets. This fact is based on the following property of c-continuations.

**Lemma 2.** *Let $c_x = H_1 \cdot H_2 \cdot \ldots \cdot H_l$. Let $r$ be an integer such that $1 \leq r \leq l$, and $H_r$ be a star-link such that $H_r = K^* = f(K)$. Then there exists a word $u$ such that: $c_x = d_u(K) \cdot K^* \cdot H_{r+1} \ldots H_l$.*

**Proposition 11.** *Consider $s$ such that $\lambda(H_i) = 1$ for all $1 \leq i < s$ and $\lambda(H_s) = 0$ (and $s = l$ if $\lambda(H_i) = 1$ for all $1 \leq i \leq l$). Let $r$, $1 \leq r \leq s$, be the greatest index such that $H_r$ is a star-link. We assume $r = 1$ if there is no star subexpression. Then, we have:*

$$T_{(x,c_x)} = \biguplus_{r \leq i \leq s} First(H_i)$$

**Proof.** By Proposition 10, for all state $(x, c_x)$ of $\mathcal{C}_E$, $T_{(x,c_x)} = \bigcup_{1 \leq i \leq s} First(H_i)$, By Lemma 2, since $H_r$ is a star-link, there exists a word $u$ such that: $c_x = d_u(K) \cdot K^* \cdot H_{r+1} \ldots H_l$. Since $First(d_u(K)) \subseteq First(K^*)$, we get $T_{(x,c_x)} = \bigcup_{r \leq i \leq s} First(H_i)$. Moreover, since $H_r$ is the "highest" star-link in $T_{c_x}$, this last union is a disjoint one.

■

Since the collection of the First sets of the subexpressions of $E$ can be computed with an overall $O(s)$ time complexity, via linkings over the syntax tree of $E$, as designed in the construction of the ZPC structure [23,20], we get an $O(s^2)$ space and time computation of the set of transitions of $\mathcal{C}_E$.

## 5.2  Transitions in $\mathcal{C}_E/\sim$

Let $[(x, c_x)]$ be a state of $\mathcal{C}_E/\sim$. We consider the set $T_{[c_x]}$ of pairs $(a, [c_y])$ such that $([(x, c_x)], a, [(y, c_y)])$ is a transition.

$$T_{[c_x]} = \{(a, [c_y]) \mid y \in Pos_E \wedge d_y(c_x) \neq 0 \wedge a = h(y)\}$$

The following Proposition is deduced from Proposition 11.

**Proposition 12.** *Consider $s$ such that $\lambda(H_i) = 1$ for all $1 \leq i < s$ and $\lambda(H_s) = 0$ (and $s = l$ if $\lambda(H_i) = 1$ for all $1 \leq i \leq l$). Let $r$, $1 \leq r \leq s$, be the greatest index such that $H_r$ is a star-link. We assume $r = 1$ if there is no star subexpression. Then, we have:*

$$T_{[c_x]} = \biguplus_{r \leq i \leq s} \{(a, [c_y]) \mid y \in First(H_i) \wedge a = h(y)\}$$

We thus get an $O(s^2)$ space and time computation of the set of transitions of $\mathcal{C}_E/\sim$.

# 6  Conclusion: Comparison with Antimirov's Construction

In Antimirov's algorithm, the successive derivations produce $O(s^2)$ expressions, each of which is compared to $O(s)$ distinct partial derivatives. Therefore there are $O(s^3)$ expression tests. Since the size of a partial derivative is $O(s^2)$ the overall complexity is $O(s^5)$. This complexity is improved by an $O(s^3)$ factor by the computation of the quotient of the c-continuation automaton for the following reasons:

(1) Only the computation of the set of states generates expression tests. The computation of the set of transitions is deduced from $O(s^2)$ procedures used in the construction of the position automaton.

(2) Each c-continuation can be directly produced over the syntax tree in $O(s^2)$ time. The set of the projections of the c-continuation can be computed in $O(s^3)$ time (each projection is compared only to one other expression after a lexicographic sort). Hence an $O(s^3)$ construction of the set of states.

(3) This construction can be refined by substituting pseudo-continuations to c-continuations. The size of a pseudo-continuation is $O(s)$. The computation of the set of pseudo-continuations, which implies a preprocessing of the starred subexpressions is in $O(s^2)$ time. Hence an $O(s^2)$ construction of the set of states.

# References

1. A.V. Aho, J.E. Hopcroft, and J.D. Ullman. *Data Structures and Algorithms.* Addison-Wesley, Reading, MA, 1983.
2. J.-M. Autebert et J.-M. Rifflet. Dérivations Formelles des Expressions Rationnelles : Un Programme de Calcul Automatique. *Rapport LITP* 78(14), 1978.

3. V. Antimirov. Partial derivatives of regular expressions and finite automaton constructions. *Theoret. Comput. Sci.*, 155:291–319, 1996.
4. D. Beauquier, J. Berstel, and P. Chrétienne. *Éléments d'Algorithmique*. Masson, Paris, 1992.
5. A. Brüggemann-Klein, Regular Expressions into Finite Automata. *Theoret. Comput. Sci.*, 120(1993), 197–213.
6. G. Berry and R. Sethi. From regular expressions to deterministic automata. *Theoret. Comput. Sci.*, 48(1):117–126, 1986.
7. J.A. Brzozowski. Derivatives of regular expressions. *J. Assoc. Comput. Mach.*, 11(4):481–494, 1964.
8. C.-H. Chang and R. Paige. From Regular Expressions to DFAs using Compressed NFAs, in Apostolico. Crochemore. Galil. and Manber. editors. *Lecture Notes in Computer Science*, 644(1992), 88-108.
9. J. -M. Champarnaud and D. Ziadi. New Finite Automaton Constructions Based on Canonical Derivatives, in CIAA'2000, *Lecture Notes in Computer Science*, S. Yu ed., Springer-Verlag, to appear.
10. M. Crochemore and C. Hancart. Automata for matching patterns. *Handbook of Formal Languages*, G. Rozenberg and A. Salomaa eds., (A.), chap. 9, 399–462, Springer-Verlag, Berlin, 1997.
11. V.M. Glushkov. The abstract theory of automata. *Russian Mathematical Surveys*, 16:1–53, 1961.
12. C. Hagenah and A. Muscholl, Computing $\varepsilon$-free NFA from Regular Expressions in $O(nlog^2(n))$ Time, in: L. Prim *et al.* (eds.), MFCS'98, *Lecture Notes in Computer Science*, 1450(1998), 277-285, Springer.
13. C. Hagenah and A. Muscholl, Computing $\varepsilon$-free NFA from Regular Expressions in $O(nlog^2(n))$ Time, *RAIRO-TIA*, 34/4(2000), 257–278.
14. J. Hromkovič, S. Seibert and T. Wilke, Translating regular expressions into small $\varepsilon$-free nondeterministic finite automata, in: R. Reischuk (ed.), STACS'97, *Lecture Notes in Computer Science*, 1200(1997), 55-66, Springer.
15. J.E. Hopcroft and J.D. Ullman. *Introduction to Automata Theory, Languages and Computation*. Addison-Wesley, Reading, MA, 1979.
16. S. Kleene. Representation of events in nerve nets and finite automata. *Automata Studies*, Ann. Math. Studies 34:3–41, 1956. Princeton U. Press.
17. R.F. McNaughton and H. Yamada. Regular expressions and state graphs for automata. *IEEE Transactions on Electronic Computers*, 9:39–57, March 1960.
18. B.G. Mirkin. An algorithm for constructing a base in a language of regular expressions. *Engineering Cybernetics*, 5:110–116, 1966.
19. R. Paige and R. E. Tarjan. Three partition refinement algorithms. *SIAM J. Comput.*, 16(6), 1987.
20. J.-L. Ponty, D. Ziadi and J.-M. Champarnaud, A new Quadratic Algorithm to convert a Regular Expression into an Automaton, In: D. Raymond and D. Wood, eds., Proc. *WIA'96, Lecture Notes in Computer Science*, Vol. 1260(1997) 109–119.
21. B. Watson, Taxonomies and Toolkits of Regular Languages Algorithms, PhD thesis, Eindhoven University of Technology, The Nederlands, 1995.
22. S. Yu. Regular languages. In G. Rozenberg and A. Salomaa, editors, *Handbook of Formal Languages*, volume I, Word, Language, Grammar, pages 41–110. Springer-Verlag, Berlin, 1997.
23. D. Ziadi, J.-L. Ponty, and J.-M. Champarnaud. Passage d'une expression rationnelle à un automate fini non-déterministe. Journées Montoises (1995), *Bull. Belg. Math. Soc.*, 4:177–203, 1997.

# On-Line Construction of Compact Directed Acyclic Word Graphs⋆

Shunsuke Inenaga[1], Hiromasa Hoshino[1], Ayumi Shinohara[1],
Masayuki Takeda[1], Setsuo Arikawa[1], Giancarlo Mauri[2], and Giulio Pavesi[2]

[1] Dept. of Informatics, Kyushu University, Japan
{s-ine,hoshino,ayumi,takeda,arikawa}@i.kyushu-u.ac.jp
[2] Dept. of Computer Science, Systems and Communication
University of Milan–Bicocca, Italy
{mauri,pavesi}@disco.unimib.it

**Abstract.** A Compact Directed Acyclic Word Graph (CDAWG) is a space–efficient text indexing structure, that can be used in several different string algorithms, especially in the analysis of biological sequences. In this paper, we present a new on–line algorithm for its construction, as well as the construction of a CDAWG for a set of strings.

## 1 Introduction

Several different string problems, like those deriving from the analysis of biological sequences, can be solved efficiently with a suitable text–indexing structure. Perhaps, the most widely used and known structure of this kind is the *suffix tree*, that can be built in linear time and permits to efficiently find and locate all the substrings of a given string. The main drawback of suffix trees is the additional space required to implement the structure. In many applications, like sequence analysis and pattern discovery in biological sequences, keeping as many data as possible in main memory might provide significant advantages. This fact has led to the introduction of more space–efficient structures, like *suffix arrays* [1], *suffix cacti* [2], and others.

In this work, we focus our attention on the *Compact Directed Acyclic Word Graph* (CDAWG), first described in [3]. The CDAWG for a string can be seen either as a compaction of the *Directed Acyclic Word Graph* (DAWG) [4], or a minimization of the suffix tree, from which it can be derived as shown in [3,5] for DAWGs and [6] for suffix trees. In the latter case, the basic idea is to merge redundant parts of the suffix tree (see Fig. 1). Experimental results [3,5] have shown how CDAWGs provide significant reductions of the memory space required by suffix trees and DAWGs when applied to genomic sequences. A linear time algorithm for the direct construction of the CDAWG of a string is presented in [5], so to avoid the additional space required by the preliminary construction of

---

⋆ The results described in this work were reached independently by the Kyushu and Milan groups, submitted simultaneously to the conference, and merged into a joint contribution.

A. Amir and G.M. Landau (Eds.): CPM 2001, LNCS 2089, pp. 169–180, 2001.
© Springer-Verlag Berlin Heidelberg 2001

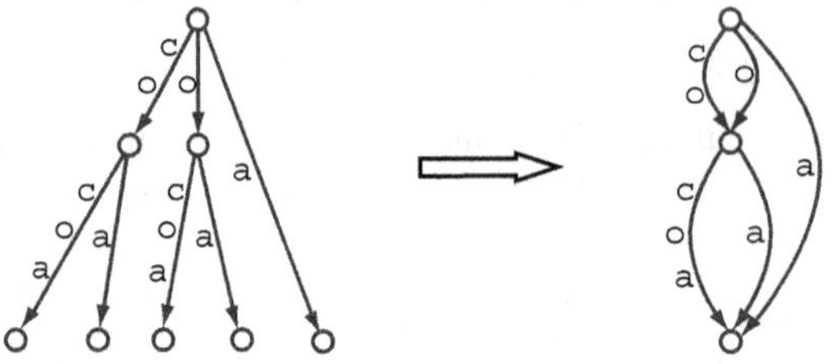

**Fig. 1.** Suffix tree and CDAWG for string *cocoa*. Substrings *co* and *o* occur as prefix of the same suffixes: the corresponding nodes are merged as well as the subtrees rooted at the nodes. Leaves are merged into a single final node.

the DAWG or the suffix tree. The algorithm is similar to McCreight's algorithm for suffix trees [7]. In this paper, we present a new algorithm for the construction of CDAWGs, based on Ukkonen's algorithm for suffix trees [8]. The algorithm is *on–line*, that is, it processes the characters of the string from left to right one by one, with no need to know the whole string beforehand. Furthermore, we show how the algorithm can be used to build a CDAWG for a set of strings, a structure first described in [3], where was derived by compacting a DAWG for a set of strings. The main drawback of this approach was the fact that, when a new string was added to the set, the DAWG had to be built again from scratch. Instead, the algorithm we present allows to add a new string directly to the compact structure.

## 2   Definitions

Let $\Sigma$ be a nonempty finite alphabet, and $\Sigma^*$ the set of strings over $\Sigma$. If $s = \alpha\beta\gamma$, with $\alpha, \beta, \gamma \in \Sigma^*$, then $\alpha$ is a prefix of $s$, $\gamma$ is a suffix of $s$, and $\alpha$, $\beta$, and $\gamma$ are substrings (factors) of $s$. If $s = s_1 \ldots s_n$ is a string in $\Sigma^*$, $|s|$ denotes its length, and $s[i..j]$ its substring $s_i \ldots s_j$. With $Suf(s)$ we will denote the set of all suffixes of $s$. Let $X$ be a subset of $\Sigma^*$. For any string $u \in \Sigma^*$, $u^{-1}X = \{x \mid ux \in X\}$. Given a string $s$, we define the syntactic congruence on $\Sigma^*$ associated with $Suf(s)$ and denoted by $\equiv_{Suf(s)}$ as:

$$u \equiv_{Suf(s)} v \iff u^{-1}Suf(s) = v^{-1}Suf(s) \quad \text{(for any } u, v \in \Sigma^*\text{)}$$

That is, $u$ and $v$ occur as prefixes of the same suffixes of $s$. In other words, the occurrences of $u$ and $v$ must end at the same positions in the string. Hence, if $u$ and $v$ occur in the string, one must be a suffix of the other. As in [3,5], we will call *classes of factors* the congruence classes of the relation $\equiv_{Suf(s)}$. The class of all strings that are not substrings of $s$ is called the *degenerate* class. The

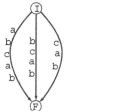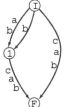

**Fig. 2.** Implicit CDAWG and CDAWG for string *abcab*.

longest string in a non–degenerate class of factors is the *representative* of the class. Given a non–degenerate class of factors $C$ of $\equiv_{Suf(s)}$, and its representative $u$, if there are at least two characters $a, b \in \Sigma$ such that $ua$ and $ub$ are substrings of $s$, then $C$ is a *strict class of factors* of $\equiv_{Suf(s)}$. From now on, we will say that two substrings are *strictly congruent* if they belong to the same strict class of factors. We are now ready to give a formal definition of a CDAWG.

**Definition 1.** *The* compact directed acyclic word graph *(CDAWG) of a string s is a directed acyclic graph, where:*

1. *two distinct nodes are marked as initial and final;*
2. *edges are labeled with non empty substrings of s;*
3. *labels of two edges leaving the same node cannot begin with the same character;*
4. *every suffix of s corresponds to a path on the graph starting from the initial node and ending at a node, such that the concatenation of the edge labels on the path exactly spells the suffix. From now on, we will call a node corresponding to a suffix of s* terminal node;
5. *substrings spelled by paths starting from the initial node and ending at the same non–terminal node of the graph belong to the same strict class of factors.*

The CDAWG of a string $s$ has at most $|s| + 1$ nodes and $2|s| - 2$ edges [3,5]. According to the definition of a strict class of factors, non–terminal nodes must have at least two outgoing edges. We will denote with $(p, \alpha, q)$ the edge $p \to q$ of the graph labeled with substring $\alpha$. The following definitions will be useful throughout the paper:

**Definition 2.** *The* implicit *CDAWG of a string s is a CDAWG where nodes with outdegree one are removed, and each edge entering a node with outdegree one is merged with the edge leaving it.*

In the implicit CDAWG of a string $s$, the suffixes of $s$ are spelled out by paths in the graph starting at the initial node, but not necessarily ending at a node. An example is shown in Fig. 2. For every node $p$, let $length_s(p)$ be the length of the longest substring spelled by a path from the initial node to $p$. Edges belonging to the spanning tree of the longest paths from the initial node are called *solid edges*. In other words, an edge $(p, \alpha, q)$ is solid iff $length_s(q) = length_s(p) + |\alpha|$. Finally, we assume that the label of each edge is implemented with a pair of integers denoting the starting and ending points in the string of the substring corresponding to the label, and every node is annotated with the length of the longest path from the initial node.

# 3    Construction of the CDAWG for a Single String

Given an alphabet $\Sigma$, let $s = s_1 \ldots s_n$ be a string on $\Sigma$. Our algorithm is divided in $n$ phases, building at each phase $i$ the implicit CDAWG $\mathcal{G}_i$ for each prefix $s[1..i]$ of $s$. More in detail, the implicit CDAWG $\mathcal{G}_{i+1}$ for $s[1..i+1]$ is constructed starting from graph $\mathcal{G}_i$ for $s[1..i]$. Each phase $i+1$ is divided in $i+1$ extensions, one for each of the $i+1$ suffixes of $s[1..i+1]$. In extension $j$ of phase $i+1$, the algorithm finds the end of the path from the initial node labeled with substring $s[j..i]$, and extends it by adding character $s_{i+1}$ to the path, unless it is already there. Therefore, in phase $i+1$, substring $s[1..i+1]$ is first put on the graph, followed by $s[2..i+1]$, $s[3..i+1]$, and so on. Extension $i+1$ of phase $i+1$ adds the single character $s_{i+1}$ after the initial node. The initial graph $\mathcal{G}_1$ has one initial node $I$ and one final node $F$, connected by an edge labeled by character $s_1$. The algorithm can be sketched as follows:

1. Construct graph $\mathcal{G}_1$
2. For $i$ from 1 to $n - 1$ do
3.     For $j$ from 1 to $i + 1$ do
4.         Find the end of the path from $I$ labeled $s[j..i]$
5.         Add character $s_{i+1}$ if needed
6.     End for
7. End for

At extension $j$ of phase $i + 1$, once the end of the path spelling $s[j..i]$ has been located, the CDAWG can be updated according to three different rules:

1. In the current graph, the path spelling $s[j..i]$ ends in $F$. To update the graph, character $s_{i+1}$ is appended to the label of the edge entering $F$.
2. The path corresponding to $s[j..i]$ does not continue with $s_{i+1}$, but continues with at least one character $c$. If the path ends at a node $p$, we create a new edge $(p, s_{i+1}, F)$. Otherwise, we create a new node $q$ at the end of the path, splitting the edge in two at the point where the path ends. Then, we create a new edge $(q, s_{i+1}, F)$.
3. Some path at the end of $s[j..i]$ continues with $s_{i+1}$. In this case, substring $s[j..i+1]$ is already in the current graph: we do nothing (hence the implicit graph).

These rules, however, do not guarantee that at the end of the phase we correctly constructed a CDAWG. In fact, the algorithm must also check whether a substring strictly congruent to another one has been encountered, or, conversely,

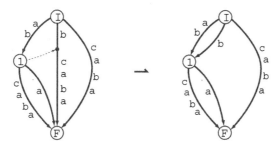

**Fig. 3.** Implicit CDAWG for string $abcaba$ before (left) and after redirection of an edge, at phase 6, extension 5. Node 1, labeled $ab$, was created at the previous extension, after the insertion of $a$ at the end of the path labeled $ab$. Now, path corresponding to $b$ is found ending in the middle of non–solid edge $(I, bcaba, F)$, that is redirected to node 1 and becomes $(I, b, 1)$.

whether a substring has to be removed from a strict class of factors, so that at the end of phase $i+1$ paths ending at the same node correspond to strict classes of factors of $s[1..i+1]$, and vice versa. Here we sketch how the algorithm has to be modified. A more detailed description of the algorithm and its implementation can be found in [9].

**Detecting Strictly Congruent Factors.** Two substrings $\alpha$ and $\beta$ belong to the same class $C$ iff they are prefixes of the same suffixes, and there are at least two characters $a, b \in \Sigma$ such that $\alpha a$, $\alpha b$, $\beta a$, and $\beta b$ occur in $s$. Moreover, $\alpha$ must be a suffix of $\beta$, or vice versa. We suppose w.l.o.g. that $\alpha = c\beta$, with $c \in \Sigma$. We also assume that $\alpha$ and $\beta$ have occurred just once, that substrings $\alpha a$ and $\beta a$ have been put in the graph in some previous phase (in two consecutive extensions), and in the current extension we have to insert $\alpha b$. The path spelling $\alpha$ ends in the middle of an edge, and the next character on the edge is $a$. A new node $p$ is created at the end of the path, as well as a new edge $(p, b, F)$. At the following extension, we have to locate $\beta$ in the graph. If $\beta$ has occurred only once (together with $\alpha$), it now belongs to the same strict class of factors, and we end in the middle of a *non–solid* edge that continues with $a$. In this case, we *redirect* the edge to $p$, labeling it with the part of the label that was contained in the path of $\beta$ (see Fig. 3). Since there can be more than two consecutive substrings to be assigned to the same class, it is possible that we again end along non–solid edges in the following extensions. In this case, we redirect the non–solid edges to $p$ as well, until we reach an extension where we end at a node or along a solid edge. Otherwise, if $\beta$ had previously occurred also by itself, either the path corresponding to $\beta$ ends at a node ($\beta$ has been followed by characters different from $a$), or the edge we end on is solid ($\beta$ had been followed only by $a$). In the former case, if there is not an edge labeled $b$ leaving the node we create a new

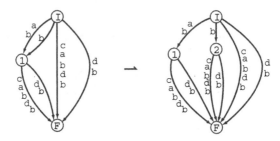

**Fig. 4.** CDAWG for string *abcabdb* at phase 7, extension 7. Character *b* is found at the end of the non–solid edge $(I, b, 1)$. At extension 6, the path spelling *db* ended at the final node. Thus, *b* has to be removed from the class associated with node 1, that is cloned into node 2. Edge $(I, b, 1)$ becomes $(I, b, 2)$.

edge labeled *b* to the final node. In the latter case, we create a new node and connect it to the final node with an edge labeled *b*. Then, there may be again non–solid edges that have to be redirected into the newly created node.

**Splitting a Strict Class of Factors.** Conversely, a substring that has been assigned to a strict class of factors has to be removed from the class if it does not occur as a suffix of the representative when a new character $s_{i+1}$ is added to the string. Let $\alpha$ and $\beta$, $\alpha = c\beta$, be the two substrings assigned to the same class in the previous example. Now, suppose that in phase $i+1$ we have to insert $\beta$ in the graph. In this case, $s_{i+1}$ is the last character of $\beta$, and we find it at the end of the edge entering node $p$, that is non–solid, since $\beta$ is not the representative of the class. Now we have two cases: $s_{i+1}$ was found at the end of an edge that entered node $p$ also at the previous extension, or we ended up somewhere else. In the former case, we had also inserted $\alpha$ at the previous extension of the same phase, therefore $\beta$ still belongs to the same class. In the latter, we have detected an occurrence of $\beta$ not preceded by $\alpha$, that is, not as a suffix of $\alpha$, and we have to remove it from the class. To reflect this in the graph, we *clone* the node $p$ into a new node $q$, and redirect the non–solid edge to $q$ keeping the same label. The redirected edge becomes solid. An example is shown in Fig. 4. If also some suffixes of $\beta$ had been previously assigned to the same class as $\beta$, in the following extensions we will again find $s_{i+1}$ at the end of a non–solid edge entering $p$. These edges are redirected to $q$. It can be proved that it suffices to check only the last edge on each path to ensure that a class has to be split. No cloning takes place if a character is found at the end of an edge entering the final node.

The two observations outlined above can be implemented in the algorithm by modifying Rules 2 and 3 accordingly. It is worth mentioning that both redirection of edges to a newly created node and node cloning can take place during the same phase. An example is shown in Fig. 5.

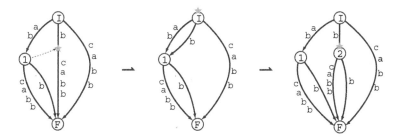

**Fig. 5.** From left to right, CDAWG for string *abcabb* at phase 6, extensions 5, 6, and 7. Character *b* is put in the graph after substring *ab*, and the path spelling *b* is found in the middle on non–solid edge $(I, bcabb, F)$ (left) that is redirected to node 1 (center). Then, at extension 7 (that adds *b* after the empty string) *b* is found at the end of a non–solid edge. Node 1 is thus cloned into node 2 (right).

### 3.1   Using Suffix Links

Naively, locating the end of $s[j..i]$ in extension $j$ of phase $i+1$ would take $O(i-j)$ time by walking from the initial node and matching the characters of $s[j..i]$ along the edges of the graph. This would lead to an overall $O(n^3)$ time complexity for the construction of the whole graph. We will now reduce it, as in [8], to $O(n)$ by introducing *suffix links* and with some remarks.

**Definition 3.** *Let $p$ be a node of the graph, different from the initial or final node. Let $\beta$ be the representative of the class associated with $p$. The suffix link of $p$, denoted by $L(p)$, is the node $q$ whose representative $\gamma$ is the longest suffix of $\beta$ whose path does not end at $p$.*

The suffix link of a node $p$ can be implemented with a pointer from $p$ to $L(p)$. If $\gamma$ is empty, then $L(p)$ is the initial node. Suffix links are not defined for the initial and the final node. Although the definition does not guarantee that every node in the graph has a suffix link, we can prove the following:

**Lemma 1.** *Any node created during phase $i + 1$ will have a suffix link from it by the end of the phase.*

*Proof.* In extension $j$ of phase $i + 1$ a new node $p$ can be created at the end of the path spelling substring $s[j..i]$ by application of Rule 2 or by cloning. In the former case, $L(p)$ will be the first node to be created or encountered at the end of the path corresponding to a suffix of $s[j..i]$ (possibly after edge redirections). Such a node always exists, since the last extension locates the empty suffix at the initial node. In the latter case, let us suppose that a node $q$ is cloned into node $p$ with path spelling $s[j..i + 1]$. Substring $s[j..i + 1]$ is the longest suffix of the representative of $q$ that does not belong to the same class. Thus, $L(q)$ is set to $p$. Suffix link $L(p)$ is left undefined until one of the suffixes of $s[j..i + 1]$ ends at a node other than $p$ (that again could be $I$).                              □

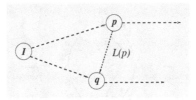

**Fig. 6.** A suffix link. Node $p$ corresponds to class $\alpha\beta$, node $q$ corresponds to $\beta$. Paths labeled with suffixes of $\alpha\beta$ longer than $\beta$ end at $p$. If at some extension $j$ character $s_{i+1}$ is added after $\alpha\beta\gamma$, then extensions from $j+1$ to $j+|\alpha|$ are implicitly performed as well.

During any phase, the only node of the graph other than the initial and the final without a suffix link from it is the last created one. Let us suppose that the algorithm has completed extension $j$ of phase $i+1$. Suffix links are used to speed up the search for the remaining suffixes of $s[j..i]$. Starting from the end of $s[j..i]$ in the graph, we walk backwards along the path corresponding to $s[j..i]$ up to either the initial node or a node $p$ that has a suffix link. This requires traversing at most one edge. Let $\gamma$ be the concatenation of the edge labels of the path from $p$ to $s[j..i]$. If $p$ is not the initial node, we move to node $L(p)$ and follow from it the path spelling $\gamma$. Otherwise, we search for $s[j+1..i]$ starting from $I$. Finally we add $s_{i+1}$ according to one of the extension rules, redirecting an edge or cloning a node if needed. Notice that, if node $p$ is the end of $l \geq 2$ different paths, the position reached after searching from $\gamma$ from $L(p)$ will be the end of path $s[j+l..i]$, that is, extensions from $j+1$ to $j+l-1$ have been implicitly performed at extension $j$.

A path spelling $\gamma$ starting from $L(p)$ always exists, since all the suffixes of $s[j..i]$ are already in the graph. Thus, to find the path spelling $\gamma$ the algorithm just matches the first characters on the edges encountered. To obtain a linear time algorithm, we need just two more "tricks".

*Remark 1.* When during any extension Rule 3 is applied, that is, a given substring $s[j..i+1]$ is already on the graph, then the same rule will apply to all further extensions, since all the suffixes of $s[j..i+1]$ are already in the graph as well. Therefore, once Rule 3 is applied (and no node has to be cloned or edges redirected), we can stop and move on to the next phase, since all the strings to be inserted are already in the graph and no adjustment is needed for the classes.

*Remark 2.* If a new edge is created entering the final node during extension $j$ of any phase $i$, then Rule 1 will always apply at extension $j$ in any successive phase. That is, new characters will always be appended at the end of the last edge in the path associated with $s[j..i]$, that will enter the final node. Thus, when a new edge is created entering the final node with label $s[j..i+1]$, we label it with integers $h$ and $e$ $(j \leq h \leq i+1)$, where $e$ denotes the current phase, that is, the current end position in the string. If we implement $e$ with a global variable, and set it to $i+1$ at the beginning of each phase $i+1$, we perform implicitly all the extensions that would end up at the final node.

Every phase $i$ starts with a series of applications of Rules 1 and 2, that put $s_i$ at the end of an edge entering the final node; when Rule 3 is applied for the first time, it will be also applied to all further extensions. Now, let $j_i$ be the first extension where Rule 3 is applied with cloning in phase $i$, and $j_i^*$ the first extension where it is applied without edge redirection to the cloned node. Extensions $j_i+1$ to $j_i^*-1$ will redirect edges to the last node created. Extensions from $j_i^*+1$ to $i$ need not to be performed, since in each of them we would not do anything. In phase $i+1$, all extensions from 1 to $j_i-1$ will apply Rule 1, therefore they are implicitly performed by setting the counter $e$ to $i+1$. Thus, we can start phase $i+1$ directly from extension $j_i^*-1$, until we find an extension where Rule 3 is applied without cloning or edge redirection. This can be done by starting phase $i+1$ from the position in the graph of the last suffix of $s[1..i]$ that had to be redirected to the cloned node. This took place at extension $j_i^*-1$. The first extension in phase $i+1$ will have to look for $s_{i+1}$ exactly at the endpoint of the last extension of phase $i$. This will also implicitly perform all extensions from $j_i$ to $j_i^*-1$. Of course, if in phase $i$ Rule 3 is first applied without cloning we can move on to phase $i+1$ as well.

The algorithm does not need to know which extension is currently performing. That is, it starts phase $i+1$ from the endpoint of phase $i$, adding $s_{i+1}$. Then it starts moving in the graph by using suffix links, and adding $s_{i+1}$ at the end of each path. If the backward walk ends at $I$, and $\gamma = \gamma_1 \ldots \gamma_k$ is the label of the path traversed, then it looks for the path labeled $\gamma_2 \ldots \gamma_k$. Phase $i+1$ ends when the algorithm applies for the first time Rule 3 without node cloning or edge redirection. Moreover, whenever we find $s_{i+1}$ at the end of a non–solid edge, we no longer have to check what happened at the previous extension, and just clone the node. In fact, if the representative of the class had been met during one of the previous extensions, we would have stopped the phase at that point, without reaching the current extension.

At the end of phase $n$, we have constructed the implicit CDAWG for string $s$. In order to obtain the actual CDAWG, we perform an additional extension phase $n+1$, extending the string to a dummy symbol \$ that does not belong to the string alphabet. Anyway, we do not increment the phase counter $e$ to $n+1$, so to avoid appending \$ to edges entering the final node. Moreover, whenever a new node $p$ has to be created, we do not add the edge $(p, \$, F)$ to the graph. Nodes created in this phase will thus have outdegree one, and will correspond to terminal nodes of the CDAWG. Notice that, whenever a path $s[j..n]$ ends along an edge, we always create a new node and mark it as terminal, while cloning of nodes and redirection of edges work as in the previous phases. When a path $s[j..n]$ ends at a node, we mark the node as terminal. At the end of the additional phase, the implicit CDAWG has been transformed into the actual CDAWG for string $s$. An example of the on–line construction of a CDAWG is shown in Fig. 7.

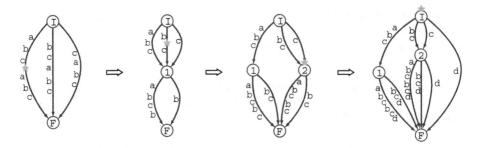

**Fig. 7.** From left to right, construction of the CDAWG for string *abcabcbcd*: at the end of phase 6 (implicit CDAWG for string *abcabc*); at the end of phase 7 (*abcabcb*, where *abc*, *bc*, and *c* belong to the same strict class of factors); at the end of phase 8 (*abcabcbc*, where *bc* and *c* have been removed from the class with representative *abc*); the final structure. Stars indicate the position in the graph reached at the end of the last explicit extension of each phase.

With arguments analogous to Ukkonen's algorithm for suffix trees, we can prove the following:

**Theorem 1.** *Given a string $s = s_1 \ldots s_n$ over a finite alphabet $\Sigma$, the algorithm implemented with suffix links and implicit extensions builds the CDAWG for $s$ in $O(n)$ time and $O(n|\Sigma|)$ space if the graph is implemented with a transition matrix, or in $O(n|\Sigma|)$ time and $O(n)$ space with adjacency lists.*

*Proof (Sketch).* The operations performed in any explicit extension (creation or cloning of nodes, edge redirections), that is, extensions that are not performed implicitly by incrementing the $e$ counter, take constant time. Let $j_i^*$ the last explicit extension performed at phase $i$, and $j_{i+1}$ the first explicit extension performed at phase $i + 1$. In the worst case, we have $j_{i+1} = j_i^* - 1$. Moreover, for each $i$, $j_i \leq j_{i+1}$. Thus, at most $3n$ explicit extensions are performed by the algorithm. At any extension $j$ of phase $i$, to locate the endpoint of $s[j..i]$ the algorithm walks back at most one edge from the endpoint of $s[j - 1..i]$, follows a suffix link, and then traverses some edges checking the first symbol on each edge. If the graph is implemented with a transition matrix, traversing an edge takes constant time. Else, it takes $O(|\Sigma|)$ time. The only thing unaccounted for is the overall number of edges traversed. For every node $p$ of the graph, let the *node depth* of $p$ be the number of nodes on the path from the root to $p$ labeled with the representative of the class associated with $p$. As in [8], the sum of the node depths counted during all the explicit extensions is reduced at most by $O(n)$, and since the maximum node–depth is $n$, the maximum number of edges traversed is bounded by $O(n)$.                                                   □

# 4    The CDAWG for a Set of Strings

The basic idea of the CDAWG for a set of strings $S = \{s^1, \ldots, s^k\}$ is the same of the single string structure. Now, the nodes of the structure correspond to patterns that occur as prefix of the same suffixes in every string of the set. In other words, given $Suf(S)$ (the set of the suffixes of the $k$ strings), the nodes of the CDAWG correspond to strict classes of factors for $\equiv_{Suf(S)}$. The only difference is that now we have $k$ final nodes $F_1 \ldots F_k$, one for each string, and we want all the suffixes of $s^i$ to end at the corresponding final node $F_i$. This result can be obtained by appending a different termination symbol, not belonging to the string alphabet, to each string of the set. More formally:

**Definition 4.** *The CDAWG for a set of strings* $s^1 \ldots s^k$ *is a directed acyclic graph, with a node marked as initial and $k$ distinct nodes $F_1 \ldots F_k$ marked as final. Edges are labeled with non empty substrings of at least one of the strings. Labels of two edges leaving the same node cannot begin with the same character. For every string $s^i$ in the set, all suffixes of $s^i$ are spelled by patterns starting at the initial node and ending at node $F_i$. Paths ending at non final nodes correspond to strict classes of factors of the congruence relation $\equiv_{Suf(S)}$.*

The CDAWG for a set of strings can be constructed with the algorithm presented in the previous section. First, we build the CDAWG for string $s^1$ (with the termination symbol) and final node $F_1$. Notice that, since the termination symbol does not occur anywhere else in $s^1$, the resulting structure is a CDAWG, with no need to perform the additional phase. Then, string $s^2$ is added to the graph, but in this case with final node $F_2$. The same will apply to every other string in the set. Node cloning and edge redirection rules ensure the correctness of the resulting structure. It can be proved that the algorithm takes $O(N)$ time to construct the structure, implemented with a transition matrix, where $N = \sum_{i=1}^{k} |s^i|$. This structure (with marginal differences) was first described in [3], where it was built by reducing a DAWG. Therefore, adding a new string to the set required the construction of a new DAWG from scratch. The algorithm presented here, instead, permits to add strings directly to the compact structure (see Fig. 8). As in [3] we can give an upper bound on the size of the structure.

**Theorem 2 (Blumer et al., [3]).** *The CDAWG for a set of strings $s^1 \ldots s^k$, has at most $N + k$ nodes, where $N = \sum_{i=1}^{k} |s^i|$.*

# 5    Conclusions

A CDAWG is a space–efficient text–indexing structure that represents all the substrings of a string. We presented a new on–line algorithm for its construction, as well as the construction of a CDAWG for a set of strings. The same structures can be computed by reduction starting from the corresponding DAWGs or suffix trees; however, the approach presented in this paper permits to save time and space simultaneously, since the CDAWGs can be built directly. Moreover, once the structure has been built for a set of strings, new strings can be added directly to the compact structure.

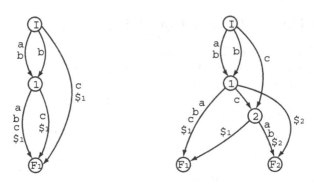

**Fig. 8.** CDAWG for strings $ababc\$_1$ and $abcab\$_2$, after the insertion of $ababc\$_1$ (left) and $abcab\$_2$ (right). Characters $\$_1$ and $\$_2$ are used as terminations. Edges $(I, \$_1, F_1)$ and $(I, \$_2, F_2)$ have been omitted.

## Acknowledgements

The Milan–Bicocca group has been supported by the Italian Ministry of University, under the project "Bioinformatics and Genomic Research".

## References

1. U. Manber and G. Myers. Suffix arrays: a new method for on–line string searches. *SIAM J. Computing*, 22(5):935–948,1993.
2. J. Kärkkäinen. Suffix cactus: a cross between suffix tree and suffix array. *Combinatorial Pattern Matching*, 937:191–204, July 1995.
3. A. Blumer, J. Blumer, D. Haussler, R. McConnell, and A. Ehrenfeucht. Complete inverted files for efficient text retrieval and analysis. *Journal of the ACM*, 34(3):578–595, 1987.
4. A. Blumer, J. Blumer, D. Haussler, A. Ehrenfeucht, M. Chen, and J. Seiferas. The smallest automaton recognizing the subwords of a text. *Theoretical Computer Science*, 40:31–55, 1985.
5. M. Crochemore and R. Verin, On compact directed acyclic word graphs, Springer Verlag LNCS 1261, pp.192–211, 1997.
6. D. Gusfield, Algorithms on Strings, Trees and Sequences: Computer Science and Computational Biology, Cambridge University Press, New York, 1997.
7. E. McCreight. A space–economical suffix tree construction algorithm. *Journal of the ACM*, 23(2):262–272, 1976.
8. E. Ukkonen. On–line construction of suffix trees. *Algorithmica*, 14(3):249–260, 1995.
9. S. Inenaga, H. Hoshino, A. Shinohara, M. Takeda, and S. Arikawa, On–line construction of compact directed acyclic word graphs. DOI Technical Report 183, Kyushu University, January 2001.

# Linear-Time Longest-Common-Prefix Computation in Suffix Arrays and Its Applications

Toru Kasai[1], Gunho Lee[2], Hiroki Arimura[1,3],
Setsuo Arikawa[1], and Kunsoo Park[2*]

[1] Department of Informatics, Kyushu University
Fukuoka 812-8581, Japan
{arim,arikawa}@i.kyushu-u.ac.jp
[2] School of Computer Science and Engineering
Seoul National University, Seoul 151-742, Korea
{ghlee,kpark}@theory.snu.ac.kr
[3] PRESTO, Japan Science and Technology Corporation, Japan

**Abstract.** We present a linear-time algorithm to compute the longest common prefix information in suffix arrays. As two applications of our algorithm, we show that our algorithm is crucial to the effective use of block-sorting compression, and we present a linear-time algorithm to simulate the bottom-up traversal of a suffix tree with a suffix array combined with the longest common prefix information.

## 1 Introduction

The suffix array [16] is a space-efficient data structure that allows efficient searching of a text for any given pattern. The suffix array is basically a sorted array *Pos* of all the suffixes of a text. A suffix array for a text of length $n$ can be built in $O(n \log n)$ time, and searching the text for a pattern of length $m$ can be done in $O(m \log n)$ time by a binary search. When a suffix array is coupled with information about the longest common prefixes (lcps) of some elements in the suffix array, string searches can be speeded up to $O(m + \log n)$ time. The *lcp* information is usually computed during the construction of suffix arrays [15,11]. In some cases, however, the *lcp* information may not be readily available.

In this paper we consider the *lcp problem in suffix arrays* that is to compute the lcp information from a text and its *Pos* array, and present a linear-time algorithm for the problem. We also describe two applications of our algorithm, i.e., block-sorting compression and the substring traversal problem.

The block-sorting algorithm [4] is a text compression method with good balance of compression ratio and speed. The original text can be decoded in linear time from block-sorting compression. An advantage of block-sorting compression is that the suffix array of the original text can also be obtained in the process of

---

* This work was supported by the Brain Korea 21 Project.

A. Amir and G.M. Landau (Eds.): CPM 2001, LNCS 2089, pp. 181–192, 2001.

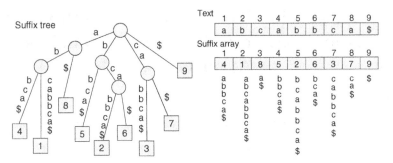

**Fig. 1.** An example of the suffix tree and the suffix array.

decoding. This means that we can compress a text and its suffix array together by simply using the block-sorting algorithm. This fact can be used for storing and transferring large full-text databases. However, the *lcp* information that is necessary for efficient searching is not obtained during the decoding of block-sorting compression. With our algorithm, block-sorting compression can be used more effectively to store a text and its suffix array.

The *substring traversal problem* is to enumerate all branching substrings appearing in a given text. Although the problem is easily solvable by a bottom-up traversal of the suffix tree, recent large scale applications in bioinformatics and data mining require a more practical and scalable solution for the problem [2].

We present a simple linear-time algorithm that simulates the bottom-up traversal of a suffix tree with a suffix array combined with the *lcp* information. Our algorithm is space-efficient and I/O-efficient, i.e., it requires only $7n$ bytes including the text while the suffix tree requires at least $15n$ bytes, and it has a good I/O complexity of $5n/B$ blocks. Furthermore, the algorithm can be modified to solve a class of problems based on the occurrence count of each branching substring, which include the longest common substring problem [12], the square/tandem repeat problem [22], and the frequent/optimal substring problem [2,3,9]. Experiments on English text data show that our proposed algorithms run efficiently in practice.

## 2    Preliminaries

Let $A = a_1 a_2 \cdots a_{n-1}\$$ be a text of length $n \geq 1$. In what follows, we assume that $A$ ends with a special end marker \$ that does not appear in other positions. Let $A_i$ denote the suffix of $A$ that starts at position $i$. For a substring $S$ of $A$, we denote by $Occ(S, A)$ the set of all occurrences of $S$ in $A$. Let $\equiv_A$ be an equivalence relation on substrings defined as follows: For any substrings $S, S'$, the relation $S \equiv_A S'$ holds if and only if $Occ(S, A) = Occ(S', A)$. A substring $S$ of $A$ is *branching* if $S$ is the longest common prefix of distinct suffixes $A_i$ and $A_j$ $(i \neq j)$.

The *suffix array* of a text $A$ [16] is a sorted array $Pos[1..n]$ of all the suffixes of $A$, i.e., $Pos[k] = i$ if $A_i$ is lexicographically the $k$-th suffix. The *suffix tree* [17] is a data structure for storing all branching substrings of $A$, which is the compacted trie $ST$ for all suffixes of $A$. The suffix tree has at most $2n - 1$ nodes and can be stored in $O(n)$ space. The suffix tree $ST$ of a text of length $n$ can be constructed in $O(n)$ time [17,23,5]. The suffix array $Pos$ of $A$ coincides with the list of the leaves of $ST$ ordered from left to right. In Fig. 1, we show the suffix tree and the suffix array of string $A = \texttt{abcabbca\$}$. We denote by $str(v)$ the substring of $A$ obtained by concatenating the labels on the path from the root to $v$. The following lemma is well known [17].

**Lemma 1.** *Let $S$ be any substring of $A$. Then, the following 1–3 are equivalent.*

1. *$S$ is branching.*
2. *$S$ is the unique longest member of the equivalence class of $S$ w.r.t. $\equiv_A$.*
3. *$S = str(v)$ for some internal node $v$ of the suffix tree of $A$.*

We denote by $lcp(A, B)$ the length of the longest common prefix between strings $A$ and $B$. The *lcps* between suffixes that are adjacent in the sorted $Pos$ array are denoted by an array $Height$: $Height[k] = lcp(A_{pos[k-1]}, A_{pos[k]})$ for $2 \leq k \leq n$. All the necessary *lcps* for $O(m + \log n)$ search (called arrays $Llcp$ and $Rlcp$ in [16]) can be computed easily in $O(n)$ time from array $Height$ [11,15]. Therefore, we define the *lcp* problem as follows.

**Definition 1.** *The lcp problem in suffix arrays is to compute the $Height$ array from a text $A$ and its suffix array $Pos$.*

For any substring $S$ of $A$, the suffix array $Pos$ gives a compact representation of all occurrences of $S$. The set of all occurrences of $S$ occupy a contiguous interval $[L, R] \subseteq \{1, \ldots, n\}$, namely, $Occ(S, A) = \{Pos[k] : L \leq k \leq R\}$. We call the pair $(L, R)$ the *rank interval* of $S$. Then, the *triple* for $S$ is the triple $(L, R, H)$ of integers, where $(L, R)$ is the rank interval of $S$ and $H = |S|$ is the length of $S$. If necessary, the substring can be immediately obtained by $S = A[Pos[L]..Pos[L] + H - 1]$.

A *bottom-up traversal* of the suffix tree is any list $L$ of its nodes such that each node appears exactly once in $L$, and a node appears in $L$ only after all of its children appear. The *post-order traversal* [1] is an example of bottom-up traversals. A *bottom-up substring traversal* of $A$ is a list $L$ of the triples $(L, R, H)$ for all branching substrings of $A$ which is generated by a bottom-up traversal of the suffix tree of $A$. Then, the substring traversal problem is stated as follows.

**Definition 2.** *The substring traversal problem is to compute the substring traversal $L$ for a text $A$.*

This problem is linear time solvable by a post-order traversal of the suffix tree $ST$. Unfortunately, it is difficult to solve this problem with the suffix array $Pos$ alone because $Pos$ has lost the information on tree topology. The array $Height$ has the information on the tree topology which is lost in the suffix array $Pos$.

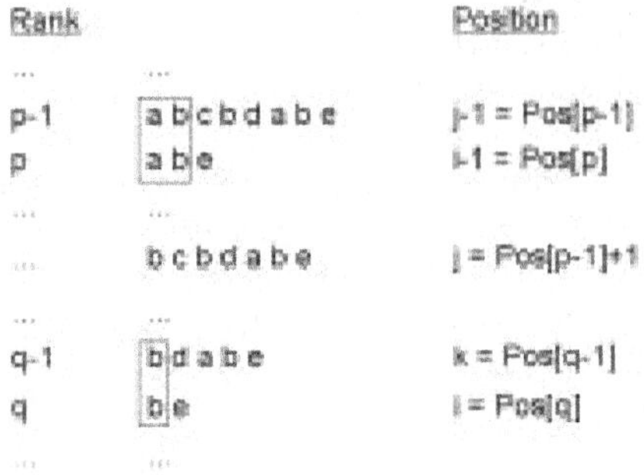

Rank                                          Position

...

p-1         | a b|c b d a b e                   |-1 = Pos[p-1]
p           | a b|e                             |-1 = Pos[p]

...

            b|c b d a b e                       | = Pos[p-1]+1

...

q-1         | b|d a b e                         k = Pos[q-1]
q           | b|e                               | = Pos[q]

...

**Fig. 2.** An example of sorted suffixes and *lcp*s.

**Lemma 2.** *A substring $S$ of a text $A$ is branching if and only if there exists some rank $1 \leq k \leq n$ such that $S$ is the longest common prefix of the adjacent suffixes $A_{Pos[k-1]}$ and $A_{Pos[k]}$.*

From Lemma 2, we can compute the list of all branching substrings associated with the *in-order traversal* of $ST$ simply by reporting $A[Pos[k]..Pos[k] + Height[k] - 1]$ for every rank $1 \leq k \leq n$. Unfortunately, the obtained list may contain duplicates since $ST$ is not a binary tree. Furthermore, there is no obvious way to compute either the associated rank intervals $(L, R)$ or the post-order traversal.

## 3    Linear-Time lcp Computation

In the *lcp* computation, we will use an intermediate array *Rank*. The array *Rank* is defined as the inverse function of *Pos*, and it can be obtained immediately when the *Pos* array is given: If $Pos[k] = i$, then $Rank[i] = k$.

### 3.1    Properties of lcp

The *lcp* between two suffixes is the minimum of the *lcp*s of all pairs of adjacent suffixes between them on the *Pos* array [16]. That is,

$$lcp(A_{Pos[x]}, A_{Pos[z]}) = \min_{x < y \leq z} \{lcp(A_{Pos[y-1]}, A_{Pos[y]})\}.$$

This implies that the *lcp* of a pair of adjacent suffixes on *Pos* is greater than or equal to the *lcp* of a pair of suffixes that surround them.

**Fact 1.** $lcp(A_{Pos[y-1]}, A_{Pos[y]}) \geq lcp(A_{Pos[x]}, A_{Pos[z]}),    x < y \leq z.$

When the *lcp* between a pair of adjacent suffixes on *Pos* is greater than 1, the lexicographical order of the suffixes is preserved when the first character of each suffix is deleted.

**Fact 2.** If $lcp(A_{Pos[x-1]}, A_{Pos[x]}) > 1$, then

$$Rank[Pos[x-1]+1] < Rank[Pos[x]+1].$$

In this case, the $lcp$ between $A_{Pos[x-1]+1}$ and $A_{Pos[x]+1}$ is one less than the $lcp$ between $A_{Pos[x-1]}$ and $A_{Pos[x]}$.

**Fact 3.** If $lcp(A_{Pos[x-1]}, A_{Pos[x]}) > 1$, then

$$lcp(A_{Pos[x-1]+1}, A_{Pos[x]+1}) = lcp(A_{Pos[x-1]}, A_{Pos[x]}) - 1.$$

Now we consider the following problem: compute the $lcp$ between a suffix $A_i$ and its adjacent suffix on $Pos$ when the $lcp$ between $A_{i-1}$ and its adjacent suffix is known. For notational convenience, let $p = Rank[i-1]$ and $q = Rank[i]$. Also let $j-1 = Pos[p-1]$ and $k = Pos[q-1]$. See Fig. 2. That is, we want to compute $Height[q]$ when $Height[p]$ is given.

**Lemma 3.** If $lcp(A_{j-1}, A_{i-1}) > 1$ then $lcp(A_k, A_i) \geq lcp(A_j, A_i)$.

*Proof.* Since $lcp(A_{j-1}, A_{i-1}) > 1$, we have $Rank[j] < Rank[i]$ by Fact 2. Since $Rank[j] \leq Rank[k] = Rank[i] - 1$, we get $lcp(A_k, A_i) \geq lcp(A_j, A_i)$ by Fact 1.

**Theorem 1.** If $Height[p] = lcp(A_{j-1}, A_{i-1}) > 1$ then

$$Height[q] = lcp(A_k, A_i) \geq Height[p] - 1.$$

*Proof.*

$$
\begin{aligned}
lcp(A_k, A_i) &\geq lcp(A_j, A_i) & \text{(by Lemma 3)} \\
&= lcp(A_{j-1}, A_{i-1}) - 1. & \text{(by Fact 3)}
\end{aligned}
$$

By Theorem 1, when the $lcp$ between suffix $A_{i-1}$ and its adjacent suffix is $h$, suffix $A_i$ and its adjacent suffix on $Pos$ has a common prefix of length at least $h - 1$. Therefore, it suffices to compare from the $h$-th characters for computing the $lcp$ between suffix $A_i$ and its adjacent suffix. If $h$ is less than or equal to 1, we will compare from the first characters.

## 3.2   Algorithm and Analysis

We now present the algorithm *GetHeight* that solves the lcp problem in suffix arrays. By Theorem 1, we do not need to compare all characters when we compute the $lcp$ between a suffix and its adjacent suffix on $Pos$. To compute all the $lcps$ of adjacent suffixes on $Pos$ efficiently, we examine the suffixes from $A_1$ to $A_n$ in order.

**Theorem 2.** *Algorithm GetHeight computes array Height in $O(n)$.*

```
Algorithm GetHeight
input: A text A and its suffix array Pos
   1  for i:=1 to n do
   2      Rank[Pos[i]] := i
   3  od
   4  h:=0
   5  for i:=1 to n do
   6      if Rank[i] > 1 then
   7          k := Pos[Rank[i]-1]
   8          while A[i+h] = A[j+h] do
   9              h := h+1
  10          od
  11          Height[Rank[i]] := h
  12          if h > 0 then h := h-1 fi
  13      fi
  14 od
```

**Fig. 3.** The linear-time algorithm for the lcp problem.

*Proof.* The correctness of *GetHeight* follows from previous discussions. The execution time of the algorithm is proportional to the number of times line 9 is executed, since line 9 is the innermost loop of *GetHeight*. The value of $h$ increases one by one in line 9, and it is always less than $n$ due to the end marker $\$$. Since the initial value of $h$ is 0 and it decreases at most $n$ times in line 12, $h$ increases at most $2n$ times. Therefore, the time complexity of Algorithm *GetHeight* is $O(n)$.

## 4   Application to Block-Sorting Compression

### 4.1   Block-Sorting Compression

The block-sorting algorithm is a text compression method with good balance of compression ratio and speed [4,8]. It achieves speed comparable to dictionary compressors, but obtains compression close to the best statistical compressor. The block-sorting algorithm is used in *bzip2* [21].

The encoder of block-sorting consists of three processes: the Burrows-Wheeler transformation, move-to-front encoding and entropy coding. The Burrows-Wheeler transformation (BWT) is the most time-consuming process. It transforms a string $A$ of length $n$ by forming the $n$ rotations (cyclic shifts) of $A$, sorting them lexicographically, and extracting the last character of each of the rotations. A string $L$ is formed from these characters, where the $i$-th character of $L$ is the last character of the $i$-th sorted rotation. In addition to $L$, the BWT computes the index $I$ of the original string $A$ in the sorted list of rotations. Fig. 4 is an example of BWT where $A$='abraca'. A move-to-front encoding encodes an instance of a character $ch$ by the count of distinct characters between itself and the previous occurrence of $ch$.

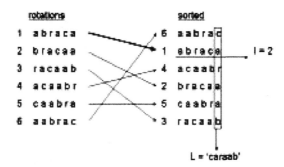

**Fig. 4.** An example of the Burrows-Wheeler transformation.

As a result of BWT, the locality of characters of $L$ goes higher than that of $A$ [4]. So, when applied to the string $L$, the output of a move-to-front encoder will be dominated by low numbers, which can be effectively encoded with Huffman coding or run-length coding.

The decoder of block-sorting is the reverse of the encoder. Decoding speed of an entropy code depends on the used method, but Huffman coding or run-length coding, which is generally used for encoding, can be reversed in linear time. A move-to-front code can be reversed in $O(n)$ time, and the original string $A$, the reverse of the BWT, can be recovered from $L$ and $I$ in $O(n)$ time. Therefore, the block-sorting decompression takes linear time in general.

## 4.2   Block-Sorting and Suffix Arrays

The first step of block-sorting, the BWT, is similar to the construction process of a suffix array. The BWT takes much time for sorting the suffixes. However, its reverse transformation from $L$ and $I$ to $A$ is quickly computed in linear time by a radix-sort-like procedure. Moreover, the suffix array $Pos$ of $A$ can be computed immediately when the compressed text is decoded.

To search for a pattern using the suffix array more efficiently, the $lcp$ information ($Llcp$ and $Rlcp$) is required. The $lcp$ information can be computed in $O(n \log n)$ time when the suffix array is constructed from the original text $A$. With our algorithm, the $lcp$ information can be computed in $O(n)$ time from the original text $A$ and its $Pos$ array. Therefore, suffix arrays can be stored and used efficiently by the block-sorting compression.

Since block-sorting has the effect of storing the compressed text and its suffix array, it can be used for storing and transferring large data. Sadakane and Imai presented a cooperative distributed text database management method unifying search and compression based on BWT [18]. Sadakane also presented a modified BWT for case-insensitive search with the suffix array [19]. Recently, Sadakane proposed a compressed text database system [20] based on the compressed suffix array [10]. Ferragina and Manzini [7,6] proposed a data structure that supports search operations without uncompressing the block-sorting compression.

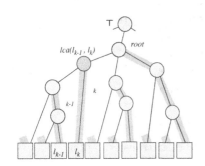

```
Algorithm BottomUpTraverse;
input: An ordered and compacted tree T
  with n ≥ 0 leaves ℓ₁,...,ℓₙ.
 1  S:={⊤} /* Initialize the stack S */
 2  for k:=1 to n+1 do /* k-th stage */
 3      v := lca(ℓₖ₋₁,ℓₖ);
 4      while (depth(top(S)) > depth(v)) do
 5          v := Pop(S) and report v; od;
 6      if (depth(top(S)) < depth(v)) then
 7          Push(v,S); fi;
 8      Push(ℓₖ,S);      /* Set Sₖ = S */
 9  od /* for-loop */
```

**Fig. 5.** An example of the right-most branch decompositions.

**Fig. 6.** The algorithm to compute the post-order traversal of an ordered tree.

## 5    Bottom-Up Traversal of Suffix Trees

### 5.1    Properties of the Post-Order Traversal

An ordered tree $T$ is *compacted* if every internal node of $T$ has at least two children. Let $T$ be an ordered and compacted tree with $n \geq 0$ leaves $\ell_1, \ldots, \ell_n$. In what follows, a *path* in $T$ is always written in the upward direction. That is, a *path* (or upward path) is a sequence $\pi = (v_0, v_1, \ldots, v_m)$ $(m \geq 0)$ of nodes in $T$ such that $v_i$ is the parent of $v_{i-1}$ for every $1 \leq i \leq m$. The length of $\pi$ is $|\pi| = m$. A path $\pi$ from the $k$-th leaf $(1 \leq k \leq n)$ to the root is called the $k$-th *branch* of $T$ and denoted by $\pi(\ell_k)$. A *node-depth* of a node $v$, denoted by $depth(v)$, is the length of the path from $v$ to the root. We write $u \preceq v$ $(u \prec v)$ if a node $u$ is an ancestor (proper ancestor) of node $v$. We denote by $lca(u, v)$ the *lowest common ancestor* of nodes $u$ and $v$ and by $\pi(\ell)$ the branch starting at a leaf $\ell$. Let $\ell$ be any leaf. A *rightmost branch* (RM branch, for short) starting with $\ell$, denoted by $\Pi(\ell)$, is the longest branch $\pi = (v_0 = \ell, v_1, \ldots, v_m)$ $(m \geq 0)$ starting at $\ell$ that consists of only *rightmost edges*, that is, $v_{i-1}$ is the rightmost child of $v_i$ for every $1 \leq i \leq m$.

$\Pi(\ell_k)$ is called the $k$-th *RM branch*. Since the set $\{\Pi(\ell_1), \ldots, \Pi(\ell_n)\}$ of all RM branches of $T$ is called the *RM branch decomposition* of $T$ since it is a partition of $T$. Fig. 5 shows an example of the RM branch decompositions, where each shadowed line indicates an RM branch (See below for the special node $\top$).

**Lemma 4.** *The post-order traversal of an ordered tree $T$ equals the concatenation $\Pi(\ell_1) \cdots \Pi(\ell_n)$ of the RM branches of $T$ from left to right.*

### 5.2    Algorithm for Bottom-Up Substring Traversal

From now on, we consider a method to compute the post-order traversal of an ordered compacted tree $T$ with $n \geq 0$ leaves $\ell_1, \ldots, \ell_n$ when the lowest common

ancestors of adjacent leaves and the depth of a node are available. Fig. 6 shows
the algorithm *BottomUpTraverse* for the problem. In the algorithm, we assume
a special *top* node$\top$ such that $\top \prec v$ for every $v$ in $T$ and special leaves $\ell_0$ and
$\ell_{n+1}$ such that $lca(\ell_0, \ell_1) = lca(\ell_n, \ell_{n+1}) = \top$ (See Fig. 5).

Scanning the height array $Height$ from left to right, the algorithm enumerates
the nodes of $T$ without duplicates by a sequence of push/pop operations to a
stack $S$ as follows. During the scan, a leaf node, say $\ell_k$, is pushed into the stack
$S$ when it is first encountered at stage $k$ and popped immediately at stage $k+1$.
The case for internal nodes is more complicated (See Fig. 5). Conceptually, a
node $v$ is pusded when it is visited from below at the first time and popped when
it is visited at the last time in the depth-first search of $T$.

An internal node $v$ is pushed into the stack when the leftmost leaf of the
second child of $v$, say $\ell_k$, is encountered at the first time in the scan, i.e., $v =
lca(\ell_{k-1}, \ell_k)$. Then, $v$ is popped from the stack $S$ when the leftmost leaf of the
next right sibling of $v$, is encountered in the scan. Then, $p = lca(\ell_{k-1}, \ell_k)$ is the
parent of $v$. Since the tree is compacted, the second leftmost leaf always exists
for every internal node. Thus from Lemma 2, the algorithm *BottomUpTraverse*
enumerates all nodes without duplicates by a scan of $Height$.

To see that the algorithm correctly computes the post-order traversal of $T$, we
need to know the precise contents of the stack during the scan. A key observation
is that if an internal node $v$ is $lca(\ell_{k-1}, \ell_k)$ for some $k$ then $v$ is on the $k$-th branch
from $\ell_k$ to the root and all nodes of $\Pi(\ell_{k-1})$ are proper descendants of $v$. We
gives the following lemma without proof due to the space limitation (See [14] for
the complete proof).

**Lemma 5.** *Let us consider the algorithm BottomUpTraverse of Fig. 6. For any
stage* $1 \leq k \leq n + 1$, *the contents of the stack* $S$ *at the beginning of the $k$-
th stage is the subsequence* $S_k = (v_{j_0}, \ldots, v_{j_k})$ *of the $k$-th branch* $\pi_k = (v_0 =
\ell_k, v_1, \ldots, v_m = \top)$ $(m \geq 0)$ *such that for every* $0 \leq j \leq m$, $v_j \in S_k$ *if and only
if the following* inclusion condition *holds at position $j$: either (i) $j = 0$ or (ii)
$v_{j-1}$ is not the leftmost child of $v_j$.*

From Lemma 5, we see that in the end of every stage $k$, the $k$-th RM branch
$\Pi(\ell_{k-1})$ is stored on the top of the stack $S$. Then, $\Pi(\ell_{k-1})$ is deleted from
the stack $S$ when $\ell_k$ is encountered in the scan. By repeating this process, the
algorithm finally outputs all RM branches $\Pi(\ell_1), \ldots, \Pi(\ell_n)$ of $T$ from left to
right. Hence, the next lemma immediately follows from Lemma 4.

**Lemma 6.** *The algorithm BottomUpTraverse of Fig. 6 computes the post-order
traversal of an ordered compacted tree with $n$ leaves in $O(n)$ time when the node-
depth for a node and the lowest common ancestor of adjacent leaves are constant
time computable.*

Now we present a linear time algorithm for the substring traversal problem
when the height array and the suffix array of $A$ is given. Fig. 7 shows the algorithm *TraverseWithArray* to compute the list of triples for text $A$ generated by
the post-order traversal of a suffix tree. In the algorithm, we encode a node $v$

```
Algorithm TraverseWithArray;
input: The height array Height and the suffix array Pos for a text A;
   1  S:= (-1, -1); n:=|T| /* Initialize the stack S */
   2  for k:=1 to n+1 do    /* k-th stage */
   3      (Llca, Hlca) := (k-1, Height[k]);
   4      (L, H) := top(S);
   5      while (H > Hlca) do
   6          (L, H) := pop(S), R := k-1; Then, report triple (L, R, H);
   7          Llca := L;    /* Update the left boundary */
   8          (L, H) := top(S);
   9      od
  10      if (H < Hlca) then
  11          Push((Llca,Hlca),S); fi;
  12      Push((k, n - Pos[k] + 1), S);    /* Set Sₖ = S */
  13 od /* for-loop */
```

**Fig. 7.** A linear time algorithm for the substring traversal problem.

by any pair $(L, H)$ such that $L$ and $H$ are the any occurrence and the length of the substring $str(v)$, respectively. The top node is encoded by $(-1, -1)$.

Recall that there were only two types of nodes processed in the algorithm *BottomUpTraverse*, a leaf and the lca of adjacent leaves. Thus for any rank $1 \le k \le n + 1$, we encode $v$ by $(L, H)$ as follows: (i) if $v$ is the leaf $\ell_k$ then $(L, H) = (k, |A_{Pos[k]}|) = (k, n - Pos[k] + 1)$ and (ii) if $v$ is the lca node $lca(\ell_{k-1}, \ell_k)$ then $(L, H) = (k - 1, Height[k])$. The depth of the node $v$ is obviously given by $H$. From Lemma 6, we know that the algorithm correctly simulates *ButtomUpTraverse*.

We then consider the computation of the rank intervals. Suppose a pair $(L, H)$ is popped from the stack $S$ at stage $k$ and it represents a node $v$. By induction on the number of nodes below $v$ on the $(k - 1)$-th path, we can show that $L$ is the rank of the leftmost leaf of $v$, where the value of $L$ is kept at the variable Llca at Line 7 of the algorithm. Since $v$ is on the $(k-1)$-th RM branch, $R = k - 1$ is obviously the rank of the rightmost leaf of $v$. Therefore, $(L, R, H)$ is the triple of $v$, and the next theorem follows from Lemma 6.

**Theorem 3.** *The algorithm TraverseWithArray of Fig. 7 computes in $O(n)$ time the list of all triples generated by the post-order traversal of the suffix tree of a text A of length n when the height array and the suffix array of A is given.*

Hence, the substring traversal problem is solvable in linear time when the height array $Height$ of a text $A$ is given. Since the algorithm *TraverseWithArray* makes only sequential I/Os and does not access the text $A$, we can also see that the algorithm is I/O efficient in the external I/O model of [24] (See [14]).

## 6  Experimental Results

We run experiments on a real dataset. For the height array construction, we implemented the naive $O(n^2)$ time algorithm (Abbreviated as NaiveHeight) and the linear time algorithm *GetHeight* (GetHeight). For the bottom-up substring traversal in Section 5, we implemented the algorithm with the suffix tree (TravTree), the naive algorithm with binary search on the suffix array (TravBinary), and the algorithm *TraverseWithArray* (TravHeight).

**Table 1.** Comparison of the computation time on English texts.

| | Height array construction | | Substring traversal | | |
|---|---|---|---|---|---|
| Algorithm | NaiveHeight | GetHeight | TravTree | TravBinary | TravHeight |
| Time (sec) | 17.59 | 7.81 | 2.07 | 13.62 | 1.94 |

In Table 1, we show the running time of the algorithms on an English text of 5.3MB [13] and a workstation (Sun UltraSPARC 300MHz, 256MB, g++ on Solaris 2.6). In the substring traversal, the preprocessing time for building the height array is not included. For the height array construction, we see from this table that GetHeight is faster than NaiveHeight more than twice on this test data. For the substring traversal, TravHeight is as fast as TravTree when the height array is precomputed, and faster than TravBinary even when the computation time of the height array is included.

# References

1. A. V. Aho, J. E. Hopcroft and U. D. Ullman, *Data Structures and Algorithms*, Addison-Wesley, 1983.
2. H. Arimura, S. Arikawa and S. Shimozono, Efficient discovery of optimal word-association patterns in large text databases, *New Generation Comput.*, 18, 49–60, 2000.
3. H. Arimura, H. Asaka, H. Sakamoto and S. Arikawa, Efficient discovery of proximity patterns with suffix arrays, In *Proc. CPM 2001*, Poster paper, LNCS, Springer-Verlag, 2001. (In this volumn).
4. M. Burrows and D. J. Wheeler, A block-sorting lossless data compression algorithm, *Digital Systems Research Center Research Report 124*, 1994.
5. M. Farach-Colton, P. Ferragina and S. Muthukrishnan, On the sorting-complexity of suffix tree construction, *Journal of the ACM*, Vol.47, No.6, 987–1011, 2000.
6. P. Ferragina and G. Manzini, Opportunistic data structures with applications, In *Proc. 41st IEEE Symposium on Foundations of Computer Science*, 390–398 2000.
7. P. Ferragina and G. Manzini, An experimental study of an opportunistic index, In *Proc. 12th ACM-SIAM Symposium on Discrete Algorithms*, 269–278 2001.
8. P. Fenwick, Block sorting text compression, In *Proc. Australian Computer Science Communications*, 18(1), 193–202, 1996.
9. R. Fujino, H. Arimura and S. Arikawa, Discovering unordered and ordered phrase association patterns for text mining, In *Proc. PAKDD2000*, LNAI 1805, 281–293, 2000.
10. R. Grossi and J. S. Vitter, Compressed suffix arrays and suffix trees with applications to text indexing and string matching, In *Proc. 32nd ACM Symposium on Theory of Computing*, 397–406, 2000.

11. D. Gusfield, An increment-by-one approach to suffix arrays and trees, *Technical Report CSE-90-39*, UC Davis, Dept. Computer Science, 1990.

12. D. Gusfield, Algorithms on Strings, Trees, and Sequences: Computer Science and Computational Biology, Cambridge University Press, New York, 1997.

13. R. Harris, Abstract Index, Monash Univ (1998).

14. T. Kasai, H. Arimura and S. Arikawa, Efficient substring traversal with suffix arrays, DOI-TR 185, Feb. 2001. (First appeared as T. Kasai, Fast algorithms for the subword statistics problems with suffix arrays, *Mc. Thesis*, Dept. Informatics, Kyushu Univ.,1999, In Japanese.)

15. S. E. Lee and K. Park, A new algorithm for constructing suffix arrays, *Journal of Korea Information Science Society (A)*, 24(7), 697–704, 1997.

16. U. Manber and G. Myers, Suffix arrays: A new method for on-line string searches, *SIAM J. Computing*, 22(5), 935–948 (1993).

17. E. M. McCreight, A space-economical suffix tree construction algorithm, *Journal of the ACM*, 23(2), 262–272, 1976.

18. K. Sadakane and H. Imai, A cooperative distributed text database management method unifying search and compression based on the Burrows-Wheeler transformation, In *Proc. International Workshop on New Database Technologies for Collaborative Work Support and Spatio-Temporal Data Management*, 434–445, 1998.

19. K. Sadakane, A modified Burrows-Wheeler transformation for case-insensitive search with application to suffix array compression, In *Proc. Data Compression Conference*, p.548, 1999.

20. K. Sadakane, Compressed text databases with efficient query algorithms based on the compressed suffix array, In *Proc. 11th Annual International Symposium on Algorithms and Computation*, 410–421, 2000.

21. J. Seward, `http://sources.redhat.com/bzip2\/`

22. J. Stoye and D. Gusfield, Simple and flexible detection of contiguous repeats using a suffix tree, In *Proc. CPM'98*, LNCS, 140–152, 1998.

23. E. Ukkonen, On-line construction of suffix trees, *Algorithmica* 14, 249–260, 1995.

24. J. S. Vitter, External memory algorithms, In *Proc. PODS'98*, 119–128 (1998).

# Multiple Pattern Matching Algorithms on Collage System

Takuya Kida, Tetsuya Matsumoto, Masayuki Takeda,
Ayumi Shinohara, and Setsuo Arikawa

Department of Informatics, Kyushu University 33
Fukuoka 812-8581, Japan
{kida,tetsuya,takeda,ayumi,arikawa}@i.kyushu-u.ac.jp

**Abstract.** Compressed pattern matching is one of the most active topics in string matching. The goal is to find all occurrences of a pattern in a compressed text without decompression. Various algorithms have been proposed depending on underlying compression methods in the last decade. Although some algorithms for multipattern searching on compressed text were also presented very recently, all of them are only for Lempel-Ziv family compressions. In this paper we propose two types of multipattern matching algorithms on collage system, which simulate the AC algorithm and a multipattern version of the BM algorithm, the most important algorithms for searching in uncompressed files. Collage system is a formal framework which is suitable to capture the essence of compressed pattern matching according to various dictionary based compressions. That is, we provide the model of multipattern matching algorithm for any compression method covered by the framework.

## 1   Introduction

The *compressed pattern matching problem* was first defined by Amir and Benson [2], and various compressed pattern matching algorithms have been proposed depending on underlying compression methods (see survey papers [19,23]).

In [7] we introduced a *collage system*, which is a formal system to represent a string by a pair of dictionary $\mathcal{D}$ and sequence $\mathcal{S}$ of phrases in $\mathcal{D}$. The basic operations are concatenation, truncation, and repetition. Collage systems give us a unifying framework of various dictionary-based compression methods, such as Lempel-Ziv family (LZ77, LZSS, LZ78, LZW), RE-PAIR [11], SEQUITUR [16], and the static dictionary based compression method. We also proposed in [7] the simple pattern matching algorithm on collage system, which simulates the move of the Knuth-Morris-Pratt automaton [10] running on the original text, by using the functions *Jump* and *Output*.

In this paper we address the *multiple pattern* matching problem on collage system. That is, given a set $\Pi$ of patterns and a collage system $\langle \mathcal{D}, \mathcal{S} \rangle$, we find all occurrences of any pattern in $\Pi$ within the text represented by $\langle \mathcal{D}, \mathcal{S} \rangle$. It is rather easy to extend *Jump* to the multipattern case. However, the extension of *Output* is not straightforward because the single pattern version utilizes some

A. Amir and G.M. Landau (Eds.): CPM 2001, LNCS 2089, pp. 193–206, 2001.

combinatorial properties on the period of the pattern. Although we have developed a multipattern searching algorithm for LZW compressed texts in [8], the same technique cannot be adopted to general collage systems. Nevertheless, we succeeded to develop an algorithm that runs in $O((\|\mathcal{D}\|+|\mathcal{S}|)\cdot height(\mathcal{D})+m^2+r)$ time with $O(\|\mathcal{D}\|+m^2)$ space, where $\|\mathcal{D}\|$ denotes the size of the dictionary $\mathcal{D}$, $height(\mathcal{D})$ denotes the maximum dependency of the operations in $\mathcal{D}$, $|\mathcal{S}|$ is the length of the sequence $\mathcal{S}$, $m$ is the total length of patterns, and $r$ is the number of pattern occurrences. Note that the time for decompressing-then-searching is linear with respect to the original text length, which can grow in proportion to $|S|\cdot 2^{\|\mathcal{D}\|}$ on the worst case. Therefore, the algorithm is more efficient than the decompress-then-search approach.

We also show an extension of the Boyer-Moore type algorithm presented in [21] to multiple patterns. The algorithm runs on the sequence $\mathcal{S}$, with skipping some tokens. It runs in $O((height(\mathcal{D})+m)|\mathcal{S}|+r)$ time after an $O(\|\mathcal{D}\|\cdot height(\mathcal{D})+m^2)$ time preprocessing with $O(\|\mathcal{D}\|+m^2)$ space. Moreover, we mention the parallel complexity of compressed pattern matching for a subclass of collage system in Section 8. Our result implies that the compressed pattern matching for regular collage system can be efficiently parallelized in principle.

## 2    Related Works

We presented in [7] a general pattern matching algorithm on collage system for a single pattern. The algorithm runs in $O((\|\mathcal{D}\|+|\mathcal{S}|)\cdot height(\mathcal{D})+m^2+r)$ time with $O(\|\mathcal{D}\|+m^2)$ space. For the subclass of collage system which contains no truncation, it runs in $O(\|\mathcal{D}\|+|\mathcal{S}|+m^2+r)$ time using $O(\|\mathcal{D}\|+m^2)$ space. We also presented a Boyer-Moore type algorithm in [21].

Independently, Navarro and Raffinot [14] developed a general technique for string matching on a text given as a sequence of blocks, which abstracts both LZ77 and LZ78 compressions, and gave bit-parallel implementations. The running time of these algorithms based on the bit-parallelism for LZW is $O(nm/w+m+r)$, where $n$ is the text length and $w$ is the length in bits of the machine word. If the pattern is short ($m<w$), these algorithms are efficient in practice. A Boyer-Moore type algorithm for a single pattern on Ziv-Lempel compressed text is also developed [15].

## 3    Preliminaries

Let $\Sigma$ be a finite set of characters, called an *alphabet*. A finite sequence of characters is called a *string*. We denote the length of a string $u$ by $|u|$. The empty string is denoted by $\varepsilon$, that is, $|\varepsilon|=0$. Let $\Sigma^*$ be the set of strings over $\Sigma$, and let $\Sigma^+=\Sigma^*\backslash\{\varepsilon\}$. Strings $x$, $y$, and $z$ are said to be a *prefix, factor,* and *suffix* of the string $u=xyz$, respectively. A prefix, factor, and suffix of a string $u$ is said to be *proper* if it is not $u$. Let $Prefix(u)$ be the set of prefixes of a string $u$, and let $Prefix(S)=\bigcup_{u\in S}Prefix(u)$ for a set $S$ of strings. We also define the sets *Suffix* and *Factor* in a similar way. For a string $u, v \in \Sigma^*$, let

$$lpf_v(u) = \text{the longest prefix of } u \text{ that is also in } Factor(v),$$
$$lsf_v(u) = \text{the longest suffix of } u \text{ that is also in } Factor(v),$$
$$lps_v(u) = \text{the longest prefix of } u \text{ that is also in } Suffix(v),$$
$$lsp_v(u) = \text{the longest suffix of } u \text{ that is also in } Prefix(v).$$

For a set $\Pi$ of strings, let $lpf_\Pi(u)$ be the longest prefix of $u$ that is also in $Factor(\Pi)$. We also define $lsf_\Pi(u)$, $lps_\Pi(u)$, and $lsp_\Pi(u)$ in a similar way.

The $i$th symbol of a string $u$ is denoted by $u[i]$ for $1 \leq i \leq |u|$, and the factor of a string $u$ that begins at position $i$ and ends at position $j$ is denoted by $u[i : j]$ for $1 \leq i \leq j \leq |u|$. Denote by $^{[i]}u$ (resp. $u^{[i]}$) the string obtained by removing the length $i$ prefix (resp. suffix) from $u$ for $0 \leq i \leq |u|$. The concatenation of $i$ copies of the same string $u$ is denoted by $u^i$. The reversed string of a string $u$ is denoted by $u^R$.

For a set $A$ of integers and an integer $k$, let $A \oplus k = \{i + k \mid i \in A\}$ and $A \ominus k = \{i - k \mid i \in A\}$. For strings $x$ and $y$, we denote the set of occurrences of $x$ in $y$ by $Occ(x, y)$. That is, $Occ(x, y) = \{i \mid |x| \leq i \leq |y|, x = y[i - |x| + 1 : i]\}$. For a set $\Pi \subset \Sigma^+$ of strings, $Occ(\Pi, y) = \bigcup_{x \in \Pi}\{\langle i, x\rangle \mid i \in Occ(x, y)\}$. Also denote by $Occ^*(x, u \bullet v)$ the set of occurrences of $x$ within the concatenation of two strings $u$ and $v$ which covers the boundary between $u$ and $v$. That is, $Occ^*(x, u \bullet v) = \{i \mid i \in Occ(x, uv), |u| < i < |u| + |x|\}$. For a set $\Pi \subset \Sigma^+$ of strings, $Occ^*(\Pi, u \bullet v) = \bigcup_{x \in \Pi}\{\langle i, x\rangle \mid i \in Occ^*(x, u \bullet v)\}$. We denote the cardinality of a set $V$ by $|V|$.

A *period* of a string $u$ is an integer $p$, $0 < p \leq |u|$, such that $x[i] = x[i + p]$ for all $i \in \{1, \ldots, |x| - p\}$. The next lemma provides an important property on periods of a string.

**Lemma 1 (Periodicity Lemma (see [3])).** *Let $p$ and $q$ be two periods of a string $x$. If $p + q - \gcd(p, q) \leq |x|$, then $\gcd(p, q)$ is also a period of $x$.*

The next lemma follows from the periodicity lemma.

**Lemma 2.** *Let $x$ and $y$ be strings. If $Occ(x, y)$ has more than two elements and the difference of the maximum and the minimum elements is at most $|x|$, then it forms an arithmetic progression, in which the step is the smallest period of $x$.*

## 4   Collage System and Text Compressions

A *collage system* [7] is a pair $\langle \mathcal{D}, \mathcal{S} \rangle$ defined as follows: $\mathcal{D}$ is a sequence of assignments $X_1 = expr_1$; $X_2 = expr_2$; $\cdots$ ; $X_\ell = expr_\ell$, where each $X_k$ is a token and $expr_k$ is any of the form

$$\begin{array}{ll}
a & \text{for } a \in \Sigma \cup \{\varepsilon\}, \quad\quad\quad (\textit{primitive assignment}) \\
X_i X_j & \text{for } i, j < k, \quad\quad\quad\quad\quad (\textit{concatenation}) \\
^{[j]}X_i & \text{for } i < k \text{ and an integer } j, \quad (\textit{prefix truncation}) \\
X_i^{[j]} & \text{for } i < k \text{ and an integer } j, \quad (\textit{suffix truncation}) \\
(X_i)^j & \text{for } i < k \text{ and an integer } j. \quad (\textit{j times repetition})
\end{array}$$

Each token represents a string obtained by evaluating the expression as it implies. The strings represented by tokens are called *phrases*. The set of phrases is called *dictionary*. Denote by $X.u$ the phrase represented by a token $X$. For example, $\mathcal{D} : X_1 = a; X_2 = b; X_3 = c; X_4 = X_1 \cdot X_2; X_5 = X_3 \cdot X_2; X_6 = X_4 \cdot X_2; X_7 = (X_4)^3; X_8 = X_7^{[2]}$, then $X_4.u$, $X_5.u$, $X_6.u$, $X_7.u$, $X_8.u$ are $ab$, $cb$, $abb$, $ababab$, and $abab$, respectively. The *size* of $\mathcal{D}$ is the number $\ell$ of assignments and denoted by $\|\mathcal{D}\|$. Also denote by $F(\mathcal{D})$ the set of tokens which are defined in $\mathcal{D}$. That is, $\|\mathcal{D}\| = |F(\mathcal{D})| = \ell$. Define the *height* of a token $X$ to be the height of the syntax tree whose root is $X$. The *height* of $\mathcal{D}$ is defined by $height(\mathcal{D}) = \max\{height(X) \mid X \text{ in } \mathcal{D}\}$. It expresses the maximum dependency of the tokens in $\mathcal{D}$.

On the other hand, $\mathcal{S} = X_{i_1}, \ldots, X_{i_n}$ is a sequence of tokens defined in $\mathcal{D}$. We denote by $|\mathcal{S}|$ the number $k$ of tokens in $\mathcal{S}$. The collage system represents a string obtained by concatenating strings $X_{i_1}.u, \cdots, X_{i_n}.u$. Most text compression methods can be viewed as mechanisms to factorize a text into a series of phrases and to store a sequence of 'representations' of the phrases. In fact, various compression methods can be translated into corresponding collage systems (see [7]). Both $\mathcal{D}$ and $\mathcal{S}$ can be encoded in various ways. The compression ratios therefore depend on the encoding sizes of $\mathcal{D}$ and $\mathcal{S}$ rather than $\|\mathcal{D}\|$ and $|\mathcal{S}|$.

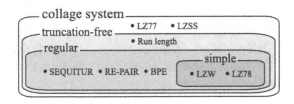

**Fig. 1.** Hierarchy of collage system.

A collage system is said to be *regular* if it contains neither repetition nor truncation. A regular collage system is said to be *simple* if, for every assignment $X = YZ$, $|Y.u| = 1$ or $|Z.u| = 1$. Through the collage systems, many dictionary-based compression methods can be categorized into some classes (see Fig. 1). Note that the collage systems for the SEQUITUR and the RE-PAIR are regular, and those for the LZW/LZ78 compressions are simple.

## 5    Main Result

Our main result is as follows.

**Theorem 1.** *The problem of compressed multiple pattern matching on a collage system* $\langle \mathcal{D}, \mathcal{S} \rangle$ *can be solved in* $O\big((\|\mathcal{D}\| + |\mathcal{S}|) \cdot height(\mathcal{D}) + m^2 + r\big)$ *time using* $O(\|\mathcal{D}\| + m^2)$ *space, where* $m$ *is the total length of patterns in* $\Pi$, *and* $r$ *is the* *number of pattern occurrences. If $\mathcal{D}$ contains no truncation, it can be solved in* $O(\|\mathcal{D}\| + |\mathcal{S}| + m^2 + r)$ *time.*

We developed in [7] an algorithm on collage system for a single pattern, which basically simulates the Knuth-Morris-Pratt (KMP) algorithm [10]. Although we devised a multipattern searching algorithm for LZ78/LZW in [9], the same technique cannot be applied directly to even the case of regular collage systems. One natural way of dealing with multiple patterns would be a simulation of the Aho-Corasick (AC) pattern matching machine. Now, we start with the definitions of $Jump_{AC}$ and $Output_{AC}$ which play a key role in our algorithm.

Let $\delta_{AC} : Q \times \Sigma \rightarrow Q$ be the deterministic state transition function of the AC machine for $\Pi$ obtained by eliminating the failure transitions (see [1]). The set $Q$ of states has a one-to-one correspondence with $Prefix(\Pi)$, and hence we identify $Q$ with $Prefix(\Pi)$ if no confusion occurs. We shall use the terms "state" and "string" interchangeably throughout the remainder of this paper. Fig. 3 is an example for $\Pi = \{aba, ababb, abca, bb\}$. Let $Jump_{AC}$ be the state transition function. For a collage system $\langle \mathcal{D}, \mathcal{S} \rangle$ and $\Pi$, define the function $Jump_{AC} : Q \times F(\mathcal{D}) \rightarrow Q$ by

$$Jump_{AC}(q, X) = \delta_{AC}(q, X.u).$$

We also define the set $Output_{AC}(q, X)$ for any pair $\langle q, X \rangle$ in $Q \times F(\mathcal{D})$ by

$$Output_{AC}(q, X) = \left\{ \langle |v|, \pi \rangle \;\middle|\; \begin{array}{l} v \text{ is a non-empty prefix of } X.u \text{ such} \\ \text{that } \pi \in \Pi \text{ is one of the outputs of} \\ \text{state } s = \delta_{AC}(q, v). \end{array} \right\}$$

That is, $Output_{AC}(q, X)$ stores all outputs emitted by the AC machine during the state transitions from the state $q$ reading the string $X.u$. The proposed algorithm can be summarized as in Fig. 2. For example, Fig. 4 shows that the move of our algorithm on $\mathcal{S}$ for $\Pi = \{aba, ababb, abca, bb\}$, where $\mathcal{D}$ is the same as the example in Section 4 and $\mathcal{S} = X_4, X_3, X_8, X_5, X_4, X_6$.

Concerning the function $Jump_{AC}(q, X)$, we can prove the next lemma in a similar way to [7] by regarding the string obtained by concatenating all patterns in $\Pi$ as a single pattern. That is, for a set $\Pi = \{\pi_1, \pi_2, \cdots, \pi_s\}$ of patterns, we make a string $P = \pi_1 \# \pi_2 \# \cdots \# \pi_s$, where $\# \notin \Sigma$ is a separate character.

**Lemma 3.** *The function $Jump_{AC}(q, X)$ can be realized in $O(\|\mathcal{D}\| \cdot height\,(\mathcal{D}) + m^2)$ time using $O(\|\mathcal{D}\| + m^2)$ space, so that it answers in $O(1)$ time. If $\mathcal{D}$ contains no truncation, the time complexity becomes $O(\|\mathcal{D}\| + m^2)$, where $m$ is the total length of patterns in $\Pi$.*

On the other hand, the realization of $Output_{AC}$ is not straightforward, and we need some additional efforts, which will be stated in the next section. Now, we have:

---

**Input.**     A set $\Pi$ of patterns and a collage system $\langle \mathcal{D}, \mathcal{S} \rangle$, where $\mathcal{S} = \mathcal{S}[1:n]$.
**Output.**     All positions at which a pattern $\pi \in \Pi$ occurs in $\mathcal{S}[1].u \cdots \mathcal{S}[n].u$.
    /* Preprocessing */
    Perform the preprocessing required for $Jump_{\mathrm{AC}}$ and $Output_{\mathrm{AC}}$
    (The complexity of this part depends on $\Pi$ and $\mathcal{D}$. See Section 6);
    /* Text scanning */
    $\ell := 0$;
    $state := 0$;
    **for** $k := 1$ **to** $n$ **do begin**
        **for each** $\langle p, \pi \rangle \in Output_{\mathrm{AC}}(state, \mathcal{S}[k])$ **do**
            Report an occurrence of $\pi$ that ends at position $\ell + p$ ;
        $state = Jump_{\mathrm{AC}}(state, \mathcal{S}[k])$;
        $\ell := \ell + |\mathcal{S}[k].u|$
    **end**

---

**Fig. 2.** Pattern matching algorithm.

**Lemma 4.** *The procedure to enumerate the set $Output_{\mathrm{AC}}$ $(q, X)$ can be realized in $O(\|\mathcal{D}\| \cdot height(\mathcal{D}) + m^2)$ time using $O(\|\mathcal{D}\| + m^2)$ space, so that it runs in $O(height(X) + \ell)$ time, where $\ell$ is the size of the set $Output_{\mathrm{AC}}(q, X)$. If $\mathcal{D}$ contains no truncation, it can be realized in $O(\|\mathcal{D}\| + m^2)$ time and space, so that it runs in $O(\ell)$ time.*

Theorem 1 follows from Lemma 3 and Lemma 4.

## 6     Realization of $Output_{\mathrm{AC}}$

Recall the definition of the set $Output_{\mathrm{AC}}(q, X)$. According to whether a pattern occurrence covers the boundary between the strings $q \in Prefix(\Pi)$ and $X.u$, we can partition the set $Output_{\mathrm{AC}}(q, X)$ into two disjoint subsets as follows.

$$Output_{\mathrm{AC}}(q, X) = Occ^\star(\Pi, q \bullet X.u) \ominus |q| \cup Occ(\Pi, X.u),$$

We consider mainly the subset $Occ(\Pi, X.u)$ below. It is easy to see that we can enumerate the set $Occ^\star(\Pi, q \bullet X.u)$ in $O(|Occ^\star(\Pi, q \bullet X.u)|)$ time if we can enumerate the set $Occ(\Pi, X.u)$ in $O(|Occ(\Pi, X.u)|)$ time, because the former is essentially the same as the problem of the concatenation case of the latter. Thus, we concentrate on proving the following lemma.

**Lemma 5.** *For a collage system $\langle \mathcal{D}, \mathcal{S} \rangle$ and a set $\Pi$ of patterns, we can enumerate the set $Occ(\Pi, X.u)$ for $X \in F(\mathcal{D})$ in $O(|Occ(\Pi, X.u)|)$ time after $O(m^2)$ time and space preprocessing, assuming that the set $Occ(\Pi, Y.u)$, $lpf_\Pi(Y.u)$, and $lsf_\Pi(Y.u)$ are already computed for all $Y$ such that $T(Y)$ is a subtree of $T(X)$ in the syntax tree.*

Now, we begin to consider the case of regular collage systems.

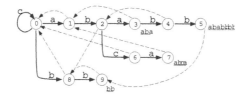

**Fig. 3.** Aho-Corasick machine for $\Pi = \{aba, ababb, abca, bb\}$. The solid and the broken arrows represent the goto and the failure functions, respectively. The underlined strings adjacent to the states mean the outputs from them.

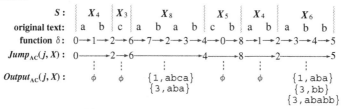

**Fig. 4.** Move of our algorithm.

## 6.1  For Regular Collage Systems

It is obvious if $X$ is a primitive assignment. If $X$ is a concatenation, i.e. $X = YZ$, we have $Occ(\Pi, X.u) = Occ(\Pi, Y.u) \cup Occ^\star(\Pi, Y.u \bullet Z.u) \cup Occ(\Pi, Z.u) \oplus |Y.u|$. Assume that $Occ(\Pi, W.u)$, $lpf_\Pi(W.u)$, and $lsf_\Pi(W.u)$ are already computed for all $W$ such that $T(W)$ is the subtree of $T(X)$ in the syntax tree. Then, we need to enumerate the set $Occ^\star(\Pi, Y.u \bullet Z.u)$ in order to enumerate the set $Occ(\Pi, X.u)$. We can reduce the above problem to the following problem since $Occ^\star(\Pi, Y.u \bullet Z.u) = Occ^\star(\Pi, lsf_\Pi(Y.u) \bullet lpf_\Pi(Z.u))$.

**Instance:** A set $\Pi$ of patterns and two factors $x$ and $y$ of $\Pi$.
**Question:** Enumerate the set $Occ^\star(\Pi, x \bullet y)$.

For the single pattern case, i.e. $\Pi = \{\pi\}$, it follows from Lemma 2 that the set $Occ^\star(\pi, x \bullet y)$ forms an arithmetic progression if it has more than two elements, where the step is the smallest period of $\pi$. Thus the $Occ^\star(\pi, x \bullet y)$ can be stored in $O(1)$ space as a pair of the minimum and the maximum values in it. The table storing those values can be computed in $O(m^2)$ time and space (see [7] for its detail).

For the multipattern case, however, we cannot apply the above technique directly to the enumeration of $Occ^\star(\Pi, x \bullet y)$. Now, we prove the next lemma.

**Lemma 6.** *For a set $\Pi$ of patterns, we can enumerate the set $Occ^\star(\Pi, x \bullet y)$ for all pairs of $x \in Prefix(\Pi)$ and $y \in Suffix(\Pi)$ in $O(|Occ^\star(\Pi, x \bullet y)|)$ time, after $O(m^2)$ time and space preprocessing.*

*Proof.* For any $x \in Prefix(\Pi)$ and $y \in Suffix(\Pi)$, we can build in $O(m^2)$ time and space a table $T$ that stores $xy$ if $xy \in \Pi$, otherwise *nil*. Then, we can enumerate

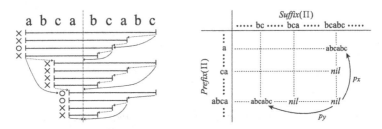

**Fig. 5.** Short-cut pointers and the table $T$ for $\Pi = \{abcabc, cabb, abca\}$.
In the left figure, $\circ$ indicates that $x'y'$ matches some pattern $\pi \in \Pi$
for $x' \in Suffix(x)$ and $y' \in Prefix(y)$, and $\times$ indicates that it does not
match.

the set $Occ^*(\Pi, x \bullet y)$ for all pairs of $x \in Prefix(\Pi)$ and $y \in Suffix(\Pi)$ by using
such table $T$ as the following manner: for each $x' \in Suffix(x) \cap Prefix(\Pi)$ and $y' \in Prefix(y) \cap Suffix(\Pi)$ in the descending order of their length, report the occurrence
of the pattern $\pi = x'y'$ if $T(x', y') \neq nil$. However, the time complexity for the
enumeration in this way becomes $O(m^2)$, not $O(|Occ^*(\Pi, x \bullet y)|)$. Then, we add
to each entry of the table $T$ a pair of two short-cut pointers $p_x$ and $p_y$ in order
to avoid increasing the time complexity, that is, for any pair of $x \in Prefix(\Pi)$
and $y \in Suffix(\Pi)$, $p_x$ and $p_y$ point to the longest proper suffix $x'$ of $x$ such that
$Occ^*(\Pi, x' \cdot y) \neq \emptyset$ or $x = \varepsilon$, and the longest proper prefix of $y$ such that $xy$ is
a pattern in $\Pi$ or $y = \varepsilon$, respectively. Fig. 5 shows an example of the pointers
and the table $T$, where $x = abca$ and $y = bcabc$ for $\Pi = \{abcabc, cabb, abca\}$.
Such pointers can be computed in $O(m^2)$ time by using the table $T$. Using
these pointers, we can get the desired sequence of pairs of $x' \in Suffix(x)$ and
$y' \in Prefix(y)$ in $O(|Occ^*(\Pi, x \bullet y)|)$ time for any pair of $x \in Prefix(\Pi)$ and
$y \in Suffix(\Pi)$. In the running example, the obtained sequence is $(abca, bcabc) \rightarrow$
$(abca, bc) \rightarrow (abca, \varepsilon) \rightarrow (a, bcabc) \rightarrow (a, bca) \rightarrow (a, \varepsilon) \rightarrow (\varepsilon, \varepsilon)$. The proof is
complete.                                                                              □

We thus finished the proof of Lemma 5 restricted to the class of regular collage
systems.

### 6.2  For Truncation-Free Collage Systems

We need to solve the following problem for dealing with repetitions.

**Instance:** A set $\Pi$ of patterns, a factor $x$ of $\Pi$, and an integer $k \geq 1$.
**Question:** Enumerate the set $Occ(\Pi, x^k)$.

For the single pattern case, i.e. $\Pi = \{\pi\}$, we presented a solution in [7] by using
Periodicity Lemma (Lemma 1). However the same technique does not work for
the multipattern case. Now, we need to prove the next lemma.

**Lemma 7.** *For a set $\Pi$ of patterns, $x \in Factor(\Pi)$, and an integer $k \geq 1$, we
can enumerate the set $Occ(\Pi, x^k)$ in $O(|Occ(\Pi, x^k)|)$ time after $O(m^2)$ time and*

*space preprocessing, assuming that $Occ(\Pi, x)$, $lpf_\Pi(x)$, and $lsf_\Pi(x)$ are already computed.*

*Proof.* It is trivial for $k \leq 2$. Suppose $k > 2$. Note that we can enumerate the set $Occ^\star(\Pi, x \bullet x)$ in $O(|Occ^\star(\Pi, x \bullet x)|)$ time from Lemma 6, and that $lps_\Pi(x) = lps_\Pi(lpf_\Pi(x))$. We use a *generalized suffix trie* [5] for a set $\Pi$ of strings ($GST_\Pi$ for short) in order to represent the set of suffixes of the strings in $\Pi$. It is an extension of the suffix trie for a single string. Note that each node of the $GST_\Pi$ corresponds to a string in $Factor(\Pi)$. The construction of the $GST_\Pi$ takes $O(m^2)$ time and space.

Now, we have two cases to consider.

Case 1: $xx \notin Factor(\Pi)$. Any pattern in $\Pi$ cannot cover more than three $x$'s. We can answer in $O(1)$ time whether $xx$ is in $Factor(\Pi)$ or not since $x \in Factor(\Pi)$ and the factor concatenation problem can be solved in $O(1)$ time. Moreover, we can obtain $Occ^\star(\Pi, x \bullet xx)$ since we can obtain $lpf_\Pi(xx)$ in $O(1)$ time (see [7]). Then, we can compute three sets $Occ(\Pi, x)$, $Occ^\star(\Pi, x \bullet x)$, and $Occ^\star(\Pi, x \bullet xx) \backslash Occ^\star(\Pi, x \bullet x)$. Therefore, the set $Occ(\Pi, x^k)$ can be enumerated in $O(|Occ(\Pi, x^k)|)$ time using these sets, $|x|$ and $k$.

Case 2: $xx \in Factor(\Pi)$. For the pattern occurrences which are within three $x$'s, we can enumerate them in the same way as Case 1. Now, we concentrate on the enumeration of the pattern occurrences that are not within three $x$'s. Suppose that a pattern $\pi$ has such an occurrence. Then $xx$ must be a factor of $\pi$. Since $|x|$ is a period of $\pi$ and $2|x| \leq |\pi|$, it follows from Lemma 1 that $|x|$ is a multiple of the smallest period $t$ of $\pi$ and therefore the set $Occ(\pi, x^k)$ forms an arithmetic progression whose step is $t$. Thus the set can be enumerated in only linear time proportional to its size. However, some occurrences in the enumeration can be included entirely within three $x$'s. In order to avoid reporting them twice, we omit $p$ in $Occ(\pi, x^k)$ satisfying the inequation $|\pi| - (p - \lfloor p/|x| \rfloor \cdot |x|) > 2|x|$ in the enumeration. So, we can enumerate all the pattern occurrences that are not within three $x$'s in time linearly proportional to the number of them, if we have the list of the patterns $\pi \in \Pi$ satisfying the conditions: (1) $xx \in Factor(\pi)$ and (2) $|x|$ is a period of $\pi$. The condition (2) can be replaced by the condition (2'): the smallest period of $xx$ equals to that of $\pi$. We add a list of the patterns $\pi$ that satisfy the conditions (1) and (2') to each node of $GST_\Pi$ that represents a string $xx = x^2$, called a *square*. It is not so hard to check up on the conditions in $O(m^2)$ time for all nodes of $GST_\Pi$. Each list added to a node of $GST_\Pi$ requires $O(|\Pi|) = O(m)$ space. The number of nodes representing squares is $O(m)$ (see [4]). Thus, the total space requirement is $O(m^2)$. Therefore, we can enumerate the set $Occ(\Pi, x^k)$ in $O(|Occ(\Pi, x^k)|)$ time with $O(m^2)$ time and space preprocessing.

The proof is complete.                                                                                 □

If $Y.u \notin Factor(\Pi)$, since any pattern in $\Pi$ cannot cover more than two $Y.u$'s, it is not hard to see that $Occ(\Pi, X)$ can be enumerated in $O(|Occ(\Pi, X)|)$ time using $Occ(\Pi, Y.u)$, $Occ^*(\Pi, Y.u \bullet Y.u)$, $|Y.u|$, and $k$. We thus finished the proof of Lemma 5 restricted to the class of truncation-free collage systems.

### 6.3   For General Collage Systems

For general collage systems, we must deal with truncation operations, in addition to concatenations and repetitions. Using the same technique of the single pattern case, we can see that Lemma 5 holds if $X = Y^{[k]}$, or $X = {}^{[k]}Y$ (see [7]), and that the time complexity increase by $height(\mathcal{D})$ as the single pattern case does. That is, the next lemma holds.

**Lemma 8.** *We can build in $O(\|\mathcal{D}\| \cdot height(\mathcal{D}) + m^2)$ time using $O(\|\mathcal{D}\| + m^2)$ space a data structure by which the enumeration of $Occ(\Pi, X.u)$ is performed in $O(height(X) + \ell)$ time, where $\ell = |Occ(\Pi, X.u)|$. If $\mathcal{D}$ contains no truncation, it can be built in $O(\|\mathcal{D}\| + m^2)$ time and space, and the enumeration requires only $O(\ell)$ time.*

Lemma 4 follows from the above. Although we need $lpf_\Pi(X.u)$ and $lsf_\Pi(X.u)$ for $X \in F(\mathcal{D})$, these can be computed in $O(\|\mathcal{D}\| \cdot height(\mathcal{D}) + m^2)$ time using $O(\|\mathcal{D}\| + m^2)$ space (see [7]).

## 7   On BM Type Algorithm for Multiple Patterns

We proposed in [21] a general, BM type algorithm for a single pattern on collage system. This algorithm is easily extensible to deal with multiple patterns if we use the techniques stated in Section 6. We give a brief sketch of the algorithm.

Recall that the BM algorithm on uncompressed texts performs the character comparisons in the right-to-left direction, and slides the pattern to the right using the so-called shift function when a mismatch occurs. Let $lpps_\Pi(w)$ denote the longest prefix of a string $w$ that is also properly in $Suffix(\Pi)$. Note that the function $\delta^{rev}_{AC}$ is the state transition function of the (partial) automaton that accepts a set $\Pi^R = \{x^R | x \in \Pi\}$ of reversed patterns. Contrary to the case of AC machine, the set $Q$ of states is $Suffix(\Pi)$. Define the functions $Jump_{MBM}$ and $Output_{MBM}$ as follows. For any state $q \in Suffix(\Pi)$ and any token $X \in F(\mathcal{D})$,

$$Jump_{MBM}(q, X) = \begin{cases} lpps_\Pi(X.u), & \text{if } q = \varepsilon \text{ and } lpps_\Pi(X.u) \neq \varepsilon; \\ \delta^{rev}_{AC}(q, X.u), & \text{if } q \neq \varepsilon; \\ \text{undefined}, & \text{otherwise.} \end{cases}$$

$$Output_{MBM}(q, X) = \{\pi \in \Pi \mid wq = \pi \text{ and } w \text{ is a proper suffix of } X.u\}.$$

The shift function is basically designed to shift the pattern to the right so as to align a text substring with its rightmost occurrence within the pattern. For a pattern $\pi$ and a string $w$, let

$$rightmost\_occ_\pi(w) = \min \left\{ \ell > 0 \;\middle|\; \begin{array}{l} \pi[|\pi| - \ell - |w| + 1 : |\pi| - \ell] = w, \\ \text{or } \pi[1 : |\pi| - \ell] \text{ is a suffix of } w \end{array} \right\}.$$

---

/\*Preprocessing for computing $Jump_{MBM}(j,t)$, $Output_{MBM}(j,t)$, and $Occ(t)$ \*/
    Preprocess the pattern $\pi$ and the dictionary $\mathcal{D}$;
/\* Main routine \*/
    $focus :=$ an appropriate value;
    $focus := \lceil m/C \rceil$;
    **while** $focus \le n$ **do begin**
Step 1:    Report all pattern occurrences that are contained in the phrase $\mathcal{S}[focus].u$
        by using $Occ(t)$;
Step 2:    Find all pattern occurrences that end within the phrase $\mathcal{S}[focus].u$
        by using $Jump_{MBM}(j,t)$ and $Output_{MBM}(j,t)$;
Step 3:    Compute a possible shift $\Delta$ based on information gathered in Step 2;
        $focus := focus + \Delta$
    **end**

---

**Fig. 6.** Overview of BM type compressed pattern matching algorithm.

For a set $\Pi$ of patterns, let $rightmost\_occ_{\Pi}(w) = \min_{\pi \in \Pi}\{rightmost\_occ_{\pi}(w)\}$, and let $Shift_{MBM}(q, X) = rightmost\_occ_{\Pi}(X.u \cdot q)$. When we encounter a mismatch against a token $X$ in state $q \in Suffix(\Pi)$, the possible shift $\Delta$ of the focus can be computed using $Shift_{MBM}$ in the same way as [21]. Figure 6 gives an overview of our algorithm.

We can prove the next lemma by using the techniques similar to those stated in Section 6, and Theorem 2 follows from Lemma 9.

**Lemma 9.** *The functions $Jump_{MBM}$, $Output_{MBM}$, and $Shift_{MBM}$ can be built in $O(height(\mathcal{D}) \cdot \|\mathcal{D}\| + m^2)$ time and $O(\|\mathcal{D}\| + m^2)$ space, so that they answer in $O(1)$ time, where $m$ is the total length of patterns in $\Pi$. The factor $height(\mathcal{D})$ can be dropped if $\mathcal{D}$ contains no truncation.*

Thus, we have the following theorem.

**Theorem 2.** *The BM type algorithm for multiple pattern searching on collage system runs in $O(height(\mathcal{D}) \cdot (\|\mathcal{D}\| + |\mathcal{S}|) + |\mathcal{S}| \cdot m + m^2 + r)$ time, using $O(\|\mathcal{D}\| + m^2)$ space, where $m$ is the total length of patterns in $\Pi$, and $r$ is the number of pattern occurrences. If $\mathcal{D}$ contains no truncation, the time complexity becomes $O(\|\mathcal{D}\| + |\mathcal{S}| \cdot m + m^2 + r)$.*

## 8   Parallel Complexity of Compressed Pattern Matching

In this section, we consider the computational complexity of the following decision problem for a class $\mathcal{C}$ of collage systems:

**Instance:** A collage system $\langle \mathcal{D}, \mathcal{S} \rangle$ in $\mathcal{C}$ over $\Sigma$ and a set $\Pi = \{\pi_1, \cdots, \pi_s\}$ of patterns.
**Question:** Is there any pattern $\pi_j \in \Pi$ that occurs in the text $T$ represented by $\langle \mathcal{D}, \mathcal{S} \rangle$? That is, are there any $i$ and $j$ such that $T[i : i + |\pi_j| - 1] = \pi_j$ or not?

*LogCFL* is the class of problems logspace-reducible to a context-free language. An auxiliary pushdown automaton (*AuxPDA*) is a nondeterministic Turing machine with a read-only input tape, a space-bounded worktape, and a pushdown store that is not subject to the space-bound. The class of languages accepted by auxiliary pushdown automata in space $s(n)$ and time $t(n)$ is denoted by $AuxPDA(s(n), t(n))$. The next lemma is quite useful.

**Lemma 10 ([22]).** *LogCFL* $= AuxPDA(\log n, n^{O(1)})$.

We now show the following theorem.

**Theorem 3.** *Compressed pattern matching problem on regular collage system is in LogCFL.*

*Proof.* We show an auxiliary pushdown automaton $M$ that accepts an input string if and only if there is some pattern $\pi_j \in \Pi$ that occurs in the text $T$ represented by $\langle \mathcal{D}, \mathcal{S} \rangle$. We note that by using pushdown store, $M$ can traverse the evaluation tree of any variable $X_k$ and 'scan' the string $X_k.u$ from left to right that is the sequence of leaves in the tree. Moreover, by utilizing the nondeterminism, $M$ can scan any substring of $X_k.u$.

$M$ represents a position $t$ of a pattern as a binary string in the worktape, and initializes it $t = 1$. For simplicity, we first consider the case that a pattern $\pi_j$ occurs within the string $X_{i_k}.u$ for some $X_{i_k}$. $M$ nondeterministically guesses such $j$ and $k$, and nondeterministically goes down the evaluation tree of $X_k$ from the root by pushing the traversed variables in the pushdown store. At a leaf, $M$ confirms that the character $X_k.u[l]$ is equal to $\pi_j[t]$. Then $M$ increments $t$ by one by using the worktape, and proceeds to the next character $X_k.u[l+1]$ by using the pushdown store. $M$ repeats this procedure until $\pi_j$ is verified to occur in $X_k$ at position $l$. Remark that $l$ is not explicitly written in the worktape: it is impossible in general since $l = O(|X_k|) = O(2^{||\mathcal{D}||})$. However, on the other hand, since $t \leq |\pi_j|$ and patterns are explicitly written in the input tape, the space required by $M$ is $O(\log |\pi_j|)$, that is logarithmic with respect to the input size. The computation time is clearly bounded by a polynomial, since the height of the evaluation tree is at most $||D||$. For a general case that a pattern $\pi_j$ spreads over a region $X_{i_k} \cdot X_{i_{k+1}} \cdots X_{i_h}$, we can show that $M$ verifies the occurrences in polynomial time using a log-space worktape in the same way. By Lemma 10, we complete the proof.                                                                                     □

Since it is known that *LogCFL* $\subseteq$ **NC**$^2$ [17,18], the above theorem implies that the compressed pattern matching for regular collage systems can be efficiently parallelized in principle. For general collage systems including repetitions and truncations, we have not succeeded to show that the problems are in **NC** nor **P**-complete yet.

## 9    Concluding Remarks

We proposed two types of multipattern matching algorithms on collage system. One is an AC-type algorithm, which runs in $O((||\mathcal{D}|| + |\mathcal{S}|) \cdot height(\mathcal{D}) + m^2 + r)$

time with $O(\|\mathcal{D}\| + m^2)$ space. Its running time becomes $O(\|\mathcal{D}\| + |\mathcal{S}| + m^2 + r)$ if a collage system contains no truncation. The other is a BM-type algorithm, which runs in $O((height(\mathcal{D}) + m)|\mathcal{S}| + r)$ time after an $O(\|\mathcal{D}\| \cdot height(\mathcal{D}) + m^2)$ time preprocessing with $O(\|\mathcal{D}\| + m^2)$ space. We also showed that compressed pattern matching on regular collage system is in $LogCFL \subseteq \mathbf{NC}^2$.

The compressed pattern matching usually aims to search in compressed files faster than a regular decompression followed by an ordinary search (Goal 1). A more ambitious goal is to perform a faster search in compressed files *in comparison with an ordinary search in the original files* (Goal 2). In this case, the aim of compression is not only to reduce disk storage requirement but also to speed up string searching task. In fact, we have achieved Goal 2 for the compression method called Byte Pair Encoding (BPE) [21,20,23].

In [6,12], approximate string matching algorithms over LZW/LZ78 compressed texts were proposed. Very recently, Navarro et al. proposed a practical solution for the LZW/LZ78 compressions and showed experimentally that it is up to three times faster than the trivial approach of uncompressing and searching [13]. The basic idea of the solution is to reduce the problem of approximate string searching to the problem of multipattern searching of a set of pattern pieces plus local decompression and direct verification of candidate text areas. Using the same technique, the result of this paper leads to speed-up of approximate string matching for various compression methods. In fact, we have verified that the suggested algorithm runs on BPE compressed texts faster than *Agrep*, known as the fastest pattern matching tool.

# References

1. A. V. Aho and M. Corasick. Efficient string matching: An aid to bibliographic search. *Comm. ACM*, 18(6):333–340, 1975.
2. A. Amir and G. Benson. Efficient two-dimensional compressed matching. In *Proc. Data Compression Conference*, page 279, 1992.
3. M. Crochemore and W. Rytter. *Text Algorithms*. Oxford University Press, New York, 1994.
4. A.S. Fraenkel and J. Simpson. How many squares can a string contain? *J. Combin. Theory Ser. A*, 82:112–120, 1998.
5. L.C.K. Hui. Color set size problem with application to string matching. In *Combinatorial Pattern Matching*, volume 644 of *Lecture Notes in Computer Science*, pages 230–243. Springer-Verlag, 1992.
6. J. Kärkkäinen, G. Navarro, and E. Ukkonen. Approximate string matching over Ziv-Lempel compressed text. In *Proc. 11th Ann. Symp. on Combinatorial Pattern Matching*, volume 1848 of *Lecture Notes in Computer Science*, pages 195–209. Springer-Verlag, 2000.
7. T. Kida, Y. Shibata, M. Takeda, A. Shinohara, and S. Arikawa. A unifying framework for compressed pattern matching. In *Proc. 6th International Symp. on String Processing and Information Retrieval*, pages 89–96. IEEE Computer Society, 1999.
8. T. Kida, M. Takeda, A. Shinohara, M. Miyazaki, and S. Arikawa. Multiple pattern matching in LZW compressed text. *Journal of Discrete Algorithms*. to appear (previous version in DCC'98 and CPM'99).

9. T. Kida, M. Takeda, A. Shinohara, M. Miyazaki, and S. Arikawa. Multiple pattern matching in LZW compressed text. In J. A. Storer and M. Cohn, editors, *Proc. Data Compression Conference '98*, pages 103–112. IEEE Computer Society, 1998.

10. D.E. Knuth, J.H. Morris, and V.R. Pratt. Fast pattern matching in strings. *SIAM J. Comput*, 6(2):323–350, 1977.

11. N.J. Larsson and A. Moffat. Offline dictionary-based compression. In *Proc. Data Compression Conference '99*, pages 296–305. IEEE Computer Society, 1999.

12. T. Matsumoto, T. Kida, M. Takeda, A. Shinohara, and S. Arikawa. Bit-parallel approach to approximate string matching in compressed texts. In *Proc. 7th International Symp. on String Processing and Information Retrieval*, pages 221–228. IEEE Computer Society, 2000.

13. G. Navarro, T. Kida, M. Takeda, A. Shinohara, and S. Arikawa. Faster approximate string matching over compressed text. In *Proc. Data Compression Conference 2001*. IEEE Computer Society, 2001. to appear.

14. G. Navarro and M. Raffinot. A general practical approach to pattern matching over Ziv-Lempel compressed text. In *Proc. 10th Ann. Symp. on Combinatorial Pattern Matching*, volume 1645 of *Lecture Notes in Computer Science*, pages 14–36. Springer-Verlag, 1999.

15. G. Navarro and J. Tarhio. Boyer-Moore string matching over Ziv-Lempel compressed text. In *Proc. 11th Ann. Symp. on Combinatorial Pattern Matching*, volume 1848 of *Lecture Notes in Computer Science*, pages 166–180. Springer-Verlag, 2000.

16. C.G. Nevill-Manning, I.H. Witten, and D.L. Maulsby. Compression by induction of hierarchical grammars. In *Proc. Data Compression Conference '94*, pages 244–253. IEEE Press, 1994.

17. W. Ruzzo. Tree-size bounded alternation. *Journal of Computer and System Sciences*, 21(2):218–235, 1980.

18. W. Ruzzo. On uniform circuit complexity. *Journal of Computer and System Sciences*, 22(3):365–383, 1981.

19. W. Rytter. Algorithms on compressed strings and arrays. In *Proc. 26th Ann. Conf. on Current Trends in Theory and Practice of Infomatics*. Springer-Verlag, 1999.

20. Y. Shibata, T. Kida, S. Fukamachi, M. Takeda, A. Shinohara, T. Shinohara, and S. Arikawa. Speeding up pattern matching by text compression. In *Proc. 4th Italian Conference on Algorithms and Complexity*, volume 1767 of *Lecture Notes in Computer Science*, pages 306–315. Springer-Verlag, 2000.

21. Y. Shibata, T. Matsumoto, M. Takeda, A. Shinohara, and S. Arikawa. A Boyer-Moore type algorithm for compressed pattern matching. In *Proc. 11th Ann. Symp. on Combinatorial Pattern Matching*, volume 1848 of *Lecture Notes in Computer Science*, pages 181–194. Springer-Verlag, 2000.

22. I. Sudborough. On the tape complexity of deterministic context-free languages. *Journal of ACM*, 25:405–414, 1978.

23. M. Takeda, Y. Shibata, T. Matsumoto, T. Kida, A. Shinohara, S. Fukamachi, T. Shinohara, and S. Arikawa. Speeding up string pattern matching by text compression: The dawn of a new era. *Transactions of Information Processing Society of Japan*, 2001. to appear.

# Finding All Common Intervals of $k$ Permutations

Steffen Heber[1,2][*] and Jens Stoye[1][**]

[1] Theoretical Bioinformatics (H0300)
[2] Functional Genome Analysis (H0800)
German Cancer Research Center (DKFZ) Heidelberg, Germany
{s.heber,j.stoye}@dkfz.de

**Abstract.** Given $k$ permutations of $n$ elements, a $k$-tuple of intervals of these permutations consisting of the same set of elements is called a *common interval*. We present an algorithm that finds in a family of $k$ permutations of $n$ elements all $K$ common intervals in optimal $O(nk+K)$ time and $O(n)$ additional space.

This extends a result by Uno and Yagiura (*Algorithmica* 26, 290–309, 2000) who present an algorithm to find all $K$ common intervals of $k = 2$ permutations in optimal $O(n + K)$ time and $O(n)$ space. To achieve our result, we introduce the set of *irreducible intervals*, a generating subset of the set of all common intervals of $k$ permutations.

## 1 Introduction

Let $\Pi = (\pi_1, \ldots, \pi_k)$ be a family of $k$ permutations of $N = \{1, 2, \ldots, n\}$. A $k$-tuple of intervals of these permutations consisting of the same set of elements is called a *common interval*.

Common intervals have applications in different fields. The *consecutive arrangement problem* is defined as follows [1,3,4]: Given a finite set $X$ and a collection $\mathcal{S}$ of subsets of $X$, find all permutations of $X$ where the members of each subset $S \in \mathcal{S}$ occur consecutively. Finding all common intervals of a set of permutations reverses this problem. Some genetic algorithms using subtour exchange crossover based on common intervals have been proposed for sequencing problems such as the traveling salesman problem or the single machine scheduling problem [2,5,7]. In a bioinformatical context, common intervals can be used to detect possible functional associations between genes. It is supposed that genes occurring in different genomes in each other's neighborhood tend to encode functionally interacting proteins [8,6,9]. If one models genomes as permutations of genes, the problem of finding co-occurring genes translates into the problem of finding common intervals.

---

[*] Present address: Department of Computer Science & Engineering, APM 3132, University of California, San Diego, La Jolla, CA 92093-0114, USA. E-mail: sheber@ucsd.edu

[**] Present address: Max Planck Institute for Molecular Genetics, Ihnestr. 73, Berlin, Germany. E-mail: stoye@molgen.mpg.de

A. Amir and G.M. Landau (Eds.): CPM 2001, LNCS 2089, pp. 207–218, 2001.

Recently, Uno and Yagiura [10] presented three algorithms for finding all common intervals of $k = 2$ permutations $\pi_1$ and $\pi_2$: two simple $O(n^2)$ time algorithms and one more complicated $O(n + K)$ time algorithm where $K \leq \binom{n}{2}$ is the number of common intervals of $\pi_1$ and $\pi_2$. Since the latter algorithm runs in time proportional to the size of the input plus the size of the output, it is optimal in the sense of worst case complexity.

An obvious extension of this algorithm to find all common intervals of a family $\Pi = (\pi_1, \ldots, \pi_k)$ of $k \geq 2$ permutations would be to compare $\pi_1$ successively with $\pi_i$ for $i = 2, \ldots, k$ and report those intervals that are common in all comparisons. This yields an $O(kn + \sum_{i=2}^{k} K_i)$ time algorithm where $K_i$ is the number of common intervals of $\pi_1$ and $\pi_i$ for $2 \leq i \leq k$. The main result of this paper is an improvement of this approach by a non-trivial extension of Uno and Yagiura's algorithm, yielding an optimal $O(kn + K)$ time and $O(n)$ space algorithm where $K$ is the number of common intervals of $\Pi$. Note that this number can be considerably smaller than any of the $K_i$.

The approach relies on restricting the set of all common intervals $C$ to a smaller subset of *irreducible* intervals $I$, from which $C$ can be easily reconstructed. While the number of common intervals can be as large as $\binom{n}{2}$, we show that $1 \leq |I| \leq n - 1$ and present an algorithm to compute $I$ in optimal $O(kn)$ time, i.e., in time proportional to the input size. Knowing $I$ we can reconstruct $C$ in $O(K)$ time, i.e., in time proportional to the output size. Both algorithms use $O(n)$ additional space and their combination yields our main result.

## 2   Permutations and Common Intervals

Given a permutation $\pi$ of (the elements of) the set $N := \{1, 2, \ldots, n\}$, we denote by $\pi(i) = j$ that the $i$th element of $\pi$ is $j$. For $x, y \in N$, $x \leq y$, $[x, y]$ denotes the set $\{x, x + 1, \ldots, y\} \subseteq N$ and $\pi([x, y]) := \{\pi(i) \mid i \in [x, y]\}$ is called an *interval* of $\pi$. Let $\Pi = (\pi_1, \ldots, \pi_k)$ be a family of $k$ permutations of $N$. W.l.o.g. we assume in the following always that $\pi_1 = id_n := (1, \ldots, n)$. A $k$-tuple $c = ([l_1, u_1], \ldots, [l_k, u_k])$ with $1 \leq l_j < u_j \leq n$ for all $1 \leq j \leq k$ is called a *common interval* of $\Pi$ if and only if

$$\pi_1([l_1, u_1]) = \pi_2([l_2, u_2]) = \ldots = \pi_k([l_k, u_k]).$$

This allows to identify a common interval $c$ with the contained elements, i.e.

$$c \equiv \pi_j([l_j, u_j]) \quad \text{for} \quad 1 \leq j \leq k.$$

Since $\pi_1 = id_n$, the above set equals the index set $[l_1, u_1]$, and we will refer to this as the *standard notation* of $c$. The set of all common intervals of $\Pi$ is denoted $C_\Pi$. Note that our definition excludes common intervals of size one.

*Example 1.* Let $N = \{1, \ldots, 9\}$ and $\Pi = (\pi_1, \pi_2, \pi_3)$ with $\pi_1 = id_9$, $\pi_2 = (9, 8, 4, 5, 6, 7, 1, 2, 3)$, and $\pi_3 = (1, 2, 3, 8, 7, 4, 5, 6, 9)$. We have

$$C_\Pi = \{[1, 2], [1, 3], [1, 8], [1, 9], [2, 3], [4, 5], [4, 6], [4, 7], [4, 8], [4, 9], [5, 6]\}.$$

## 3   Finding All Common Intervals of Two Permutations

In order to keep this paper self-contained, here we briefly recall the algorithm RC (short for *Reduce Candidate*) of Uno and Yagiura [10] that finds all $K$ common intervals of $k = 2$ permutations $\pi_1 = id_n$ and $\pi_2$ of $N$ in $O(n + K)$ time and $O(n)$ space. For the correctness and analysis of the algorithm we refer to [10].

An easy test if an interval $\pi_2([x, y])$, $1 \leq x < y \leq n$, is a common interval of $\Pi = (\pi_1, \pi_2)$ is based on the following functions:

$$l(x, y) := \min \pi_2([x, y])$$
$$u(x, y) := \max \pi_2([x, y])$$
$$f(x, y) := u(x, y) - l(x, y) - (y - x).$$

Since $f(x, y)$ counts the number of elements in $[l(x, y), u(x, y)] \setminus \pi_2([x, y])$, an interval $\pi_2([x, y])$ is a common interval of $\Pi$ if and only if $f(x, y) = 0$. A simple algorithm to find $C_\Pi$ is to test for each pair of indices $(x, y)$ with $1 \leq x < y \leq n$ if $f(x, y) = 0$, yielding a naive $O(n^3)$ time or, using running minima and maxima, a slightly more involved $O(n^2)$ time algorithm.

The main idea of Algorithm RC is to save the time to test $f(x, y) = 0$ for some pairs $(x, y)$ by eliminating *wasteful* candidates for $y$.

**Definition 1.** *For a fixed $x$, a right interval end $y > x$ is called* wasteful *if it satisfies $f(x', y) > 0$ for all $x' \leq x$.*

In Algorithm RC (Algorithm 1), the common intervals are found using a data structure $Y$ consisting of a doubly-linked list *ylist* for indices of non-wasteful right interval end candidates and, storing intervals of *ylist*, two further doubly-linked lists *llist* and *ulist* that implement the functions $l$ and $u$ in order to compute $f$ efficiently. They are also essential for an efficient update of *ylist*. In our pseudocode we use the standard list operations *L.head* for the first element of list $L$, *L.succ(e)* for the successor and *L.pred(e)* for the predecessor of element $e$ in $L$.

---

**Algorithm 1** (Reduce Candidate, RC)

---

**Input:** A family $\Pi = (\pi_1 = id_n, \pi_2)$ of two permutations of $N = \{1, \ldots, n\}$.
**Output:** $C_\Pi$ in standard notation.
 1: initialize $Y$
 2: **for** $x = n - 1, \ldots, 1$ **do**
 3:    update $Y$    // (see Algorithm 2)
 4:    $y \leftarrow x$
 5:    **while** $(y \leftarrow ylist.succ(y))$ defined **and** $f(x, y) = 0$ **do**
 6:       output $[l(x, y), u(x, y)]$
 7:    **end while**
 8: **end for**

---

In the first step of Algorithm 1, $ylist$ is initialized containing one element that stores the index $n$, and $llist$ and $ulist$ are both initialized with the one-element interval $[n, n]$ consisting of the last/only element of $ylist$.

Then $Y$ is updated iteratively. A counter $x$ (corresponding to the currently investigated left interval end) runs from $n - 1$ down to 1. For any fixed $x$, the elements of $llist$ are maximal intervals of $ylist$ such that for an interval $[y, y']$ we have $l(x, y) = l(x, ylist.succ(y)) = \cdots = l(x, y')$; similar for $ulist$. For an interval $[y, y']$ in $llist$ or $ulist$, we define its $value$ by $val([y, y']) := \pi_2(y)$ and its $end$ by $end([y, y']) := y'$. Algorithm 2 shows the update procedure for $\pi_2(x) > \pi_2(x+1)$. The case $\pi_2(x) < \pi_2(x + 1)$ is treated in a symmetric way.

---

**Algorithm 2** (Update of data structure $Y$ in line 3 of Algorithm 1)

---

1: prepend $x$ at the head of $ylist$
2: prepend $[x, x]$ at the head of $llist$
3: **while** $(u^* \leftarrow ulist.head)$ has a successor $u$ and $val(u) < \pi_2(x)$ **do**
4:     delete $u^*$ from $ulist$ and the corresponding elements from $ylist$
5: **end while**
6: $y^* \leftarrow end(u^*)$
7: **if** $(\tilde{y} \leftarrow ylist.succ(y^*))$ is defined **then**
8:     **while** $f(x, y^*) > f(x, \tilde{y})$ **do**
9:         delete $y^*$ from $ylist$
10:         $y^* \leftarrow ylist.pred(\tilde{y})$
11:     **end while**
12: **end if**
13: update the left and right end of $u^* \leftarrow [x, y^*]$

---

First, index $x$ is prepended at the head of $ylist$ and $[x, x]$ is prepended at the head of $llist$. Then $ylist$ is trimmed by deleting all elements $y$ $(> x)$ that can be concluded to be wasteful (lines 3–12). This is called TRIMMING_YLIST in [10]. Simultaneously, $ulist$ is trimmed in line 3. Finally, the interval ends of the new head of $ulist$, $u^*$, are updated.

Coming back to Algorithm 1, Uno and Yagiura show that in iteration step $x$, after the update of $Y$, the function $f(x, y)$ is monotonically increasing for the elements $y$ remaining in $ylist$. This allows in lines 5–7 to find efficiently all common intervals with left end $x$ by evaluating $f(x, y)$ running left-to-right through $ylist$ until an index $y$ is encountered with $f(x, y) > 0$.

## 4     Irreducible Intervals

In this section we define the set of *irreducible intervals* and show how they can be used to reconstruct all common intervals. We start by characterizing the structure of the set of common intervals.

**Lemma 1.** *Let $\Pi$ be a family of permutations. For $c_1, c_2 \in C_\Pi$ we have*

$$|c_1 \cap c_2| \geq 2 \quad \Leftrightarrow \quad c_1 \cap c_2 \in C_\Pi,$$
$$c_1 \cap c_2 \neq \emptyset \quad \Rightarrow \quad c_1 \cup c_2 \in C_\Pi.$$

*Proof.* This follows immediately from the definition of common intervals.     □

Two common intervals $c_1, c_2 \in C_\Pi$ have a *non-trivial overlap* if $c_1 \cap c_2 \neq \emptyset$ and they do not include each other. A list $p = (c_1, \ldots, c_{\ell(p)})$ of common intervals $c_1, \ldots, c_{\ell(p)} \in C_\Pi$ is a *chain* (of length $\ell(p)$) if every two successive intervals in $p$ have a non-trivial overlap. A chain of length one is called a *trivial chain*, all other chains are called *non-trivial chains*. A chain that can not be extended to its left or right is a *maximal chain*. By Lemma 1, every chain $p$ *generates* a common interval $c = \tau(p) := \bigcup_{c' \in p} c'$.

**Definition 2.** *A common interval $c$ is called* reducible *if there is a non-trivial chain that generates $c$, otherwise it is called* irreducible.

This definition partitions the set of common intervals $C_\Pi$ into the set of reducible intervals and the set of irreducible intervals, denoted $I_\Pi$. Obviously, $1 \leq |I_\Pi| \leq |C_\Pi| \leq \binom{n}{2}$. For a common interval $c \in C_\Pi$ we count the number of irreducible intervals that properly contain $c$ and call this number the *nesting level* of $c$.

**Lemma 2.** *Let $\Pi$ be a family of permutations, $c \in C_\Pi$ a common interval, and $(b_1, \ldots, b_\ell)$ a chain of irreducible intervals generating $c$. The nesting levels of $c$ and all the $b_i$ for $i = 1, \ldots, \ell$ are equal.*

*Proof (Sketch).* Let $n_c$ be the nesting level of $c$ and $n_i$ the nesting level of $b_i$ for $i = 1, \ldots, \ell$. Since $b_i \subseteq c$ we have $n_i \geq n_c$ for $i = 1, \ldots, \ell$. If $n_i > n_c$, there exists an irreducible interval $c^* \not\supseteq c$ with $b_i \subset c^*$ and $\ell > 1$. Now we distinguish between internal and terminal intervals $b_i$ in the chain. In both cases one can easily see that $c^*$ can be generated by smaller common intervals, contradicting the assumption that $c^*$ is irreducible.     □

We can further partition $I_\Pi$ into maximal chains. This partitioning is unique. For a maximal chain $p = (c_1, \ldots, c_{\ell(p)})$ and $1 \leq i \leq j \leq \ell(p)$, we call $p[i, j] := (c_i, \ldots, c_j)$ a *subchain* of $p$.

**Lemma 3.** *The set of common intervals that is generated from the subchains of the maximal chains of $I_\Pi$ equals $C_\Pi$.*

*Proof.* This follows directly from the definition of the partition.     □

*Example 1 (cont'd).* For $\Pi = (\pi_1, \pi_2, \pi_3)$ as above, the irreducible intervals are

$$I_\Pi \;=\; \{[1,2], [1,8], [2,3], [4,5], [4,7], [4,8], [4,9], [5,6]\}.$$

The reducible intervals are generated as follows:

$$[1,3] = [1,2] \cup [2,3],$$
$$[1,9] = [1,8] \cup [4,9],$$
$$[4,6] = [4,5] \cup [5,6].$$

A sketch of the structure of maximal chains of irreducible intervals and their nesting levels is shown in Figure 1.

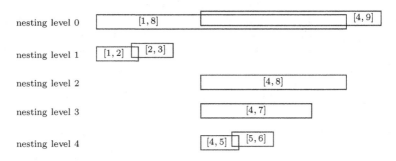

**Fig. 1.** Visualization of the irreducible intervals in $I_\Pi$ and their nesting levels.

**Lemma 4.** *Given two different maximal chains $p_1$ and $p_2$, exactly one of the following alternatives is true:*

- *$\tau(p_1)$ and $\tau(p_2)$ are disjoint,*
- *$\tau(p_1)$ is contained in a single element of $p_2$, or*
- *$\tau(p_2)$ is contained in a single element of $p_1$.*

*Proof.* $\tau(p_1)$ and $\tau(p_2)$ are either disjoint or have a non-empty intersection. In the latter case, $\tau(p_1)$ and $\tau(p_2)$ cannot overlap non-trivially, because of the maximality of $p_1$ and $p_2$. Therefore, w.l.o.g. suppose $\tau(p_2) \subseteq \tau(p_1)$. No element of $p_2$ can overlap non-trivially with any element of $p_1$, otherwise one could find an element of $p_1$ or $p_2$ that is generated by smaller intervals, contradicting its irreducibility. This yields the existence of exactly one irreducible interval $c$ of $p_1$ that includes $\tau(p_2)$ completely, while no other element of $p_1$ overlaps with $\tau(p_2)$. □

Based on the above lemmas, we describe a linear time algorithm to reconstruct the set $C_\Pi$ of common intervals of a family of permutations $\Pi$ from its set $I_\Pi$ of irreducible intervals (Algorithm 3). The algorithm partitions $I_\Pi$ into maximal chains (line 1). This can be done, for example, by the following three steps. First, $I_\Pi$ is partitioned according to the nesting level. This is possible in $O(|I_\Pi|)$ time by applying a sweep line technique to all interval start and end points. Then the intervals in the resulting classes are sorted by their left end. Using radix sort, this can also be done in $O(|I_\Pi|)$ time. Finally, the classes are further refined at non-overlapping consecutive intervals, yielding the maximal chains of irreducible intervals. This again takes $O(|I_\Pi|)$ time. Using Lemma 3, we create $C_\Pi$ by generating all subchains of the maximal chains (lines 2–4). This takes $O(|C_\Pi|)$ time. Since $|I_\Pi| \leq |C_\Pi|$, Algorithm 3 takes $O(|C_\Pi|)$ time in total.

---

**Algorithm 3** (Reconstruct $C_\Pi$ from $I_\Pi$)

---
**Input:** $I_\Pi$ in standard notation
**Output:** $C_\Pi$ in standard notation
1: partition $I_\Pi$ into maximal chains $p_1, p_2, \ldots$
2: **for each** $p_m = (b_1, \ldots, b_{\ell(p_m)})$ **do**
3:     output $\tau(p_m[i, j])$ in standard notation **for all** $1 \leq i \leq j \leq \ell(p_m)$
4: **end for**

---

The following theorem is the basis for the complexity analysis of our algorithm in the following section.

**Theorem 1.** *Given a family* $\Pi = (\pi_1, \ldots, \pi_k)$ *of permutations of* $N = \{1, 2, \ldots, n\}$, *we have* $1 \leq |I_\Pi| \leq n - 1$.

*Proof.* For each interval $[j, j+1]$, $j = 1, \ldots, n-1$, of $\pi_1$ denote by $b_{[j,j+1]} \in I_\Pi$ the irreducible interval of smallest cardinality containing $[j, j+1]$. It is easy to see that $b_{[j,j+1]}$ is uniquely defined. For any $c = [x, y] \in C_\Pi$, a subset of $\{b_{[x,x+1]}, \ldots, b_{[y-1,y]}\}$ generates $c$. This yields $\{b_{[j,j+1]} \mid j = 1, \ldots, n-1\} = I_\Pi$. □

*Example 2.* The limits given in Theorem 1 are actually achieved. For $\Pi = (id_{2k}, (1, k+1, 2, k+2, \ldots, k, 2k))$ we have $C_\Pi = I_\Pi = \{[1, n]\}$. For $\Pi = (id_n, id_n)$ we have $C_\Pi = \{[i, j] \mid 1 \leq i < j \leq n\}$ and $I_\Pi = \{[i, i+1] \mid 1 \leq i < n\}$.

## 5   Finding All Irreducible Intervals of $k$ Permutations

In this section we present our algorithm that finds all irreducible intervals of a family $\Pi = (\pi_1, \pi_2, \ldots, \pi_k)$ of $k \geq 2$ permutations of $N = \{1, \ldots, n\}$ in $O(kn)$ time. Together with Algorithm 3 this allows to find all $K$ common intervals of $\Pi$ in optimal $O(kn + K)$ time.

### 5.1   Outline of the Algorithm

For $1 \leq i \leq k$, set $\Pi_i := (\pi_1, \ldots, \pi_i)$. Starting with $I_{\Pi_1} = \{[j, j+1] \mid 1 \leq j < n\}$, the algorithm successively computes $I_{\Pi_i}$ from $I_{\Pi_{i-1}}$ for $i = 2, \ldots, k$ (see Algorithm 4). To construct $I_{\Pi_i}$ from $I_{\Pi_{i-1}}$, we define the mapping

$$\varphi_i : I_{\Pi_{i-1}} \to I_{\Pi_i}$$

where for $c \in I_{\Pi_{i-1}}$, $\varphi_i(c)$ is the smallest common interval $c' \in C_{\Pi_i}$ that contains $c$. Since $I_{\Pi_i} \subseteq C_{\Pi_i} \subseteq C_{\Pi_{i-1}}$ and, by Lemma 3, $I_{\Pi_{i-1}}$ generates the elements of $C_{\Pi_{i-1}}$, $I_{\Pi_{i-1}}$ also generates $I_{\Pi_i}$. One can easily see that $c' \in I_{\Pi_i}$ and that $\varphi_i$ is surjective, i.e. $I_{\Pi_i} = \{\varphi_i(c) \mid c \in I_{\Pi_{i-1}}\}$. This implies the correctness of Algorithm 4. In Section 5.2 we will show how $\varphi_i(I_{\Pi_{i-1}})$ can be computed in $O(n)$ time and space, yielding the $O(kn)$ time complexity to compute $I_\Pi$ $(= I_{\Pi_k})$.

---

**Algorithm 4** (Computation of $I_{\Pi_k}$)

---

**Input:** A family $\Pi = (\pi_1 = id_n, \pi_2, \ldots, \pi_k)$ of $k$ permutations of $N = \{1, \ldots, n\}$.
**Output:** $I_\Pi$ in standard notation.
1: $I_{\Pi_1} \leftarrow ([1,2], [2,3], \ldots, [n-1, n])$
2: **for** $i = 2, \ldots, k$ **do**
3:     $I_{\Pi_i} \leftarrow \{\varphi_i(c) \mid c \in I_{\Pi_{i-1}}\}$     // (see Algorithm 5)
4: **end for**
5: output $I_{\Pi_k}$ in standard notation

---

## 5.2   Computing $I_{\Pi_i}$ from $I_{\Pi_{i-1}}$

For the computation of $\varphi_i(I_{\Pi_{i-1}})$ we use a modified version of Algorithm RC where the data structure $Y$ is supplemented by a data structure $S$ that is derived from $I_{\Pi_{i-1}}$. $S$ consists of several doubly-linked lists of intervals of $ylist$, one for each maximal chain of $I_{\Pi_{i-1}}$.

Using $\pi_1$ and $\pi_i$, as in Algorithm RC, the $ylist$ of $Y$ allows for a given $x$ to access all non-wasteful right interval end candidates $y$ of $C_{(\pi_1, \pi_i)}$. The aim of $S$ is to further reduce these candidates to only those indices $y$ for which simultaneously $[x, y] \in C_{\Pi_{i-1}}$ (ensuring $[x, y] \in C_{\Pi_i}$) and $[x, y]$ contains an interval $c \in I_{\Pi_{i-1}}$ that is not contained in any smaller interval from $C_{\Pi_i}$. Together this ensures that exactly the irreducible intervals $[x, y] \in I_{\Pi_i}$ are reported.

An outline of our modified version of Algorithm RC is shown in Algorithm 5. Since the first permutation handed to the algorithm has to be the identity and $S$ (derived from $I_{\Pi_{i-1}}$) is compatible only with the index set of $\pi_1$, we supply the algorithm with $id_n = \pi_i^{-1} \circ \pi_i$ and $\pi_i^{-1} \circ \pi_1$ instead of $\pi_1(= id_n)$ and $\pi_i$. (As usual, $\pi_i^{-1}$ denotes the inverse of permutation $\pi_i$.) This does not change the index set of the computed irreducible intervals.

In line 1 of Algorithm 5, $Y$ is initialized as in Algorithm RC. To initialize $S$, $I_{\Pi_{i-1}}$ is partitioned into maximal chains of non-trivially overlapping irreducible intervals as in line 1 of Algorithm 3. For each such chain, $S$ contains a doubly-linked $clist$ that initially holds the intervals of that chain in left-to-right order. Moreover, intervals from different $clists$ with the same left end are connected by *vertical pointers* yielding for each index $x \in N$ a doubly-linked *vertical list*. It is not difficult to add the vertical pointers during the construction of the $clists$ such that the intervals in each vertical list are ordered by increasing length (decreasing nesting level).

**Algorithm 5** (Extended Algorithm RC)

**Input:** Two permutations $\pi_1 = id_n$ and $\pi_2 = \pi_i^{-1}$ of $N = \{1, \ldots, n\}$; $I_{\Pi_{i-1}}$ in standard notation.

**Output:** $I_{\Pi_i}$ in standard notation.

1: initialize $Y$ and $S$
2: **for** $x = n - 1, \ldots, 1$ **do**
3:     update $Y$ and $S$    // (see text)
4:     **while** $([x', y] \leftarrow S.\textit{first\_active\_interval}(x))$ defined **and** $f(x, y) = 0$ **do**
5:         output $[l(x, y), u(x, y)]$
6:         remove $[x', y]$ from its active sublist    // (the interval is satisfied)
7:     **end while**
8: **end for**

To describe the update of $S$ in line 3 of the algorithm, we introduce the notion of *sleeping*, *active*, and *satisfied* intervals. Initially all intervals of the *clist*s are sleeping. In iteration step $x$, all intervals with left end $x$ become active and are included at the head of an (initially empty) *active sublist* of their *clist*. An interval remains active until it is satisfied or deleted. A *clist* $L$ and the contained intervals are deleted whenever $x$ becomes smaller than the left interval end of $L.head$. It might be that the right end $y$ of an interval $[x, y]$ at the time of activation is already deleted from the *ylist*. In this case, the interval is merged with the successing interval $[x', y']$ in its *clist*, i.e. the corresponding two elements of *clist* are replaced by a new one, containing the interval $[x, y']$. If no successor exists, the interval $[x, y]$ is deleted.

Concerning the function $\varphi_i$, sleeping or active intervals correspond to irreducible intervals from $I_{\Pi_{i-1}}$ whose images have not yet been determined. The status changes to satisfied when the image is known.

The update of $Y$ in line 3 is the same as in Algorithm RC, the only difference being that whenever an element $y$ is removed from *ylist* and $y$ is the right end of some active or satisfied interval, this interval is merged with its successor in its *clist* if such a successor exists, otherwise it is deleted. The resulting interval inherits the active status if one of the merged intervals was active, otherwise it is satisfied. (If both merged intervals are active, this reflects the case that $\varphi_i$ maps both intervals of $I_{\Pi_{i-1}}$ to the same (larger) interval of $I_{\Pi_i}$.) Note that even though $y$ can be the right end of many irreducible intervals, at any point of the algorithm $y$ can be the right end of at most one active or satisfied interval. This is due to the fact that no two intervals of a maximal chain can have the same right end, and whenever two intervals from different chains have the same right end, the chain of the shorter interval is deleted before the longer interval is made active (cf. Lemma 4). Hence it suffices to keep for each index $y > x$ a pointer to the (only) active or satisfied interval with right end $y$. This *right end pointer* is set when the interval is made active and is deleted when the interval's *clist* is deleted.

In contrast to the simple traversal of the *ylist* in Algorithm RC, here the generation of right interval end candidates in lines 4–7 is slightly more complicated.

Function $S.first\_active\_interval(x)$ returns the first active interval $[x', y]$ in the *clist* of the interval at the head of the vertical list at index $x$. If the right end $y$ of this interval gives rise to a common interval $[l(x, y), u(x, y)]$, i.e., if $f(x, y) = 0$, a common interval of smallest size containing an active interval is encountered. Hence we have found an element of $\varphi_i(I_{\Pi_{i-1}})$ which then is reported in line 5. Therefore $[x', y]$ becomes satisfied and is removed from its active sublist in line 6. In case this interval was the last active interval of its *clist*, the pointer to the head of the vertical list at index $x$ is redirected to the successor of the current head, such that in the next iteration $S.first\_active\_interval(x)$ returns the leftmost active interval from the *clist* with the next lower nesting level (if such a list containing an interval with left end $x$ exists).

This way we only look at elements of *ylist* that are candidates for right ends of minimal common intervals with left end $x$ and that contain an active interval. $S.first\_active\_interval(x)$ generates these candidates in left-to-right order such that, since $f(x, y)$ is monotonically increasing for the elements $y$ of *ylist* and hence also for the elements of any sublist of *ylist*, by evaluating $f(x, y)$ until an index $y$ is encountered with $f(x, y) > 0$, all irreducible intervals from $I_{\Pi_i}$ with left end $x$ are found. This implies the correctness of our implementation of $\varphi_i$.

The complete data structure $S$ for $\Pi = (\pi_1, \pi_2, \pi_3)$ as in Example 1 while processing index $x = 4$ of permutation $\pi_3$ is shown in Figure 2.

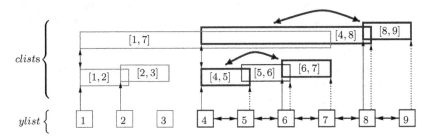

**Fig. 2.** Sketch of *ylist* and the *clists* while processing element $x = 4$ of $\pi_3$ for $\Pi = (\pi_1, \pi_2, \pi_3)$ as in Example 1. Shaded boxes represent sleeping intervals, boxes with thick solid lines represent active intervals, and boxes with thin lines represent satisfied intervals. Thick arrows connect the elements of the active sublists, solid vertical arrows denote vertical lists (the vertical pointer of index 5 was deleted after reporting interval $[5, 6]$ in iteration step $x = 5$), and dotted vertical arrows are the right end pointers.

## 5.3   Analysis of Algorithm 5

Since all operations modifying $Y$ are the same as in Algorithm RC, this part of the analysis carries over from [10], and we can restrict our analysis to the initialization and update of $S$.

The initialization of $S$ in line 1, including the creation of the vertical lists, can easily be implemented in linear time in a way similar to the first step of Algorithm 3.

In line 3, the intervals with left end $x$ are easily found using the vertical lists, marked active, and prepended to the active sublists in constant time per interval. Since $I_{\Pi_{i-1}}$ contains $O(n)$ intervals and since each interval is activated exactly once, this step takes overall $O(n)$ time. Moreover, each index $y$ that is deleted from the $ylist$ can cause the merge of two intervals. Since merging two neighbors in a doubly-linked list takes constant time, and since each of the in total $n$ elements of $ylist$ is deleted at most once, this part takes overall $O(n)$ time as well.

As in Algorithm RC, the time required for reporting the output is proportional to the size of the output, here $|I_{\Pi_i}| < n$. Using vertical list and active sublist, the first active interval is found in constant time. Hence, and since the removal of interval $[x', y]$ from the active sublist in line 6 is a constant-time operation as well, the loop in lines 4–7 takes overall $O(n)$ time.

Putting things together, Algorithm 5 takes $O(n)$ time and space. Since at any point of Algorithm 4 we need to store only two permutations $\pi_1$ and $\pi_i$ and the current $I_{\Pi_i}$, we have

**Theorem 2.** *The irreducible intervals of $k$ permutations of $n$ elements can be found in optimal $O(kn)$ time and $O(n)$ additional space.*

Combining this result with Algorithm 3, we get

**Corollary 1.** *The $K$ common intervals of $k$ permutations of $n$ elements can be found in optimal $O(kn + K)$ time and $O(n)$ additional space.*

## Acknowledgments

We wish to thank Richard Desper, Dan Gusfield, and Christian N.S. Pedersen for helpful comments.

## References

1. K.S. Booth and G.S. Lueker. Testing for the consecutive ones property, interval graphs and graph planarity using $PQ$-tree algorithms. *J. Comput. Syst. Sci.*, 13(3):335–379, 1976.
2. R.M. Brady. Optimization strategies gleaned from biological evolution. *Nature*, 317:804–806, 1985.
3. D. Fulkerson and O. Gross. Incidence matrices with the consecutive 1s property. *Bull. Am. Math. Soc.*, 70:681–684, 1964.
4. M.C. Golumbic. *Algorithmic Graph Theory and Perfect Graphs*. Academic Press, New York, 1980.
5. S. Kobayashi, I. Ono, and M. Yamamura. An efficient genetic algorithm for job shop scheduling problems. In *Proc. of the 6th International Conference on Genetic Algorithms*, pages 506–511. Morgan Kaufmann, 1995.

6. E.M. Marcotte, M. Pellegrini, H.L. Ng, D.W. Rice, T.O. Yeates, and D. Eisenberg. Detecting protein function and protein-protein interactions from genome sequences. *Science*, 285:751–753, 1999.

7. H. Mühlenbein, M. Gorges-Schleuter, and O. Krämer. Evolution algorithms in combinatorial optimization. *Parallel Comput.*, 7:65–85, 1988.

8. R. Overbeek, M. Fonstein, M. D'Souza, G.D. Pusch, and N. Maltsev. The use of gene clusters to infer functional coupling. *Proc. Natl. Acad. Sci. USA*, 96(6):2896–2901, 1999.

9. B. Snel, G. Lehmann, P. Bork, and M.A. Huynen. STRING: A web-server to retrieve and display the repeatedly occurring neigbourhood of a gene. *Nucleic Acids Res.*, 28(18):3443–3444, 2000.

10. T. Uno and M. Yagiura. Fast algorithms to enumerate all common intervals of two permutations. *Algorithmica*, 26(2):290–309, 2000.

# Generalized Pattern Matching and the Complexity of Unavoidability Testing

Christine E. Heitsch[1]

Department of Computer Science
University of British Columbia
201-2366 Main Mall, Vancouver, B. C. V6T 1Z4
heitsch@cs.ubc.ca

**Abstract.** We formulate the GENERALIZED PATTERN MATCHING problem, a natural extension of string searching capturing regularities across scale. The special case of UNAVOIDABILITY TESTING is obtained for pure generalized patterns by fixing an appropriate family of text strings – the Zimin words. We investigate the complexity of this restricted decision problem. Although the efficiency of standard string searching is well-known, determining the occurrence of generalized patterns in Zimin words does not appear so tractable. We provide an exponential lower bound on any algorithmic decision procedure relying exclusively on the equivalent deletion sequence characterization of unavoidable patterns. We also demonstrate that the four other known necessary conditions are not sufficient to decide pattern unavoidability.

## 1   Introduction

Numerous efficient algorithms have been developed to handle pattern matching in the exact or approximate cases [1], when potential matchings are considered only between a symbol of the pattern and a symbol of the text. However, excepting the well-known problem of finding consecutively repeated substrings [4] [7], the question of detecting string regularities across different scales has remained largely unaddressed. Generalized patterns capture this wider focus by expanding the target of some potential matches beyond individual symbols of the text to include all nonempty substrings. Furthermore, the complexity questions arising with this generalization represent an intriguing departure from the known polynomial time algorithms in the standard cases.

Pure GENERALIZED PATTERN MATCHING (GPM), as alluded to by Cassaigne in [6] and explicitly defined in section 2, is clearly in *NP*. A given correspondence between a generalized pattern consisting purely of variables and a substring of the text can be efficiently verified by standard string searching techniques. However, the current pattern matching methods appear unlikely to yield an efficient deterministic solution to the pure GPM problem. The success of such sequential algorithms, fundamentally dependent on preprocessing of pattern, text, or associated structure, does not obviously generalize. In fact, the

A. Amir and G.M. Landau (Eds.): CPM 2001, LNCS 2089, pp. 219–230, 2001.

results of this paper are intended to suggest an intrinsic computational difficulty at the heart of pure GPM.

We focus on investigating the complexity of UNAVOIDABILITY TESTING (UT). As outlined in section 3, the link between string unavoidability and generalized pattern matching follows from the decidability results of Zimin [8]. As a special case of GPM, the problem instance depends solely on the pattern, with the associated text string chosen from the fixed family of Zimin words. Further results of Zimin [8] and Bean et al. [3] yield a method for constructing a matching correspondence which exists if and only if the pattern is unavoidable. The method depends on the existence of a sequence of deletions reducing a pattern to the empty string. We show that, interpreted as the appropriate algorithm, this reductive deletion method cannot decide pattern unavoidability without encountering an inescapable exponential initial branching of the computation tree.

To prove such an exponential lower bound on the deletion approach, we begin in section 4 by explicitly defining the bipartite graph induced by the deletion criteria. Combinatorial analysis of the connected components yields theorem 3. This significant result permits a further restriction of the problem to patterns having a unique first step in the deletion sequence. The difficulty of deterministically locating the correct initial deletion forms the basis of our complexity conclusions; we show that the unique choice must be made from among exponentially many valid possibilities.

More precisely, section 5 introduces a notion of size which reflects, to a degree, the number of choices a deletion reduction algorithm might face. With the appropriate idea of related patterns, we prove theorem 4, a fundamental result about the possibility of creating specific combinatorial combinations of certain types of patterns. Applying theorem 4 and its immediate consequences, beginning with experimentally located base cases, yields a progression of patterns with increasing difficulty in determining the correct deletion.

Section 6 fully addresses the complexity of UT under the deletion criteria algorithmic approach, by formulating the results in terms of the input pattern length and confirming the exponential distribution among valid deletion choices. Given these results, we conclude that an algorithm employing the deletion sequence criteria cannot determine the unavoidability of arbitrarily many patterns without considering an exponential number of possibilities. This inescapable exponential initial branching of the computational tree implies that UNAVOIDABILITY TESTING can not be solved in polynomial time by such a deletion algorithm.

Finally, section 7 addresses some alternate algorithmic approaches, based on the four other known necessary conditions. An optimal extension of a combination of these four criteria can be interpreted as a binary tree structure. However, theorem 10 shows that there exist arbitrarily many avoidable patterns also encoded by this tree structure and satisfying the necessary conditions. Hence, the conditions are not sufficient to produce a polynomial time algorithm deciding pattern unavoidability.

## 2  Generalized Pattern Matching

The collection of all possible non-empty strings over a finite (and non-empty) set of symbols $S$ will be denoted $S^+$. An overbar distinguishes elements of $S^+$ from individual string symbols. $\varepsilon$ denotes the empty string, and $S^+ \cup \{\varepsilon\} = S^*$. Note that $S^*$ is the set of reduced strings, since $s\varepsilon = \varepsilon s = s$ for all $s \in S$ and by extension for all $\bar{t} \in S^*$. Finally, $\leq$ will refer to the substring relation on $S^+ \times S^+$, that is $\bar{t} \leq \bar{s}$ if there exist strings $\bar{u}, \bar{v} \in S^*$ so that $\bar{u}\bar{t}\bar{v} = \bar{s}$.

Although pattern matching is typically phrased as a constructive problem, we are primarily interested in complexity issues. Hence, the decision problem associated with exact pattern matching can be stated:

**Problem:** PATTERN MATCHING (PM)
**Instance:** $\bar{p} \in A^+$ and $\bar{w} \in A^+$
**Question:** Is $\bar{p} \leq \bar{w}$?

As remarked in section 1, we wish to expand the concept of matching patterns to capture string regularities across scales. PM can be viewed as asking whether there exists an identity mapping between a pattern $\bar{p}$ and a subword of $\bar{w}$. We broaden this notion of a matching correspondence by introducing a new set $V$ of "variable" pattern symbols, which may map to non-empty words in $A^+$, while the symbols in $A$ remain "constant."

**Definition 1 (Generalized Pattern Occurrence).** *For $\bar{p} \in (V \cup A)^+$ and $\bar{w} \in A^+$, $\mathbf{\bar{p}} \mid \mathbf{\bar{w}}$ if there exists a map $\phi : (V \cup A) \to A^+$ such that $\phi(a) = a$ for all $a \in A$ and $\phi(\bar{p}) \leq \bar{w}$ under the induced homomorphism.*

Henceforth, any pattern or its occurrence should be considered in the context defined above. Note that $\phi$ is a non-erasing homomorphism, restricted to the identity on $A$. As such, PM is the special case of GPM with only constant pattern symbols. "Pure" GPM will refer to the case when $\bar{p} \in V^+$, while a pattern that includes both variables and constants will be called "mixed" GPM.

**Problem:** GENERALIZED PATTERN MATCHING (GPM)
**Instance:** $\bar{p} \in (V \cup A)^+$ and $\bar{w} \in A^+$
**Question:** Does $\bar{p} \mid \bar{w}$?

The rest of this paper considers the complexity of UNAVOIDABILITY TESTING, a special case of pure GPM depending only on the given pattern.

## 3  Unavoidability Testing

**Definition 2 (Unavoidable).** *[3] [8] A string $\bar{s}$ is called **unavoidable** if every infinite word on $n$ letters has an occurrence of $\bar{s}$ as a pure generalized pattern.*

The complementary characteristic of avoidability, often phrased in terms of sets of finite words, has been the primary research focus of this area. (A summary of current results can be found in [6].) The above definition, itself a statement about asymptotic pure GPM, is preferred for the complexity questions of this paper. Decidability results on unavoidable patterns utilize the Zimin words, which will be chosen as our family of text words, indexed by the number of distinct symbols appearing in each string[1].

**Definition 3 (Zimin Words).** *[8] Let $Z_1 = a_1$ and recursively define $\mathbf{Z_n}$ on $A_n = \{a_1, a_2, \ldots, a_n\}$ as $Z_n = Z_{n-1} a_n Z_{n-1}$. Equivalently, define a mapping $\theta : A_{n-1} \to A_n$ where $\theta(a_i) = a_{i+1} a_1$. Then $Z_n = a_1 \theta(Z_{n-1})$.*

For any string $\bar{s}$, let $\boldsymbol{\alpha}(\bar{s})$ be the number of distinct symbols occurring in $\bar{s}$. Also, for $\bar{p} \in V^+$, it will be assumed that $|V| = \alpha(\bar{p})$.

**Theorem 1.** *[8] $\bar{p}$ is an unavoidable pattern if and only if $\bar{p}$ occurs in $Z_{\alpha(\bar{p})}$.*

Thus, theorem 1 states that the problem of determining pattern unavoidability is a special case of pure GPM depending only on the pattern with the associated Zimin word as text.

**Problem:** UNAVOIDABILITY TESTING (UT)
**Instance:** $\bar{p} \in V^+$
**Question:** Does $\bar{p} \mid Z_{\alpha(\bar{p})}$?

One direction in theorem 1 follows immediately from the unavoidability of Zimin words themselves and the transitivity of generalized pattern occurrence. However, the other direction depends on an equivalent characterization of unavoidable patterns in terms of the $\sigma$-deletion of free sets. Although the two characterizations are polynomially equivalent, $\sigma$-deletions and free sets offer a decided advantage in terms of algorithmic analysis.

**Definition 4 (Free Set).** *[3] [8] $F \subseteq V$ is **free** for $\bar{p} \in V^+$ if and only if there exist sets $A, B \subseteq V$ such that $F \subseteq B \setminus A$ where, for all $xy \leq \bar{p}$, $x \in A$ if and only if $y \in B$.*

The requirement on all substrings $xy$ is called the "two-window" criteria. Note that the definition could equivalently require $F \subseteq A \setminus B$.

**Definition 5 ($\sigma$-Deletion).** *[3] [8] The mapping $\sigma_F$ is a $\boldsymbol{\sigma}$-**deletion** of $\bar{p} \in V^+$ if and only if $F \subseteq V$ is a free set for $\bar{p}$ and $\sigma_F : V \to V \cup \{\varepsilon\}$ is defined by*

$$\sigma_F(x) = \begin{cases} x \text{ if } x \notin F \\ \varepsilon \text{ if } x \in F \end{cases}$$

---

[1] The decision results published independently in [3] and [8] use different terminology and notation. For the purposes of this paper, whichever seemed the most appropriate was chosen.

Note that $\sigma_F(\bar{p})$ always refers to the reduced string in $V^*$. Moreover, a matching correspondence lifts through a $\sigma$-deletion. Intuitively, the $\sigma$-deletion criteria permits variables to be "squeezed" into the mapping as needed. Hence, there is also the following unavoidability characterization.

**Theorem 2.** *[3] [8] $\bar{p}$ is an unavoidable pattern if and only if $\bar{p}$ can be reduced to $\varepsilon$ by a sequence of $\sigma$-deletions.*

Since $\sigma$-deletions are designed specifically to permit a "bottom-up" construction of the matching correspondence, there appears to be an immediate algorithmic advantage to this method of resolving UNAVOIDABILITY TESTING. Rather than searching through all possible matching correspondences, recursively generate a complete $\sigma$-deletion sequence, which is easily converted into the desired mapping. Moreover, the $\sigma$-deletion characterization is the more combinatorially accessible, as the next section demonstrates.

# 4    Minimality and Uniqueness

A $\sigma$-deletion requires a free set $F \subseteq B \setminus A$, where $A$ and $B$ satisfy the two-window criteria. The minimal such sets can be simultaneously constructed by considering a pattern's adjacency graph (also found in [2]).

**Definition 6 (Minimal $\sigma$-Graph).** *Let $\mathcal{G}(\bar{p})$ be the bipartite graph with vertices $[a, x]$ and $[x, b]$ for every variable $x$ in $\bar{p}$ and $a, b \notin V$. $([a, x], [y, b])$ is an edge in $\mathcal{G}(\bar{p})$ if and only if $xy \leq \bar{p}$.*

For each connected component of $\mathcal{G}(\bar{p})$, the projections of the left and right sides back down onto subsets of $V$ yield sets $A$ and $B$ minimally satisfying the two-window criteria for $\bar{p}$. A **trivial** $\mathcal{G}(\bar{p})$ has only one connected component, and no possible $\sigma$-deletions. However, there certainly may be more than two pairs of minimal $A$ and $B$ sets. Furthermore, the necessity of considering unions of these minimal sets has not yet been ruled out.

Call a $\sigma$-deletion and its associated free set $F$ **worthwhile** if $\sigma_F(\bar{p})$ is unavoidable, that is if a $\sigma$-deletion sequence beginning with $\sigma_F$ reduces $\bar{p}$ to the empty string. Also, let $\equiv$ be the equivalence relation on the vertices of $\mathcal{G}(\bar{p})$ where two vertices are equivalent if and only if they are in the same connected component.

**Lemma 1.** *If $\bar{p}$ is unavoidable, then there exists a worthwhile $\sigma$-deletion $\sigma_F$ of $\bar{p}$ such that, for all $y_1, y_2 \in F$, $[y_1, b] \equiv [y_2, b]$ in $\mathcal{G}(\bar{p})$.*

The result follows from a proof that any worthwhile $\sigma$-deletion of $\bar{p}$ can be separated into free sets of equivalent variables, which may then be $\sigma$-deleted in sequence. A detailed justification of it, and all subsequent results, is given in [5]. Additionally, the symmetry possible in the free set definition, where $B \setminus A$ was arbitrarily chosen over $A \setminus B$, should not be overlooked. Either by considering the implications of this dual definition or by investigating the relationship between the $\sigma$-deletions of a pattern $\bar{p}$ and its reverse $(\bar{p})^R$, the following result is obtained.

**Corollary 1.** *If $\bar{p}$ is unavoidable, then there exists a worthwhile $\sigma$-deletion $\sigma_F$ of $\bar{p}$ such that, for all $y_1, y_2 \in F$, $[a, y_1] \equiv [a, y_2]$ in $\mathcal{G}(\bar{p})$.*

Hence, $\mathcal{G}(\bar{p})$ is justifiably considered the minimal $\sigma$-graph of $\bar{p}$; unavoidability can be determined by considering only the free sets arising from the connected components of $\mathcal{G}(\bar{p})$. Although an unavoidable pattern must have such a **minimal** $\sigma$-deletion, it is by no means unique.

**Theorem 3.** *If $\bar{p}$ is unavoidable and $\mathcal{G}(\bar{p})$ has at least three connected components, then there exists more than one minimal worthwhile $\sigma$-deletion of $\bar{p}$.*

The proof rests on showing that, under the given conditions, the order of the first two minimal $\sigma$-deletions can be reversed. For this paper, the result's primary application is in restricting to unavoidable patterns with a unique worthwhile $\sigma$-deletion, insuring that $\mathcal{G}(\bar{p})$ has exactly two connected components.

## 5    Unavoidable Combinations

A pattern with a unique worthwhile $\sigma$-deletion has exactly two pairs of sets minimally satisfying the two-window criteria, which shall be denoted $A$, $B$, and $A^c$, $B^c$ with $F \subseteq B \setminus A$. The size of a unique $\sigma$-deletion is introduced as a measure of the difficulty in choosing the corresponding free set.

**Definition 7 ($\sigma$-Deletion Size).** *Suppose $\bar{p}$ has a unique worthwhile $\sigma$-deletion $\sigma_F$. Let $|\sigma_{\mathbf{F}}| = k_1/k_2$ where $k_1 = |F|$ and $k_2 = |B \setminus A|$.*

Implicit in the notation is that $|\sigma_F|$ is measured with respect to a particular pattern $\bar{p}$ and its minimal $\sigma$-graph $\mathcal{G}(\bar{p})$. In terms of the most general bounds, clearly $1 \leq k_1 \leq k_2$. See theorem 7 for a statement of the precise relationship among possible $k_1$, $k_2$, and $\alpha(\bar{p})$.

The difficulty of deciding pattern unavoidability on the basis of the $\sigma$-deletion criteria will rest on demonstrating two facts. The first is that $k_2$ can increase exponentially as a function of $|V|$ without necessitating an exponential increase in the lengths of the patterns involved. Secondly, for many such $k_2$, there exist patterns where $k_1$ can take on all possible values between 1 and $k_2$.

These results are achieved by looking at specific combinatorial combinations of unavoidable patterns. As motivation, recall the recursive Zimin word definition, noting that every $Z_n$ has a unique $\sigma$-deletion of size $1/1$. It is necessary to generalize only slightly this method of combining unavoidable patterns "buffered" by newly introduced variables by defining an appropriate restriction on equivalence classes of isomorphic patterns.

**Definition 8 ($\sigma$-Isomorphic).** *Suppose $\bar{p} \in V^+$ and $\bar{q} \in U^+$. Say $\bar{\mathbf{q}} = \bar{\mathbf{p}}'$ if there exists a one-to-one mapping $\phi : V \to U$ such that $\phi(x) = x$ for $x \in V \cap U$ and under the induced isomorphism $\phi(\bar{p}) = \bar{q}$.*

The prime notation is carried through structures related under $\sigma$-isomorphism. Although pattern unavoidability is preserved under unrestricted isomorphisms, $\sigma$-isomorphisms constrain the possible relabeling of variables so that unavoidability is preserved under the Zimin-type construction, $\bar{p}z\bar{p}'$. Moreover, the following fundamental result states that, for specific kinds of unavoidable patterns and certain of their $\sigma$-isomorphisms, the construction yields new patterns with unique worthwhile $\sigma$-deletions.

**Theorem 4.** *Let $\bar{p} \in V^+$ with $\alpha(\bar{p}) \geq 3$. Assume $\bar{p}$ has a unique worthwhile $\sigma$-deletion with associated sets $A$, $B$, $A^c$, $B^c$. Suppose further that $\bar{p} = s\bar{u}t$ and that either $t \notin A$ or $s \notin B^c$. Consider a $\sigma$-isomorphism $\bar{p}'$ with $(V \setminus V') \subseteq (B \setminus A)$. Then, for $z \notin V \cup V'$, $\bar{p}z\bar{p}'$ has a unique worthwhile $\sigma$-deletion as well.*

The proof proceeds in several stages, beginning with consideration of $\mathcal{G}(\bar{p}z\bar{p}')$. The conditions of the theorem are sufficient to show, by exhausting all other possibilities, that the union of the original free set and its $\sigma$-isomorphic image form the only worthwhile $\sigma$-deletion. According to those conditions, only variables of $B \setminus A$ may differ between $\bar{p}$ and $\bar{p}'$. However, any subset of $B \setminus A$ may be relabeled by such a $\sigma$-isomorphism, yielding an immediate quantitative corollary.

**Corollary 2.** *Suppose $\bar{p}$ satisfies the requirements of theorem 4 and $|\sigma_F| = k_1/k_2$. Then, for $0 \leq i \leq k_1$ and for $0 \leq j \leq k_2 - k_1$, there exists an unavoidable pattern $\bar{q}$ with $\alpha(\bar{q}) = \alpha(\bar{p}) + 1 + i + j$ such that the unique $\sigma$-deletion of $\bar{q}$ has size $(k_1 + i)/(k_2 + i + j)$.*

Beyond individual patterns, theorem 4 can be inductively applied to yield a progression of non-empty equivalence classes of patterns.

**Definition 9.** *Let $(\mathbf{k_1/k_2})_n$ be the set of all unavoidable patterns having $n$ distinct variables and unique $\sigma$-deletions of size $k_1/k_2$.*

**Theorem 5.** *Given $\bar{p} \in (k_1/k_2)_n$ with $\bar{p}$ satisfying the conditions of theorem 4, then for $0 \leq l_1 \leq (2^i - 1)k_1$ and $0 \leq l_2 \leq (2^i - 1)(k_2 - k_1)$ with $i \geq 0$,*

$$(k_1 + l_1/k_2 + l_1 + l_2)_{n+i+l_1+l_2} \neq \emptyset$$

Of course, these results are contingent on the initial existence of specific patterns, which has not yet been provided. However, when $|V| = 3$, $4$, and $5$, it is practically feasible to enumerate all nonisomorphic patterns and decide their unavoidability. Tables in the Appendix summarize the numerical results of such enumeration and decision programs, with the conclusion that the required patterns do indeed exist.

**Theorem 6.** *There exist patterns $\bar{p}$ satisfying theorem 4 for $\alpha(\bar{p}) = 3, 4, 5$.*

Hence, progressions of collections of pattern classes $(k_1/k_2)_n$ are known to exist, where $k_2$ increases linearly with $n$ and $k_1$ takes on entire ranges of values. In fact, using results and techniques well beyond the scope of this paper, there are even worse implications for the algorithmic complexity of deciding unavoidability by a $\sigma$-deletion reduction. The constructive proof techniques do not cover the exceptional cases, whose emptiness has been exhaustively verified.

**Theorem 7.** *With the exception of $(2/2)_4$ and $(4/4)_7$, for $1 \leq k_1 \leq k_2 \leq n - 2$, $(k_1/k_2)_n \neq \emptyset$ if and only if $1 + \sum_{i=k_2-k_1}^{n-2-k_1} 2^i \geq k_1$.*

## 6   Computational Complexity

Thus far, the difficulty of constructing a unique worthwhile $\sigma$-deletion has been expressed in terms of the ratio of $|F|$ to $|B \setminus A|$. The previous results have demonstrated that, with increasing numbers of variables, ever greater ranges of free set cardinalities must be considered. However, the complexity of an algorithm solving UNAVOIDABILITY TESTING is measured with respect to its input size, the length of a given pattern. The following theorem confirms the existence of patterns $\bar{p}$ with $|\bar{p}| = \mathcal{O}(\alpha(\bar{p}))$ and unique worthwhile free sets with sizes ranging from 1 to $\mathcal{O}(\alpha(\bar{p}))$.

**Theorem 8.** *Suppose $i \geq 0$ and $k_2 = 2^i + 1$. For all $1 \leq k_1 \leq k_2$, there exist $\bar{p} \in (k_1/k_2)_{2^i+i+4}$ such that $|\bar{p}| \leq 2^{i+4} - 1$.*

Although the constructive proof depends on an inductive application of theorem 4, most patterns satisfying the length bound of $|\bar{p}| < 16 \cdot \alpha(\bar{p})$ in theorem 8 will not be of the form $\bar{p}z\bar{p}'$. Furthermore, $k_2 > |\bar{p}|/16$ so that with longer patterns (and a greater number of variables) the size of $B \setminus A$ increases arbitrarily while the unique worthwhile free set may be of any size from $1, \ldots, k_2$. However, because this growth has been shown only for set cardinalities, demonstrating a truly exponential branching of the $\sigma$-deletion decision tree requires considering the actual distribution of free sets among unavoidable patterns.

Consider these patterns satisfying the conditions of theorem 4 with $\alpha(\bar{p}) = 5$, $|\bar{p}| \leq 15$, and $B \setminus A = \{x, v\}$.

$$xyxzuzyvyx \quad \text{and} \quad F = \{x\}$$
$$xyzxuzvuv \quad \text{and} \quad F = \{v\}$$
$$xyxzuvyvzy \quad \text{and} \quad F = \{x, v\}$$

Replacing either $x$ or $v$ with $t$ and combining $\bar{p}s\bar{p}'$ under theorem 4 yields patterns with $B \setminus A = \{x, v, t\}$ where $F$ may be any one of $\{x\}$, $\{v\}$, $\{x, t\}$, $\{v, t\}$, $\{x, v, t\}$. Generalizing this technique demonstrates that there exist patterns $\bar{p}$ with the same associated $B \setminus A$ sets where the unique choice for a worthwhile $\sigma$-deletion may be any one of an exponential, in $|\bar{p}|/16$, number of free sets $F$. Hence, the criteria $F \subseteq B \setminus A$ is genuine; a deterministic algorithm attempting to decide unavoidability on the basis of the $\sigma$-deletion criteria faces patterns with exponentially many valid choices for the initial free set. Thus there is an exponential lower bound on deciding UT by means of $\sigma$-deletions. While this does not rule out other better algorithms, further results show that the other known necessary conditions are not sufficient.

## 7    Insufficiency

The following four necessary conditions can be extracted from the definition and decision characterization of unavoidable patterns. Note that the last three criteria are clearly verifiable in polynomial time.

1. $\bar{q}$ is unavoidable for any $\bar{q} \leq \bar{p}$,
2. $|\bar{p}| \leq 2^{\alpha(\bar{p})} - 1$,
3. $\bar{p}$ has an isolated variable,
4. $\mathcal{G}(\bar{p})$ is nontrivial.

The first condition follows immediately from the definition of string avoidability, the second from the characterization that $\bar{p} \mid Z_{\alpha(\bar{p})}$, while the third comes most clearly from the $\sigma$-deletion reduction of $\bar{p}$ to a single variable, and the fourth from the $\sigma$-deletion definition. Conditions 1 and 3 strongly suggest a "divide-and-conquer" heuristic for an efficient decision algorithm. Having located the necessary isolated variable, the pattern may be broken into two substrings. Each of which must also have an isolated variable, if nonempty and likewise unavoidable, and so can be further subdivided. Two different algorithms are possible, depending on whether the second branching occurs at the same variable in the right and left substrings or not. It can easily be seen that requiring the subdivision to occur at the same variable would lead to false negatives, while unconstrained division ultimately leads to arbitrarily many false positives for reasons similar to those outlined below.

A third and optimal algorithmic approach would require that substrings be divided at the same variable, but permitted to remain intact if this is not possible, so long as at least one division does occur for each recursion. This algorithm can be embodied by a rooted binary tree structure, with a level for each distinct variable. In addition to the nodes labeled by variables, each of which has two children, there are $\varepsilon$ nodes which can be the parent of only one node and hence are considered "placeholders" for an eventual variable node. Let $\mathcal{T}(V)$ be the collection of all such trees for some variable set $V$. For $T \in \mathcal{T}(V)$, let $\pi(T)$ be the pattern which can be reconstructed from $T$ by reversing the splitting algorithm. Note that the number of nodes of $T \in \mathcal{T}(V)$ can not exceed $2^{|V|} - 1$, so $\pi(T)$ will automatically satisfy the length bound of condition 2.

**Theorem 9.** *For unavoidable $\bar{p} \in V^{+}$, there exists $T \in \mathcal{T}(V)$ with $\pi(T) = \bar{p}$.*

Since $\mathcal{T}(V)$ were motivated by combining and extending conditions 1 and 3 necessary for unavoidability, the result is to be expected. However, the proof depends on an inductive construction using a $\sigma$-deletion sequence for $\bar{p}$, and so implies the difficulty of uniquely associating a $T \in \mathcal{T}(V)$ with a given $\bar{p}$. If the converse of theorem 9 were also true, then pattern unavoidability would be decided efficiently by the recursive algorithm under consideration. And, in conjunction with verifying the nontriviality of $\mathcal{G}(\bar{p})$, the correspondence between $T$ and unavoidable $\bar{p}$ is exact (although non-unique) when $|V| = 3$ and 4. However, the situation deteriorates rapidly after this.

When $|V| = 5$, there are now known to be 55,572 nonisomorphic unavoidable patterns. But there are a grand total of 2,384,331 *other* nonisomorphic $\bar{p}$ with $\bar{p} = \pi(T)$ for some $T \in \mathcal{T}(V)$. Of those, 60,702 have nontrivial $\mathcal{G}(\bar{p})$ of which 3720 have no avoidable proper subpatterns. Thus, there exist thousands of avoidable patterns with five variables which satisfy the four conditions listed above. Moreover, given even one, it is possible to generate arbitrarily more.

**Theorem 10.** *Let $\bar{q} \in V^+$ be an avoidable pattern. Suppose that*

1. *there exists $T \in \mathcal{T}(V)$ with $\pi(T) = \bar{q}$,*
2. *$\bar{q}$ has nontrivial $\mathcal{G}(\bar{q})$,*
3. *and all proper subpatterns of $\bar{q}$ are unavoidable.*

*Then there exist infinitely many $\bar{r}$ such that $\bar{r}$ also satisfy the three conditions above and $\bar{r}$ is not unavoidable either.*

One of the implications of the result is that $\bar{q} \not\leq \bar{r}$, which prevents the potential decision algorithm from being adjusted to take finitely many such $\bar{q}$ into consideration. It also indicates that the Zimin-type construction, which was used to successfully in the results of section 5, will not be the basis for the proof of this theorem. Instead, the concept of "expanding" selected nodes of the tree structure will be exploited by replacing the instances of a variable by another pattern.

**Definition 10.** *Let $\bar{p} \in V^+$ and $\bar{q} \in U^+$. Suppose that $y$ is a variable of $\bar{q}$ and that $(U \setminus \{y\}) \cap V = \emptyset$. Define $\rho(\bar{q}, y, \bar{p})$ to be the pattern obtained by replacing every instance of $y$ in $\bar{q}$ by $\bar{p}$.*

The restriction that $(U \setminus \{y\}) \cap V = \emptyset$ optimally insures the consistency of the associated tree structure.

**Theorem 11.** *$\rho(\bar{q}, y, \bar{p})$ is unavoidable if and only if both $\bar{q}$ and $\bar{p}$ are.*

The proof in the more complicated direction depends on showing that the $\sigma$-deletion sequence for $\bar{p}$ can be inserted appropriately into the $\sigma$-deletion sequence for $\bar{q}$ to produce one which reduces $\rho(\bar{q}, y, \bar{p})$ to the empty string. Hence, the proof of theorem 10 follows almost immediately from theorem 11. Note that although $\bar{q}$ does exist as a subsequence of $\bar{r} = \rho(\bar{q}, y, \bar{p})$, it can be shown that only subsequences generated by $\sigma$-deletions are useful in deciding pattern unavoidability.

Consequently, the other known necessary conditions are not sufficient for deciding pattern unavoidability, while the $\sigma$-deletion decision procedure can not be effected in subexponential time. Additionally, there are no clearly easier means of determining whether $\bar{p} \mid Z_{\alpha(\bar{p})}$. Hence, this special case of GPM is known to be in *NP* and, in the absence of a more efficient characterization, is not expected to succumb to any polynomial time algorithm.

# References

[1] A. Apostolico and Z. Galil, editors. *Pattern Matching Algorithms*. Oxford University Press, New York, 1997.

[2] K.A. Baker, G.F. McNulty, and W. Taylor. Growth problems for avoidable words. *Theoret. Comput. Sci.*, 69(3):319–345, 1989.

[3] D.R. Bean, A. Ehrenfeucht, and G.F. McNulty. Avoidable patterns in strings of symbols. *Pacific J. Math.*, 85(2):261–294, 1979.

[4] M. Crochemore. An optimal algorithm for computing the repetitions in a word. *Inform. Process. Lett.*, 12(5):244–250, 1981.

[5] C.E. Heitsch. *Computational Complexity of Generalized Pattern Matching*. PhD thesis, Department of Mathematics, University of California at Berkeley, 2000.

[6] M. Lothaire. *Algebraic Combinatorics on Words*, chapter Unavoidable Patterns (Julien Cassaigne). http://www-igm.univ-mlv.fr./~berstel/Lothaire/, 2001.

[7] M.G. Main and R.J. Lorentz. An $O(n \log n)$ algorithm for finding all repetitions in a string. *J. Algorithms*, 5(3):422–432, 1984.

[8] A.I. Zimin. Blocking sets of terms. *Mat. Sb. (N.S.)*, 119(161)(3):363–375, 447, 1982.

# A   Appendix

**Table 1.** Decomposition of $(k_1/k_2)_n$ sets according to the four possibilities for $\bar{p} = s\bar{u}t$ relevant to theorem 4. Only the second column is excluded under the theorem's assumptions. The symmetry between the first and fourth cases is due to the dual nature of the $\sigma$-deletion definition.

| $(k_1/k_2)_n$ | $t \in A, s \in B$ | $t \in A, s \in B^c$ | $t \in A^c, s \in B$ | $t \in A^c, s \in B^c$ |
|---|---|---|---|---|
| $(1/1)_3$ | 2 | 0 | 4 | 2 |
| $(1/1)_4$ | 63 | 34 | 116 | 63 |
| $(1/2)_4$ | 9 | 0 | 20 | 9 |
| $(1/1)_5$ | 7571 | 5072 | 12,156 | 7571 |
| $(1/2)_5$ | 2558 | 1432 | 4544 | 2558 |
| $(1/3)_5$ | 117 | 0 | 362 | 117 |
| $(2/2)_5$ | 153 | 41 | 411 | 153 |
| $(2/3)_5$ | 6 | 0 | 29 | 6 |

**Table 2.** The complete numeration of nonisomorphic unavoidable patterns with 3, 4, and 5 distinct variables, respectively. Each group of results is broken down by pattern length, and then further refined by considering only those patterns with a unique $\sigma$-deletion of size $k_1/k_2$.

$$\alpha(\bar{p}) = 3$$

| $|\bar{p}|$ | # of $\bar{p}$ | 1/1 |
|---|---|---|
| 3 | 1 | |
| 4 | 3 | |
| 5 | 5 | 3 |
| 6 | 4 | 4 |
| 7 | 1 | 1 |
| | 14 | 8 |

$$\alpha(\bar{p}) = 4$$

| $|\bar{p}|$ | # of $\bar{p}$ | 1/1 | 1/2 |
|---|---|---|---|
| 4 | 1 | | |
| 5 | 6 | | |
| 6 | 22 | | |
| 7 | 61 | 20 | 4 |
| 8 | 89 | 52 | 8 |
| 9 | 91 | 63 | 9 |
| 10 | 74 | 58 | 8 |
| 11 | 49 | 41 | 6 |
| 12 | 28 | 26 | 2 |
| 13 | 12 | 11 | 1 |
| 14 | 4 | 4 | |
| 15 | 1 | 1 | |
| | 438 | 276 | 38 |

$$\alpha(\bar{p}) = 5$$

| $|\bar{p}|$ | # of $\bar{p}$ | 1/1 | 1/2 | 1/3 | 2/2 | 2/3 |
|---|---|---|---|---|---|---|
| 5 | 1 | | | | | |
| 6 | 10 | | | | | |
| 7 | 61 | | | | | |
| 8 | 292 | | | | | |
| 9 | 1076 | 199 | 46 | 2 | | |
| 10 | 2529 | 878 | 228 | 14 | 8 | 4 |
| 11 | 4292 | 1842 | 606 | 32 | 103 | 14 |
| 12 | 5773 | 2832 | 986 | 46 | 220 | 8 |
| 13 | 6590 | 3512 | 1295 | 89 | 203 | 15 |
| 14 | 6667 | 3862 | 1456 | 84 | 144 | |
| 15 | 6240 | 3811 | 1430 | 92 | 80 | |
| 16 | 5429 | 3506 | 1306 | 66 | | |
| 17 | 4536 | 3047 | 1074 | 58 | | |
| 18 | 3600 | 2496 | 852 | 40 | | |
| 19 | 2741 | 1965 | 632 | 29 | | |
| 20 | 1998 | 1476 | 448 | 18 | | |
| 21 | 1397 | 1061 | 299 | 13 | | |
| 22 | 938 | 734 | 190 | 6 | | |
| 23 | 601 | 477 | 118 | 4 | | |
| 24 | 368 | 304 | 62 | 2 | | |
| 25 | 212 | 174 | 37 | 1 | | |
| 26 | 116 | 100 | 16 | | | |
| 27 | 60 | 52 | 8 | | | |
| 28 | 28 | 26 | 2 | | | |
| 29 | 12 | 11 | 1 | | | |
| 30 | 4 | 4 | | | | |
| 31 | 1 | 1 | | | | |
| | 55,572 | 32,370 | 11,092 | 596 | 758 | 41 |

# Balanced Suffix Trees
## (Invited Lecture)

S. Rao Kosaraju[*]

Department of Computer Science
The Johns Hopkins University
Baltimore, Maryland 21218
kosaraju@cs.jhu.edu

**Abstract.** We consider the problem of developing an efficient tree-based data structure for storing and searching large data sets. It is assumed that the data set is stored in secondary storage, and hence the goal is to reduce the number of accesses to the storage. The number of accesses is measured by the number of edges of the tree that get accessed while processing a query.

The data consists of a large number of words, each word being a string of characters. Each query consists of searching for a given word (*member*). Our goal is to design a simple data structure that permits efficient execution of the queries. The special case when the words are the suffixes of a string is the classic suffix tree data structure.

We discuss different approaches for reducing the depth of as suffix tree.

---

[*] Supported by NSF grant CCR-9821058, ARO grant DAAH04-96-1-0013, and a NASA Contract.

A. Amir and G.M. Landau (Eds.): CPM 2001, LNCS 2089, pp. 231–231, 2001.

# A Fast Algorithm for Optimal Alignment between Similar Ordered Trees

Jesper Jansson and Andrzej Lingas

Dept. of Computer Science, Lund University
Box 118, 221 00 Lund, Sweden
{Jesper.Jansson,Andrzej.Lingas}@cs.lth.se

**Abstract.** We present a fast algorithm for optimal alignment between two similar ordered trees with node labels. Let $S$ and $T$ be two such trees with $|S|$ and $|T|$ nodes, respectively. An optimal alignment between $S$ and $T$ which uses at most $d$ blank symbols can be constructed in $O(n \log n \cdot (maxdeg)^4 \cdot d^2)$ time, where $n = \max\{|S|, |T|\}$ and $maxdeg$ is the maximum degree of a node in $S$ or $T$. In particular, if the input trees are of bounded degree, the running time is $O(n \log n \cdot d^2)$.

## 1 Introduction

Let $R$ be a rooted tree. $R$ is called a *labeled tree* if each node of $R$ is labeled by a symbol from a fixed finite set $\Sigma$. $R$ is an *ordered tree* if the left-to-right order among siblings in $R$ is given.

The problem of determining the similarity between two labeled trees occurs in several different areas of computer science. For example, in computational biology, methods for measuring the similarity between ordered labeled trees of bounded degree can be used in the comparison of RNA secondary structures [2,4,9]. The problem also occurs in evolutionary trees comparison, organic chemistry, pattern recognition, and image clustering [2,4,8,12].

The similarity between two labeled trees can be defined in various ways analogous to the ways of defining the similarity between two sequences [5,7,8]. For example, one can look for the largest maximum agreement subtree, the largest common subgraph, the smallest common supertree, the minimum tree edit distance *etc.* [2,3,4,7,10,12].

In [2], Jiang *et al.* generalized the concept of an alignment between sequences to include labeled trees as follows. An *insert operation* on a labeled tree adds a new node $u$ which is labeled by a blank symbol $\lambda$ (space) not belonging to $\Sigma$. The operation either (1) turns the current root of the tree into a child of $u$ and lets $u$ become the new root, or (2) makes $u$ the parent of a subset of (if the tree is unordered) or consecutive subsequence of (if the tree is ordered) children of an existing node $v$, and $u$ a child of $v$. An *alignment between two labeled trees* is obtained by performing insert operations on the two trees so they become isomorphic when labels are ignored, and then overlaying the first augmented tree on the other one. The *score* of the alignment is the sum of the scores of

A. Amir and G.M. Landau (Eds.): CPM 2001, LNCS 2089, pp. 232–240, 2001.

all matched pairs of labels, where the score of a pair of labels is defined by a given function $\mu : (\Sigma \cup \{\lambda\}) \times (\Sigma \cup \{\lambda\}) \rightarrow \mathbb{Z}$. An *optimal alignment* between a pair of labeled trees is an alignment between them achieving the highest possible score[1]. See Fig. 1 for an example.

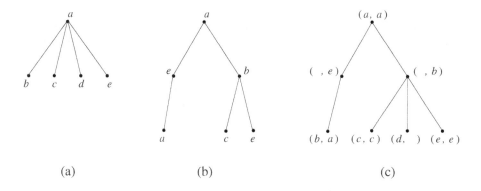

(a)                              (b)                              (c)

**Fig. 1.** Example: Let $\Sigma = \{a, b, c, d, e\}$ and define the scoring function $\mu$ as $\mu(x, x) = 3$, $\mu(x, y) = -1$, $\mu(x, \lambda) = \mu(\lambda, x) = -2$, $\mu(\lambda, \lambda) = -2$ for all $x, y \in \Sigma$ with $x \neq y$. Then the score of the alignment in (c) of the two ordered trees shown in (a) and (b) is equal to 2.

Jiang *et al.* presented an $O(n^2 (maxdeg)^2)$-time algorithm for computing an optimal alignment between two ordered trees with node labels, where $n$ stands for the maximum number of nodes in one of the input trees, and *maxdeg* for the maximum degree of a node in the input trees. They also provided a polynomial time algorithm for finding an optimal alignment of two unordered trees in case $maxdeg = O(1)$, and showed the latter problem to be MAX SNP-hard in general.

Inspired by the known fast method for an optimal alignment between similar sequences (see Section 3.3.4 in [8]), we give a fast algorithm for optimal alignment between two similar ordered trees with node labels. If there is an optimal alignment between the two input ordered trees which uses at most $d$ blank symbols then our algorithms runs in $O(n \log n \cdot (maxdeg)^4 \cdot d^2)$ time. Hence, under a natural assumption on the scoring, if the maximum possible score of an alignment between the two trees is $O(d)$ apart from the score of a perfect alignment of the first tree with itself then the algorithm runs in $O(n \log n \cdot (maxdeg)^4 \cdot d^2)$ time. In particular, if both trees are of bounded degree the running time reduces to $O(n \log n \cdot d^2)$.

---

[1] In fact, Jiang *et al.* consider the arithmetically complementary distance measure of an alignment which is the subject of minimization [2].

## 2   $d$-Relevant Pairs

The general idea of our algorithm is to modify the dynamic programming algorithm of Jiang *et al.* to only consider what we call $d$-relevant pairs of subtrees or subforests. In order to introduce our slightly technical concept of $d$-relevance we need the following definition.

**Definition 1.** *For an ordered tree $T$ and a node $u$ of $T$, $T[u]$ denotes the ordered subtree of $T$ rooted at $u$. When $u$ is not the root of $T$, $\overline{T[u]}$ stands for the ordered subtree of $T$ resulting from removing $T[u]$ and the edge between $u$ and the parent of $u$ from $T$. Next, $L(T, u)$ denotes the set of leaves in $T$ that are to the left of the leaves of $T[u]$. The number of nodes in $T$ is denoted by $|T|$ and the cardinality of $L(T, u)$ by $|L(T, u)|$.*

Now, we are ready to introduce the concept of $d$-relevant pairs of subtrees as well as those of $d$-descendant and $d$-ancestor.

**Definition 2.** *Let $d$ be a positive integer. For two ordered trees $S, T$ containing nodes $u$ and $v$ respectively, the pair of subtrees $(S[u], T[v])$ is called $d$-relevant if $||S[u]| - |T[v]|| \leq d$ and $||L(S, u)| - |L(T, v)|| \leq d$. For a node $w$ of $T$, $T[w]$ is called a $d$-descendant of $T[v]$ if $w$ is a descendant of $v$ in $T$ and $|T[v]| - |T[w]| \leq d$. Symmetrically, $T[w]$ is called a $d$-ancestor of $T[v]$ if $w$ is an ancestor of $v$ in $T$ and $|T[w]| - |T[v]| \leq d$.*

The definition of $d$-relevance immediately yields the following lemma.

**Lemma 1.** *Let $S$, $T$ be two labeled ordered trees, and let $u$, $v$ be two nodes in $S$ and $T$ respectively. If there is an alignment between $S$ and $T$ which uses at most $d$ blank symbols (spaces) and consists of an alignment between $S[u]$ and $T[v]$ and an alignment between $\overline{S[u]}$ and $\overline{T[v]}$ then $(S[u], T[v])$ is $d$-relevant for $S$ and $T$.*

The next three lemmas will be useful for bounding the number of $d$-relevant pairs from above.

**Lemma 2.** *If the pairs $(S[u], T[v])$ and $(S[u], T[w])$ are $d$-relevant for two ordered trees $S$ and $T$, and $w$ is a descendant (or, ancestor) of $v$ in $T$ then $T[w]$ is a $2d$-descendant (or, $2d$-ancestor) of $T[v]$.*

*Proof.* Since $(S[u], T[v])$ is $d$-relevant, it holds that $||S[u]| - |T[v]|| \leq d$. Suppose that $T[w]$ is not a $2d$-descendant of $T[v]$ in $T$, i.e., $|T[v]| - |T[w]| > 2d$. Then we have $|S[u]| - |T[w]| = |S[u]| - |T[v]| + |T[v]| - |T[w]| > -d + 2d = d$, which contradicts the $d$-relevance of $(S[u], T[w])$.     □

**Lemma 3.** *For a node $u$ of an ordered tree $S$, the number of $d$-ancestors of $S[u]$ is at most $d$.*

*Proof.* Assume that the number of $d$-ancestors of $S[u]$ is greater than $d$. By the pigeonhole principle there exists a $d$-ancestor $S[u']$ whose root $u'$ is located at distance greater than $d$ from $u$. But then $|S[u']| - |S[u]| > d$, which is a contradiction.     □

**Lemma 4.** *Let* $\{(S[u], T[v_i])\}_{i=0}^{l}$ *be a sequence of distinct d-relevant pairs in two ordered trees* $S, T$ *such that for any* $0 \leq i, j \leq l$, $v_i$ *is not a descendant of* $v_j$. *Then,* $l \leq 2d$ *holds.*

*Proof.* We may assume w.l.o.g. that the sequence is ordered according to the left-right order in $T$. Since $(S[u], T[v_l])$ is $d$-relevant, $||L(S, u)| - |L(T, v_l)|| \leq d$ holds. On the other hand, we have $|L(T, v_l)| - |L(T, v_0)| \geq l$. Hence, if $l > 2d$ then $|L(S, u)| - |L(T, v_0)| > d$, which contradicts the $d$-relevance of $(S[u], T[v_0])$. □

By combining the three lemmas above, we obtain an upper bound on the number of $d$-relevant pairs of subtrees.

**Theorem 1.** *For two ordered trees* $S, T$ *and a node* $u$ *of* $S$, *the number of distinct d-relevant pairs of subtrees in which* $u$ *participates is* $O(d^2)$. *Consequently, there are* $O(|S| \cdot d^2)$ *d-relevant pairs of subtrees for* $S$, $T$.

*Proof.* Let $\{(S[u], T[v_i])\}_{i=0}^{l}$ be a maximal sequence of distinct $d$-relevant pairs of subtrees for two ordered trees $S, T$ such that for each $0 \leq i \leq l$ there is no $d$-relevant pair $(S[u], T[v])$, where $v$ is a descendant of $v_i$. It follows from Lemma 2 that for each $d$-relevant pair $(S[u], T[w])$, it either belongs to the sequence or $T[w]$ is a $2d$-ancestor of a member in the sequence. Hence, the number of $d$-relevant pairs in which $u$ participates is at most $(2d + 1) \cdot (l + 1)$ by Lemma 3. Now, it is sufficient to observe that $l$ cannot exceed $2d$ by Lemma 4. □

## 2.1   *d*-Relevant Pairs of Subforests

The dynamic programming algorithm of Jiang *et al.* recursively computes scores not only between pairs of subtrees of the input trees but also between some pairs of subforests of the trees. Therefore, in order to modify this algorithm, we need to generalize the concept of $d$-relevance for pairs of subtrees to include the aforementioned pairs of subforests. For this purpose, we introduce the following technical notations.

**Definition 3.** *For an ordered tree* $S$ *and a node* $u$ *of* $S$, *let* $d_u$ *be the degree of* $u$ *and denote the children of* $u$ *by* $u_1, ..., u_{d_u}$, *according to their left-to-right order.* $S(u, i, j)$ *refers to the ordered forest* $S[u_i], ..., S[u_j]$, *and* $S(u)$ *is short for* $S(u, 1, d_u)$.

Thus, $S(u)$ is the *complete ordered forest* obtained by removing $u$ and all edges incident to $u$ from $S[u]$. Also note that $S(u, i, i) = S[u_i]$.

**Definition 4.** *Let* $S(u, i, j)$ *be an ordered forest in an ordered tree* $S$. $\overline{S(u, i, j)}$ *stands for the ordered subtree of* $S$ *obtained by removing* $S(u, i, j)$ *and all edges incident to* $S(u, i, j)$ *from* $S$. $L(S(u, i, j))$ *denotes the set of leaves in* $S$ *that are to the left of the leaves of* $S(u, i, j)$. *The number of nodes in* $S(u, i, j)$ *is denoted by* $|S(u, i, j)|$ *and the cardinality of* $L(S(u, i, j))$ *by* $|L(S(u, i, j))|$.

Now, we are ready to generalize the concept of $d$-relevance as well as those of $d$-descendant and $d$-ancestor for pairs of nodes inducing full subtrees to include pairs of subforests of the form $(S(u, i, j), T(v, k, l))$.

**Definition 5.** *Let $d$ be a positive integer. For two ordered trees $S, T$ containing nodes $u$ and $v$ respectively, the pair of ordered subforests $(S(u, i, j), T(v, k, l))$ is called $d$-relevant if $||S(u, i, j)| - |T(v, k, l)|| \leq d$ and $||L(S(u, i, j))| - |L(T(v, k, l))|| \leq d$. For a node $w$ of $T$, $T(w, k', l')$ is called a $d$-descendant of $T(v, k, l)$ if $w$ is a descendant of $v$ in $T$ and $||T(w, k', l')| - |T(v, k, l)|| \leq d$. Symmetrically, $T(w, k', l')$ is called a $d$-ancestor of $T(v, k, l)$ if $w$ is an ancestor of $v$ in $T$ and $||T(w, k', l')| - |T(v, k, l)|| \leq d$.*

The definition of $d$-relevance of subforests immediately yields the following lemma analogous to Lemma 1.

**Lemma 5.** *Let $S, T$ be two labeled ordered trees, and let $S(u, i, j)$ and $T(v, k, l)$ be ordered forests in $S$ and $T$ respectively. If there is an alignment between $S$ and $T$ which uses at most $d$ blank symbols (spaces) and consists of an alignment between $S(u, i, j)$ and $T(v, k, l)$ and an alignment between $\overline{S(u, i, j)}$ and $\overline{T(v, k, l)}$ then $(S(u, i, j), T(v, k, l))$ is $d$-relevant for $S$ and $T$.*

The next three lemmata will be useful for bounding the number of $d$-relevant pairs of subforests from above. Their proofs are analogous to the corresponding proofs of Lemmata 2–4.

**Lemma 6.** *If the pairs $(S(u, i, j), T(v))$ and $(S(u, i, j), T(w))$ are $d$-relevant for two ordered trees $S, T$ and $w$ is a descendant (or, ancestor) of $v$ in $T$ then $T(w)$ is a $2d$-descendant (or, $2d$-ancestor) of $T(v)$.*

**Lemma 7.** *For a node $v$ of an ordered tree $T$, the number of $d$-ancestors of the form $T(w)$ of a forest $T(v)$ is at most $d$.*

**Lemma 8.** *Let $\{(S(u, i, j), T(v_q)\}_{q=0}^l$ be a sequence of distinct $d$-relevant pairs in two ordered trees $S, T$ such that for any $0 \leq q', q'' \leq l$, $v_{q'}$ is not a descendant of $v_{q''}$. Then, $l \leq 2d$ holds.*

By combining the three lemmas above, we obtain an upper bound on the number of $d$-relevant pairs $(S(u), T(v, k, l))$ and $(S(u, i, j), T(v))$ as in Theorem 1.

**Theorem 2.** *For two ordered trees $S, T$ and a node $u$ of $S$, the number of distinct $d$-relevant pairs of the form $(S(u, i, j), T(v))$ is $O(d^2(\deg(S))^2)$. Symmetrically, for a node $v$ of $T$, the number of distinct $d$-relevant pairs of the form $(S(u), T(v, k, l))$ is $O(d^2(\deg(T))^2)$. Consequently, there are $O(n \cdot d^2(maxdeg)^2)$ $d$-relevant pairs of the form $(S(u), T(v, k, l))$ or $(S(u, i, j), T(v))$ for $S, T$, where $n = \max\{|S|, |T|\}$.*

## 3  Constructing the $d$-Relevant Pairs

The test for $d$-relevance for a pair of subtrees can easily be accomplished in constant time after appropriate preprocessing. However, in order to speed up the at least quadratic algorithm of Jiang *at al.*, we cannot afford testing each possible subtree pair for $d$-relevance. Instead, we proceed as follows.

First, we compute all vectors $(|T[v]|, |L(T, v)|)$, where $v \in T$. This can be done in linear total time by using the Eulerian tour technique from [11]. We insert the vectors into a standard range search data structure, e.g., a layered range tree [6]. The construction of the data structure takes $O(|T| \cdot \log |T|)$ time. Then, for all $u$ in $S$ we compute the vectors $(|S[u]|, |L(S, u)|)$ in linear total time in the same way as above. For each $u$ in $S$, we query the data structure with the square centered at $(|S[u]|, |L(S, u)|)$ having side length $2d$. Each query takes $O(\log |S| + r)$ time, where $r$ is the number of reported vectors. Since each of the returned vectors is in one-to-one correspondence with a node $v$ such that the pair $(u, v)$ is $d$-relevant, $r = O(d^2)$ holds by Theorem 1.

Putting everything together, we obtain the following theorem.

**Theorem 3.** *For two ordered trees on at most $n$ nodes each and a non-negative integer $d$, all $d$-relevant pairs of subtrees can be reported in $O(n(\log n + d^2))$ time.*

We can use the same technique to precompute all pairs of $d$-relevant subforests. In fact, for our purposes it is sufficient to report all pairs of $d$-relevant subforests where at least one of the subforests is complete, i.e., is of the form $S(u)$ or $T(v)$. To report all $d$-relevant pairs of the form $(S(u), T(v, k, l))$, the number of vectors to insert into the layered range tree is $O(|T|(maxdeg)^2)$ since $O((maxdeg)^2)$ ordered forests of the form $T(v, k, l)$ originate from each node $v$ in $T$. Thus, the construction time becomes $O(|T| \cdot (maxdeg)^2 \cdot \log(|T| \cdot (maxdeg)^2)) = O(n \cdot (maxdeg)^2 \cdot \log n)$. The number of queries to the data structure is $O(|S|)$, and the query time is $O(\log(|T| \cdot (maxdeg)^2) + r) = O(\log n + r)$ time, where the sum of the $r$'s over $S$ is $O(nd^2 \cdot (maxdeg)^2)$ by Theorem 2. The reporting of $d$-relevant pairs of the form $(S(u, i, j), T(v))$ can be done symmetrically within the same (in terms of $n$) preprocessing and query time bounds.

Summing up, we obtain:

**Theorem 4.** *For two ordered trees on at most $n$ nodes each and a non-negative integer $d$, all $d$-relevant pairs of subforests, where at least one subforest is complete, can be reported in $O(n \cdot (maxdeg)^2 \cdot (\log n + d^2))$ time.*

## 4  The Fast Algorithm

Our fast algorithm for an optimal alignment between two ordered trees works under the assumption that there is an optimal alignment between the trees $S$, $T$ which uses at most $d$ blank symbols (spaces). First, we compute all $d$-relevant pairs of subtrees of $S, T$ as described in Section 3. As each $d$-relevant pair is reported, we insert it into a balanced binary search tree $\mathcal{B}_1$. Next, all

$d$-relevant pairs of subforests in which at least one subforest is complete are computed and inserted into a balanced binary search tree $\mathcal{B}_2$. According to Theorems 1 and 2, there are $O(nd^2(maxdeg)^2)$ $d$-relevant pairs of subtrees or subforests where at least one subforest is complete, so this preprocessing takes $O(n \cdot (maxdeg)^2 \cdot (\log n + d^2) + n \cdot d^2(maxdeg)^2) \log(n \cdot d^2(maxdeg)^2)) = O(n \log n \cdot (maxdeg)^2 \cdot d^2)$ time by Theorems 3 and 4. Then, we modify the algorithm of Jiang $et$ $al.$ recursively evaluating the score values (see [2]) solely for $d$-relevant pairs of subtrees or pairs of subtrees where one of the subtrees is empty.

This evaluation involves also recursive evaluation of the score values for $d$-relevant pairs of subforests where one of the forests is complete or empty. In fact, the recursive procedures in [2] include also intermediate terms with the scores values for pairs of subforests when none of the forests is complete or empty. However, these intermediate terms are eliminated by the composition of the aforementioned procedures, resulting in recursive formulas for the score values expressed in the form of maximum of some sums of score values for pairs of smaller subtrees or subforests where at least one of the subforests is complete or empty. Whenever the left handside is $d$-relevant in the application of such a formula, the components of the sum on the right hand side yielding the maximum, with the exception of the scores for the pairs including an empty subtree or subforest, also have to be $d$-relevant. Therefore, before an application of such a formula to an evaluation of a $d$-relevant pair, we simply test each of the components of the sums on the right handside, which is not a score for pair containing an empty subtree or subforest, for membership in $\mathcal{B}_1$ or $\mathcal{B}_2$. Such a membership query takes $O(\log n)$ time. If the test is positive we fetch the score value for the argument pair which should be evaluated by this time, otherwise we set that score value to minus infinity. The score values for pairs containing an empty subtree or subforest can be trivially precomputed in time $O(|S| + |T|)$. We conclude that the cost of determining the score for a $d$-relevant pair on the left handside of such a recursive formula on the basis of the scores for $d$-relevant pairs occuring on its right handside does not exceed the cost of determining the scores for this pair on the basis of the scores of pairs occuring on the right handside in the algorithm of Jiang $et$ $al.$ multiplied by $O(\log n)$.

Jiang $et$ $al.$ show that the cost of determining the score for a pair of subtrees or subforests by using the aforementioned formulas and already computed scores for pairs of smaller subtrees or subforests is $O(deg(z) \cdot (maxdeg)^2)$, where $z$ is a node in $S$ or $T$ which is either the root of the first subtree or the second subtree, or the parent of the roots of the trees in the first forest or the second forest. Hence, the corresponding cost for $d$-relevant pairs in our modification of this algorithm is $O(deg(z) \cdot (maxdeg)^2 \cdot \log n)$. By Theorems 1 and 2, for a given node $z$ in $S$ or $T$, there are $O(d^2(maxdeg)^2)$ $d$-relevant pairs of subtrees of the form $(S[z], T[v])$ or $(S[u], T[z])$, or subforests of the form $(S(z,i,j), T[v])$ or $(S(u), T(z,l,k))$. Hence, our modified algorithm runs in $O(\sum_{z \in S \cup T} deg(z) \cdot (maxdeg)^2 \cdot \log n \cdot d^2(maxdeg)^2)$ time, i.e., $O(nd^2(maxdeg)^4 \log n)$ assuming the preprocessing has been done.

**Theorem 5.** *An optimal alignment of two ordered trees which uses at most d blank symbols can be constructed in $O(n \log n \cdot (maxdeg)^4 \cdot d^2)$ time.*

Under the natural assumption that the score of a pair including at least one blank symbol is negative and by $\Omega(1)$ smaller than that of a pair consisting of two identical symbols, we immediately obtain the following lemma.

**Corollary 1.** *An optimal alignment of two ordered trees whose score is $O(d)$ apart from the score of the perfect alignment between the first tree and its copy can be constructed in $O(n \log n \cdot (maxdeg)^4 \cdot d^2)$ time.*

**Corollary 2.** *An optimal alignment of two ordered trees of bounded degree whose score is $O(d)$ apart from the score of the perfect alignment between the first tree and its copy can be constructed in $O(n \log n \cdot d^2)$ time.*

## 5   Final Remarks

An optimal alignment between two sequences whose score is at most $d$ apart from that of a perfect alignment between the first sequence and its copy can be constructed in $O(nd)$ time [8]. Since a sequence can be interpreted as a line ordered tree with node labels, a natural question arises: is it possible to lower the time complexity of our method, especially the exponent 2 of $d$?

Our method does not seem to generalize to include unordered trees directly. Simply, the proof of Lemma 4 relies on the ordering of the trees (i.e., on the sets $L(\ ,\ )$). It is an interesting open problem whether a substantial speed-up in the construction of an optimal alignment between similar unordered trees of bounded degree is achievable.

In the construction of the $d$-relevant pairs we could use more sophisticated and more asymptotically efficient data structures for two dimensional range search on an integer grid [1]. However, this would not lead to an improvement of the overwhole asymptotic time complexity of our alignment algorithm.

## References

1. S. Alstrup, G.S. Brodal, T. Rauhe. New Data Structures for Orthogonal Range Searching. Proc. of 41st Annual Symposium on Foundations of Computer Science (FOCS 2000), 2000, pp. 198–207.
2. T. Jiang, L. Wang, and K. Zhang. Alignment of Trees - An Alternative to Tree Edit. Theoretical Computer Science, 143 (1995), pp. 137–148. (A preliminary version in Proc. of 5th Annual Symposium on Combinatorial Pattern Matching (CPM'94), Lecture Notes in Computer Science, Vol. 807, Springer, 1994, pp. 75–86.)
3. D. Keselman and A. Amir. Maximum agreement subtree in a set of evolutionary trees – metrics and efficient algorithms. Proc. of 35th Annual IEEE Symposium on the Foundations of Computer Science (FOCS'94), 1994, pp. 758–769.
4. S.-Y. Le, R. Nussinov, and J.V. Maizel. Tree graphs of RNA secondary structures and their comparisons. Computers and Biomedical Research, 22 (1989), pp. 461–473.

5. P.A. Pevzner. Computational Molecular Biology: An Algorithmic Approach. The MIT Press, Cambridge, Massachusetts, 2000.

6. F. Preparata and M.I. Shamos. *Computational Geometry*. Springer-Verlag, New York, 1985.

7. D. Sankoff and J. Kruskal (Eds). *Time Warps, String Edits, and Macromolecules, the Theory and Practice of Sequence Comparison.* Addison Wesley, Reading Mass., 1983.

8. J.C. Setubal and J. Meidanis. *Introduction to Computational Molecular Biology.* PWS Publishing Company, Boston, 1997.

9. B. Shapiro. An algorithm for comparing multiple RNA secondary structures. Comput. Appl. Biosci. (1988) pp. 387–393.

10. K.C. Tai. The tree-to-tree correction problem. J. ACM 26 (1979), pp. 422-433.

11. R.E. Tarjan and U. Vishkin. Finding biconnected components and computing tree functions in logarithmic parallel time. SIAM Journal of Computing 14, 4 (1985), pp. 862–874.

12. K. Zhang and D. Shasha. Simple fast algorithms for the editing distance between trees and related problems. SIAM Journal of Computing 18, 6 (1989), pp. 1245–1262.

# Minimum Quartet Inconsistency
# Is Fixed Parameter Tractable

Jens Gramm* and Rolf Niedermeier

Wilhelm-Schickard-Institut für Informatik, Universität Tübingen
Sand 13, D-72076 Tübingen, Fed. Rep. of Germany
{gramm,niedermr}@informatik.uni-tuebingen.de

**Abstract.** We study the parameterized complexity of the problem to reconstruct a binary (evolutionary) tree from a complete set of quartet topologies in the case of a limited number of errors. More precisely, we are given $n$ taxa, exactly one topology for every subset of 4 taxa, and a positive integer $k$ (the parameter). Then, the *Minimum Quartet Inconsistency* (MQI) problem is the question of whether we can find an evolutionary tree inducing a set of quartet topologies that differs from the given set in only $k$ quartet topologies. MQI is NP-complete. However, we can compute the required tree in worst case time $O(4^k \cdot n + n^4)$—the problem is fixed parameter tractable. Our experimental results show that in practice, also based on heuristic improvements proposed by us, even a much smaller exponential growth can be achieved. We extend the fixed parameter tractability result to weighted versions of the problem. In particular, our algorithm can produce *all* solutions that resolve at most $k$ errors.

## 1 Introduction

In recent years, quartet methods for reconstructing evolutionary trees have received considerable attention in the computational biology community [6,11]. In comparison with other phylogenetic methods, an advantage of quartet methods is, e.g., that they can overcome the data disparity problem (see [6] for details). The approach is based on the fact that an evolutionary tree is uniquely characterized by its set of induced quartet topologies [5]. Herein, we consider an *evolutionary tree* to be an unrooted binary tree $T$ in which the leaves are bijectively labeled by a set of taxa $S$. A quartet, then, is a size four subset $\{a, b, c, d\}$ of $S$, and the topology for $\{a, b, c, d\}$ induced by $T$ simply is the four leaf subtree of $T$ induced by $\{a, b, c, d\}$. The three possible quartet topologies for $\{a, b, c, d\}$ are $[ab|cd]$, $[ac|bd]$, and $[ad|bc]$.[1] E.g., the topology is $[ab|cd]$ when, in $T$, the paths from $a$ to $b$ and from $c$ to $d$ are disjoint. The fundamental goal of quartet

---

* Work supported by the DFG projects "KOMET," LA 618/3-3, and "OPAL" (optimal solutions for hard problems in computational biology), NI-369/2-1.
[1] The fourth possible topology would be the star topology, which is not considered here because it is not binary.

A. Amir and G.M. Landau (Eds.): CPM 2001, LNCS 2089, pp. 241–256, 2001.
© Springer-Verlag Berlin Heidelberg 2001

methods is, given a set of quartet topologies, to reconstruct the corresponding evolutionary tree. The computational interest in this paradigm derives from the fact that the given set of quartet topologies usually is fault-prone.

In this paper, we focus on the following, perhaps most often studied optimization problem in the context of quartet methods.

MINIMUM QUARTET INCONSISTENCY (MQI)
**Input:** A set $S$ of $n$ taxa and a set $Q_S$ of quartet topologies such that there is exactly one topology for *every* quartet set[2] corresponding to $S$ and a positive integer $k$.
**Question:** Is there an evolutionary tree $T$ where the leaves are bijectively labeled by the elements from $S$ such that the set of quartet topologies induced by $T$ differs from $Q_S$ in at most $k$ quartet topologies?

MQI is NP-complete [12]. Concerning the approximability of MQI, it is known that it is polynomial time approximable with a factor $n^2$ [11,12]. It is an open question of [11] whether MQI can be approximated with a factor at most $n$ or even with a constant factor. The parameterized complexity [7] of MQI, however, so far, has apparently been neglected—we close this gap here. Assuming that the number $k$ of "wrong" quartet topologies is small in comparison with the total number of given quartet topologies, we show that MQI is *fixed parameter tractable*; that is, MQI can be solved exactly in worst case time $O(4^k n + n^4)$. Observe that the input size is $O(n^4)$. It is worth noting here that the variant of MQI where the set $Q_S$ is not required to contain a topology for every quartet is NP-complete, even if $k = 0$ [16]. Hence, this excludes parameterized complexity studies and also implies inapproximability (with any factor).

To develop our algorithm, we exhibit some nice combinatorial properties of MQI. For instance, we point out that "global conflicts" due to erroneous quartet topologies can be reduced to "local conflicts." The basis for this was laid by Bandelt and Dress [2]. This is the basic observation in order to show fixed parameter tractability of MQI. Our approach makes it possible to construct *all* evolutionary trees that can be (uniquely) obtained from the given input by changing at most $k$ quartet topologies. This puts the user of the algorithm in the position to pick (e.g., based on additional biological knowledge) the probably best, most reasonable solution or to construct a consensus tree from all solutions. Moreover, our method also generalizes to weighted quartets.

We performed several experiments on artificial and real (fungi) data and, thereby, showed that our algorithm (due to several tuning tricks) in practice runs much faster than its theoretical (worst case) analysis predicts. For instance, with a small $k$ (e.g., $k = 100$), we can solve relatively large ($n = 50$ taxa) instances optimally in around 40 minutes on a LINUX PC with a Pentium III 750 MHz processor and 192 MB main memory.

A full version (containing all proofs) is available [10].

---

[2] Note that given $n$ species, there are $\binom{n}{4} = O(n^4)$ corresponding quartet topologies.

## 2    Preliminaries

**Minimum Quartet Inconsistency.** In order to find the "best" binary tree for a given set of quartet topologies, we can ask for a tree that violates a minimum number of topologies. In case we are given exactly one quartet topology for every set of four taxa, this question gives the MQI problem. If there is not a quartet topology for necessarily every set of four taxa, Ben-Dor *et al.* [3] propose two solutions, namely, a heuristic approach and an exact algorithm. The heuristic solution is based on semidefinite programming and does not guarantee to produce the optimal solution, but has a polynomial running time. The exact algorithm uses dynamic programming for finding the optimal solution and has exponential running time, namely, $O(m3^n)$, where $n$ is the number of species and $m$ is the number of given quartet topologies. Note that Ben-Dor *et al.* run all their experiments on MQI instances, i.e., there was exactly one quartet topology for every set of four taxa. In that case, we have $m = O(n^4)$. The memory requirement of their exact solution is $\Theta(2^n)$. According to Jiang *et al.* [11] there is a factor $n^2$-approximation, and, at the same time, they asked about better approximation results. Note that the complement problem of MQI, where one tries to maximize $|Q_T \cap Q|$ ($Q_T$ being the set of quartet topologies induced by a tree $T$), possesses a polynomial time approximation scheme [12].

**Some Notation.** Assume that we are given a set of $n$ taxa $S$. For a quartet $\{a, b, c, d\} \subseteq S$, we refer to its possible quartet topologies by $[ab|cd]$, $[ac|bd]$, and $[ad|bc]$. These are the only possible topologies up to isomorphism. A set of quartet topologies is *complete* if it contains exactly one topology for every quartet of $S$. A complete set of quartet topologies over $S$ we denote by $Q_S$. A set of quartet topologies $Q$ is *tree-consistent* [2] if there exists a tree $T$ such that for the set $Q_T$ of quartet topologies induced by $T$, we have $Q \subseteq Q_T$. Set $Q$ is *tree-like* [2] if there exists a tree with $Q = Q_T$. Since an evolutionary tree is uniquely characterized by the topologies for all its quartets [5], a complete set of topologies is tree-consistent iff it is tree-like. A set of topologies has a "conflict" whenever it is not tree-consistent. We will call a conflict "global" when a complete set of topologies is not tree-consistent. We call it "local" when a size three set of topologies, which necessarily is incomplete, is not tree-consistent.

## 3    Global Conflicts Are Local

Given a complete set of quartet topologies which is not tree-consistent, the results of Bandelt and Dress [2] imply that there already is a subset of only three quartet topologies which is not tree-consistent. This is the key to developing a fixed parameter solution for the problem: It is sufficient to examine the size three sets of quartet topologies and to recursively branch on those sets which are not tree-consistent, as will be explained in Section 5.

**Proposition 1.** *(Proposition 2 in [2]) Given a set of taxa $S$ and a complete set of quartet topologies $Q_S$ over these taxa, $Q_S$ is tree-like iff the following so-called substitution property holds for every five distinct taxa $a, b, c, d, e \in S$:*

$$[ab|cd] \in Q_S \text{ implies } [ab|ce] \in Q_S \text{ or } [ae|cd] \in Q_S.$$

In the following, we show that in Proposition 1, we can replace the substitution property introduced by Bandelt and Dress with the more common term of tree-consistency. This is because, for an incomplete set of only three topologies, the substitution property is tightly connected to the tree-consistency of the topologies. We will state this in the following technical Lemmas 1 and 2 (proofs omitted, see [10]) and later use it to give, in Theorem 1, another interpretation of Proposition 1.

**Lemma 1.** *Three topologies involving more than five taxa are tree-consistent.*

When searching for local conflicts, Lemma 1 makes it possible to focus on the case of three topologies involving only five taxa. If the substitution property, as given in Proposition 1, is not satisfied, we say that the topologies for the quartets $\{a, b, c, d\}$, $\{a, b, c, e\}$, and $\{a, c, d, e\}$ *contradict* the substitution property.

**Lemma 2.** *For a given a set of taxa $S$, three topologies consisting of taxa from $S$ are tree-consistent iff they do not contradict the substitution property.*

Note that Lemma 2 involving a necessarily incomplete set of three topologies does not generalize from size three to an incomplete set of arbitrary size, as exhibited in the following example. For taxa $\{a, b, c, d, e, f\}$, consider the incomplete set of topologies $[ab|cd]$, $[ab|ce]$, $[bc|de]$, $[cd|ef]$, and $[af|de]$. Without going into the details, we only state here that these topologies are not tree-consistent, although there are no three topologies which contradict the substitution property.

Theorem 1 now will make it clearer that "global" tree-consistency of a complete set of topologies reflects in "local" tree-consistency of every three topologies taken from this set.

**Theorem 1.** *Given a set of taxa $S$ and a complete set of quartet topologies $Q_S$ over $S$, $Q_S$ is tree-like (and, thus, tree-consistent) iff every set of three topologies from $Q_S$ is tree-consistent.*

*Proof.* Due to Lemma 2 we may replace the substitution property in Proposition 1 with tree constistency. This gives the result. □

When we have a complete set of topologies $Q_S$ for a set of taxa $S$, we do not necessarily know whether the set is tree-like or not. If it is not, we can, according to Theorem 1, track down a subset of three topologies that is not tree-consistent. Our goal will be to detect all these local conflicts. This will be the preprocessing stage of the algorithm that will be described in Section 5, in order to (try to) "repair" the conflicts in a succeeding stage of the algorithm. We can find all these local conflicts in time $O(n^5)$ as follows. Since, following Lemma 1, only three topologies involving five taxa can form a local conflict, it suffices to consider all size five sets of taxa $\{a, b, c, d, e\} \subseteq S$. There are five quartets over this size five set of taxa, namely, $\{a, b, c, d\}$, $\{a, b, c, e\}$, $\{a, b, d, e\}$, $\{a, c, d, e\}$, and $\{b, c, d, e\}$. For the topologies of these quartets, we can test, in constant time, whether there are three among them that are not tree-consistent. Doing so for every size five set, we will, if $Q_S$ is not tree-consistent, certainly obtain a size three subset of $Q_S$ which is not tree-consistent. Moreover, from Lemma 2 we know that we find *all* these local conflicts in time $O(n^5)$.

We can improve this time bound for the preprocessing stage of the algorithm to be described in Section 5 with the following result by Bandelt and Dress [2]. They show that it is sufficient to restrict our attention to the size five sets containing some arbitrarily fixed taxon $f$.

**Proposition 2.** *(Proposition 6 in [2]) Given a set of taxa $S$, a complete set of quartet topologies $Q_S$, and some taxon $f \in S$, then $Q_S$ is tree-like iff every size five set of taxa which contains $f$ satisfies the substitution property.*

Following Proposition 2, we can select some arbitrary $f \in S$ and examine only the size five sets involving $f$. Similar to our procedure described above, we consider every such size five set containing $f$ separately. Among the topologies over this size five set, we search the size three sets which are not tree-consistent. If the set of quartet topologies $Q_S$ is not tree-consistent, we will find a size three set of quartet topologies which is not tree-consistent. Finding these local conflicts which involve $f$ can be done in time $O(n^4)$.

## 4   Combinatorial Characterization of Local Conflicts

Given three topologies, we need to decide whether they are tree-consistent or not. Directly using the definition of tree-consistency turns out to be a rather technical, troublesome task, since we have to reason whether or not a tree topology exists that induces the topologies. Similarly, it can be difficult to test, for the topologies, whether or not they contradict the substitution property. To make things less technical and easier to grasp, we subsequently give a useful combinatorial characterization of local conflicts, i.e., three topologies which are not tree-consistent. Note that in the following definition, we distinguish two possible *orientations* of a quartet topology $[ab|cd]$, namely, $[ab|cd]$, with $a, b$ on its left hand side and $c, d$ on its right hand side, and $[cd|ab]$, with the sides interchanged.

**Definition 1.** *Given a set of topologies where each of the topologies is assigned an orientation, let $l$ be the number of different taxa occurring in the left hand sides of the topologies and let $r$ be the number of different taxa occurring in the right hand sides of the topologies. The signature, then, is the pair $(l, r)$ that, over all possible orientations for these topologies, minimizes $l$.*

**Theorem 2.** *Three quartet topologies are not tree-consistent iff they involve five taxa and their signature is $(3, 4)$ or $(4, 4)$.*

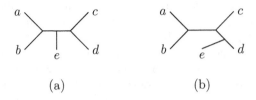

**Fig. 1.** Possible trees for $[ab|cd]$ and taxon $e$ in the proof of Theorem 2.

*Proof.* ($\Rightarrow$) We show that, given three topologies $t_1$, $t_2$, and $t_3$ which are not tree-consistent, they involve five taxa and have signature $(3, 4)$ or $(4, 4)$. From Lemma 2, we know that three topologies are not tree-consistent iff they contradict the substitution property. To recall, three topologies contradict the substitution property if, for one of these topologies, w.l.o.g., $t_1 = [ab|cd]$, neither the topology $t_2$ for quartet $\{a, b, c, e\}$ is $[ab|ce]$ nor the topology $t_3$ for quartet $\{a, c, d, e\}$ is $[ae|cd]$. Therefore, the topology $t_2$ is either $[ac|be]$ or $[ae|bc]$, and the topology $t_3$ is either $[ac|de]$ or $[ad|ce]$. By exhaustively checking the possible combinations, we can find that the topologies involve five taxa and their signature is $(3, 4)$ (e.g., for $t_2 = [ac|be]$ and $t_3 = [ac|de]$) or $(4, 4)$ (e.g., for $t_2 = [ac|be]$ and $t_3 = [ad|ce]$).

($\Leftarrow$) We are given three topologies, $t_1$, $t_2$, and $t_3$, involving five taxa and having signature $(3, 4)$ or $(4, 4)$. Assume that they are tree-consistent. Showing that this implies signature $(2, 3)$ or $(3, 3)$, we prove that the assumption is wrong. For tree-consistent $t_1, t_2$, and $t_3$, we can find a tree inducing them. With, w.l.o.g., taxa $\{a, b, c, d, e\}$ and $t_1 = [ab|cd]$, we mainly have two possibilities: we can attach the leaf $e$ on the middle edge of topology $t_1$, as shown in Figure 1(a), or we can attach $e$ on one of the four side branches of $t_1$, as exemplarily shown in Figure 1(b). Considering the sets of quartet topologies induced by these trees, we find, in each case, that the set has signature $(3, 3)$. For instance, the topologies induced by the tree in Figure 1(a) are, besides $t_1$, $[ab|ce]$, $[ab|de]$, $[ae|cd]$, and $[be|cd]$. Three topologies selected from these have signature $(3, 3)$ (e.g., $[ab|cd]$, $[ab|ce]$, and $[ae|cd]$) or $(2, 3)$ (e.g., $[ab|cd]$, $[ab|ce]$, and $[ab|de]$). $\qquad\square$

Using Theorem 2, we can determine whether three topologies are conflicting by simply counting the involved taxa and computing their signature.

## 5    Fixed Parameter Algorithm for MQI

In this section, we present a recursive algorithm solving MQI with parameter $k$. Before calling the recursive part for the first time, one has to build the list of size three sets of quartets whose topologies are not tree-consistent. The preparation of this *conflict list* is explained in Section 3. After that, we call the recursive procedure of the algorithm with argument $k$.

The recursive procedure selects a local conflict to branch on from the conflict list. This branching is done by changing one topology from the selected local conflict, updating the conflict list, and calling the recursive procedure with argument $k - 1$ on the thereby created subcases. We will later explain how to select and change the topologies when branching. After a topology $t$ has changed, the algorithm updates the conflict list as follows: It (1) removes the size three sets of quartets in the list whose topologies are now tree-consistent, and (2) adds the size three sets of quartets not in the list whose topologies now form a local conflict.

The recursion stops if no conflicts are left in the conflict list (we have found a solution), or if $k = 0$ (in case the conflict list is not empty, we did not find a solution in this branch of the search tree). When a solution is found, the algorithm outputs the current set of topologies, i.e., a complete set of quartet topologies that is tree-like and that can be obtained by altering at most $k$ topologies in the given set of topologies. From this tree-like set of quartet topologies, it is possible to derive the evolutionary tree in time $O(n^4)$ [4]. Thus scanning the whole search tree, we find *all* solutions that we can obtain by altering at most $k$ topologies.

**Running Time.** For establishing an upper bound on the running time, we consider the preprocessing, the update procedure, and the size of the search tree. The preprocessing can be done in time $O(n^4)$, as explained in Section 3.

Updating the conflict list can be done in time $O(n)$: Following Lemma 1, local conflicts can only occur among three topologies consisting of no more than five taxa. Therefore, having changed the topology of one quartet $\{a, b, c, d\}$, we only have to examine the "neighborhood" of the quartet, i.e., those sets of five taxa containing $a, b, c, d$. For every such set of five taxa, it can be examined in constant time whether for three topologies over the five taxa, a new conflict emerged, or whether an existing conflict has been resolved. Given taxa $a, b, c, d$, we have $n - 4$ choices for a fifth taxon. Thus, $O(n)$ is an upper bound for the update procedure.[3]

Now, we consider the search tree size. By a careful selection of subcases to branch into, we can find a way to make at most four recursive calls on an arbitrarily selected local conflict, i.e., for every three topologies which are not tree-consistent. Let $t_1$, $t_2$, and $t_3$ be three topologies which are not tree-consistent, and let, w.l.o.g., $t_1 = [ab|cd]$. Following Lemma 1, the topologies involve only one additional taxon, say $e$. Following Lemma 2, $t_1, t_2, t_3$ contradict the substitution property. Given $t_1 = [ab|cd]$, the substitution property requires topology $[ab|ce]$ or topology $[ae|cd]$. Therefore, we can, w.l.o.g., assume the following setting for three quartets contradicting the substitution property: Topology $t_1 = [ab|cd]$, topology $t_2$ is the topology for quartet $\{a, b, c, e\}$ different from $[ab|ce]$, and

---

[3] In fact, as explained in Section 3, we only consider sets of five species containing a designated taxon $f$. Therefore, if we change the topology of a quartet $\{a, b, c, d\}$ which does not contain the designated taxon $f$, then we only have to consider *one* set of five topologies, namely, $\{a, b, c, d, f\}$. In this special case, the update procedure can be done in time $O(1)$.

topology $t_3$ is a topology for quartet $\{a, c, d, e\}$ different from $[ae|cd]$. In order to change the three topologies to satisfy the substitution property, we have the following possibilities. We can change $t_1$; either (1) we change $t_1$ to $[ac|bd]$, or (2) we change $t_1$ to $[ad|bc]$. Otherwise, we can assume that $t_1$ is not changed. Then, we have to (3) change $t_2$ to $[ab|ce]$ or (4) change $t_3$ to $[ae|cd]$, because these are the only remaining possibilities to satisfy the substitution property. Since the height of the search tree is at most $k$, the preceding considerations justify an upper bound of $4^k$ on the exponential growth and yield the following theorem, which summarizes our findings.

**Theorem 3.** *The* MQI *problem can be solved in time* $O(4^k \cdot n + n^4)$.

Note that this running time is not only true for the algorithm reporting *one* solution, but also for reporting *all* evolutionary trees satisfying the requirement. Our algorithm has $O(kn^4)$ memory requirement, where the input size is already $O(n^4)$. The correctness of the algorithm follows easily from Theorem 1.

## 6    Improving the Running Time in Practice

Besides improving the worst case bounds on the algorithm's running time, we can also extend the algorithm in order to improve the running time in practice without affecting the upper bounds. In this section, we collect some ideas for such heuristic improvements.

**Fixing Topologies.** It does not make sense to change a topology which, at some previous level of recursion, has been altered, or for which we explicitly decided not to alter it. If we decide not to alter a topology in a later stage of recursion, we call this *fixing* the topology. This avoids redundant branchings in the search tree.

**Forcing Topologies to Change.** It might be possible to identify topologies which necessarily have to be altered in order to find a solution. We call this *forcing* a topology to change. The ideas described here are similar to those used in the so-called reduction to problem kernel for the 3-Hitting Set problem [13].

**Lemma 3.** *Consider an instance of the* MQI *problem in which quartet $q$ has topology $t$. If there are more than $3k$ distinct local conflicts which contain $t$ then, in a solution for this instance, the topology for $q$ is different from $t$.*

*Proof.* In Section 3, we showed that three topologies only can form a local conflict if there are not more than five taxa occurring in them (see Lemma 1). For five taxa, there are five quartets consisting of these taxa, e.g., for taxa $\{a, b, c, d, e\}$ the quartets are $\{a, b, c, d\}$, $\{a, b, c, e\}$, $\{a, b, d, e\}$, $\{a, c, d, e\}$, and $\{b, c, d, e\}$. Therefore, when given two quartet topologies $t_1$ and $t_2$, we make the following observations. If there are more than five taxa occurring in $t_1$ and $t_2$, they cannot form a conflict with a third topology. If there are exactly five taxa occurring in $t_1$ and $t_2$, then there are five quartets consisting of these five taxa, two of which are the quartets for $t_1$ and $t_2$. The remaining three topologies are the only possibilities for a topology $t_3$ that could form a conflict with $t_1$ and $t_2$.

Now, consider the situation in which, for a quartet topology $t$, we have more than $3k$ distinct local conflicts which contain $t$. From the preceding discussion, we know that for any $t'$, there are at most three topologies such that $t$ and $t'$ can form a conflict with it. Consequently, there must be more than $k$ distinct topologies $t'$ that occur in a local conflict with $t$. We show by contradiction that we have to alter topology $t$ to find a solution. Assume that we can find a solution while not altering $t$. By changing a topology $t'$, we can cover at most three conflicts, since there are at most three local conflicts containing both $t$ and $t'$. Therefore, by changing $k$ topologies, we can resolve at most $3k$ local conflicts. This contradicts our assumption and shows that we have to alter $t$ to find a solution.                                                                                    $\square$

**Recognizing Hopeless Situations.** Now, we describe situations in which, at some level in the search tree when we are allowed to alter at most $k$ topologies, we can recognize that we cannot find a solution. Thus, we can "cut off," i.e., omit, complete subtrees of the search tree.

Having a local conflict consisting only of fixed topologies, we obviously cannot resolve this conflict while not changing one of the fixed topologies. As another observation, we know that for a solution, we have to change the forced topologies. If after identifying these forced topologies, there are more than $k$ of them, it is obvious that a solution is not possible—already by changing these topologies, we would change more topologies than we are allowed to.

The following two lemmas contain more involved observations. Their proofs use similar ideas as used in the proof of Lemma 3 (see [10]). If a local conflict does not contain a topology which is forced to change, then we call it an *unforced local conflict*.

**Lemma 4.** *Let us have an instance of the* MQI *problem in which we have identified $p$ conflicts which are forced to change. If the number of unforced local conflicts is greater than $3(k - p)k$, then the instance has no solution.*

**Lemma 5.** *An instance of the* MQI *problem in which the number of local conflicts is greater than $6(n - 4)k$ has no solution.*

**Clever Branching.** Applying the rules described above will also significantly improve our situation when branching. For the general branching situation on a local conflict, we have shown in Section 5 that it is sufficient to branch into four subcases. Regarding topologies forced to change, we can, however, reduce the number of subcases. When we have identified a topology $t$ which is forced to change, it is sufficient to branch into two subcases: one for each alternative topology of $t$. Regarding fixed topologies, we can take advantage of local conflicts which contain fixed topologies. Having a local conflict with one or two fixed topologies, we omit the subcases which change a fixed topology. This will reduce the number of subcases to three, two, or even one subcase.

**Preprocessing by the Q\*-Method.** The algorithmic improvements described above do not sacrifice the guarantee to find the optimal solutions. Using these improvements, we will find every solution that we would find without them. This is not true for the following idea. We propose to use the Q\*-method described by Berry and Gascuel [4] as a preprocessing for our algorithm. The Q\*-method produces the maximum subset of the given quartet topologies that is tree-like. In the combined use with our algorithm, we fix these quartet topologies from the beginning. Therefore, our algorithm will compute the minimum number of quartet topologies we have to change in order to obtain a tree-like set of topologies that contains the topologies fixed by the Q\*-method. The tree we obtain will be a refinement of the tree reported by the Q\*-method which may contain unresolved branches. Thus, we cannot guarantee that the reported tree is the optimal solution for the MQI problem. On real data, however, it is the optimal tree with high certainty: Suppose it is not. Then there are four taxa $a,b,c$, and $d$ that are arranged in another way by the Q\*-method than they would be arranged in the optimal solution for the MQI problem. As we are working on a complete set of topologies, this would imply that there are at least $n-3$ quartets that would make the same wrong prediction for the arrangement of $a, b, c, d$: the quartet $\{a, b, c, d\}$ and, for all $e \in S - \{a, b, c, d\}$, one quartet over $\{a, b, c, d, e\}$ that involves $e$. On real data, this is very unlikely. Our experiments described in Section 8 support the conjecture that with the preprocessing by the Q\*-method, we find every solution that the MQI algorithm would find. Moreover, the experiments show that this enhancement allows us to process much larger instances than we could without using it.

## 7   Related Problems

We now come to some variants and generalizations of the basic MQI problem and their fixed parameter tractability. These variations arise in practice due to the fact that often quartet inference methods cannot non-ambiguously predict a topology for every quartet. Perhaps the most natural generalization of MQI is to consider weighted quartet topologies.

WEIGHTED MQI. Weights arise since a quartet inference method can predict the topology for a quartet with more or less certainty. Therefore, we can assign weights to the quartet topologies reflecting the certainty they are predicted with. Given a complete set of weighted topologies $Q_S$ and a positive integer $k$, we distinguish two different questions.

1. Assume that we are given a complete set of weighted topologies $Q_S$, with positive real weights, and a positive integer $k$. A binary tree is a candidate for a solution if the set of quartet topologies induced by this tree differs from

$Q_S$ in the topologies for at most $k$ quartets. Can we, among all candidate trees satisfying this property, find the one such that the topologies in $Q_S$ which are not induced by the tree have minimum total weight?

The algorithm in Section 5 can compute *all* solution trees. So, we can, without sacrificing the given time bounds, find this tree among the solution trees for which the "wrong" quartet topologies have minimal total weight.

2. Assume that we are given a complete set of weighted topologies $Q_S$, each topology having a real weight $\geq 1$, and a positive real $K$. Is there a binary tree such that the quartet topologies induced by the tree differ from the given topologies only for topologies having total weight less than $K$?

Again, we can use the algorithm presented in Section 5. When branching into different subcases, the time analysis of the algorithm relied on the fact that in each subcase at least one quartet topology is changed, i.e., added to the "wrong" topologies. In the current situation of weighted topologies with weights $\geq 1$, each subcase changes quartet topologies having a total weight of at least 1. The time analysis of our algorithm is, therefore, still valid and the time bounds remain the same.

Allowing arbitrarily small weights in question 2, the problem cannot be fixed parameter tractable, unless P = NP. To see this, take an instance of unweighted MQI with parameter $k$. We can turn this instance into an instance of weighted MQI by assigning all topologies weight $1/k$ and setting the parameter to 1. A fixed parameter algorithm for the problem with arbitrary weights $> 0$ would thus give a polynomial time solution for MQI, which contradicts the NP-completeness of MQI unless P = NP. Having, however, weights of size at least $\epsilon$ for some positive real $\epsilon$, the problem is fixed parameter tractable as we described here for the special case that $\epsilon = 1$ (similar to WEIGHTED VERTEX COVER in [14]).

*Underspecified* MQI. Due to lack of information or due to ambiguous results, a quartet inference method may not be able to compute a topology for every quartet, so there may be quartets for which no topology is given. Assuming a bounded number of quartets with missing topology, we formulate the problem as follows. Given a set $S$ of taxa, integers $k$ and $k'$, and a set of topologies $Q_S$, such that $Q_S$ contains quartet topologies for all quartets over $S$ except for $k'$ many. Then, we ask whether there is a binary tree such that the quartet topologies induced by the tree differ from the given topologies only for $k$ topologies.

The set of topologies is "underspecified" by $k'$ topologies. We can solve the problem as follows. Having three possible topologies for each quartet, we can, for a quartet without given topology, branch into three subcases, one for each of its three possible topologies. Having selected a topology for each such quartet, we run the algorithm from Section 5. The resulting algorithm has time complexity $O(3^{k'} \cdot 4^k \cdot n + n^5)$ and shows that the problem is fixed parameter tractable for parameters $k$ and $k'$. Note that for unbounded $k'$ this problem is NP-complete even for $k = 0$ [16] and, therefore, is not fixed parameter tractable.

We only briefly mention another variant of MQI, *Overspecified* MQI: In that problem, we are, compared to MQI, given an additional integer $k''$ and two topologies instead of one for $k''$ many quartets. For these quartets, we are free to choose one of the given topologies. In a similar way as for underspecified MQI, we can show that overspecified MQI is fixed parameter tractable for parameters $k$ and $k''$.

# 8    Experimental Evaluation

To investigate the usefulness and practical relevance of the algorithm for unweighted MQI, we performed experiments on artificial as well as on real data from fungi. The implementation of the algorithm was done using the programming language C. The algorithm contains the enhancements described in Section 6. The combined use with the Q*-method was, however, only applied when processing the fungi data, not when processing the artificial data. The reported tests were done on a LINUX PC with a Pentium III 750 MHz processor and 192 MB main memory.

## 8.1    Artificial Data

We performed experiments on artificially generated data in order to find out which kind of data sets our algorithm can be especially useful for. For a given number $n$ of taxa and parameter $k$, we produce a data file as follows. We generate a random evolutionary tree for $n$ taxa and derive the quartet topologies from that tree. Then, we change $k$ distinct, arbitrarily selected topologies in a randomly chosen way. This results in an MQI instance that certainly can be solved with parameter $k$. For each pair of values for $n$ and $k$, ten different data sets were created. The reported results are the average for test runs on ten data sets.

We experimented with different values of $n$ and $k$. As a measure of performance, we use two values: We report the processing time and, since processing time is heavily influenced by system conditions, e.g., memory access time in case of cache faults, also the search tree size. The search tree size is the number of the search trees nodes, both inner nodes and leaves, and it reflects the exponential growth of the algorithm's running time.

Figure 2(a) gives a table of results for different values of $n$ and $k$. Regarding the processing time, we note, on the one hand, the increasing time for fixed $n$ and growing $k$. On the other hand, we observe that for moderate values of $k$, we can process large instances of the problem, e.g., $n = 50$ and $k = 100$ in 40 minutes. For comparison of the algorithm's performance, consider the results reported by Ben-Dor $et\ al.$ [3], who solve MQI instances also giving guaranteed optimal results. They only report about processing up to 20 taxa and list, admittedly for a high number of erroneous topologies, a running time of 128 hours for this case (on a SUN Ultra-4 with 300 MHz).

In Figure 2(b) we compare, on a logarithmic scale, the theoretical upper bound of $4^k$ to the real size of the search tree. For each fixed number of taxa $n$, we give a graph displaying the growth of search tree size for increasing $k$. The search trees are, by far, smaller than the $4^k$ bound. This is mainly due to the practical improvements of the algorithm (see Section 6). We also note that for equal value of $k$, a higher number $n$ of taxa often results in a smaller search tree.

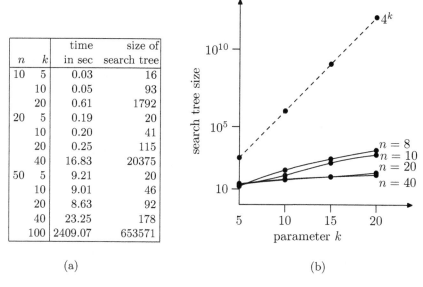

| $n$ | $k$ | time in sec | size of search tree |
|-----|-----|-------------|---------------------|
| 10  | 5   | 0.03        | 16                  |
|     | 10  | 0.05        | 93                  |
|     | 20  | 0.61        | 1792                |
| 20  | 5   | 0.19        | 20                  |
|     | 10  | 0.20        | 41                  |
|     | 20  | 0.25        | 115                 |
|     | 40  | 16.83       | 20375               |
| 50  | 5   | 9.21        | 20                  |
|     | 10  | 9.01        | 46                  |
|     | 20  | 8.63        | 92                  |
|     | 40  | 23.25       | 178                 |
|     | 100 | 2409.07     | 653571              |

(a)                                              (b)

**Fig. 2.** Comparing running time and search tree size for different values of $n$ and $k$.

## 8.2   Real Data

Using our algorithm, we analyzed the evolutionary relationships of species from the mushroom genus *Amanita*, a group that includes well-known species like the Fly Agaric and the Death Cap. The underlying data are an alignment of nuclear DNA sequences coding for the D1/D2 region of the ribosomal large subunit (alignment length 576) from *Amanita* species and one outgroup taxon, as used by Weiß *et al.* [18]. We inferred the quartet topologies by (1) using dnadist from the Phylip package [9] to compute pairwise distances with the maximum likelihood metric, and (2) using distquart from the Phyloquart package [4] to infer quartet topologies based on the distances.

The analysis was done by a preprocessing of the data using the Q*-method, also taken from the Phyloquart package. Experiments on small instances, e.g., 10 taxa, show that all solutions we find without using the Q*-method are also found when using it. Using the Q*-method, however, results in a significant speed-up of the processing. Figure 3(a) shows this impact for small numbers of *Amanita* species. Note, however, that the speed-up heavily depends on the data. In Figure 3(a) and in the following, we neglect the time needed for the preprocessing by the Q*-method, which is, e.g., 0.11 seconds for $n = 12$.

We processed a set of $n = 22$ taxa in 35 minutes. The resulting tree was rooted using the outgroup taxon *Limacella glioderma* and is displayed in Figure 3(b). We found the best solution for $k = 979$ for the given 7315 quartet topologies. The Q*-method had fixed 41 percent of the quartet topologies in advance. Considering the tree, the grouping of taxa is consistent with the grouping

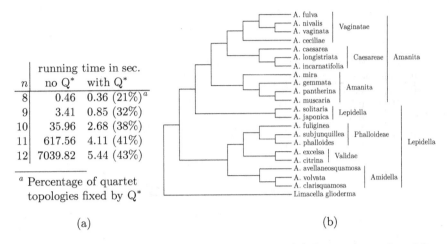

| | running time in sec. | |
|---|---|---|
| $n$ | no $Q^*$ | with $Q^*$ |
| 8 | 0.46 | 0.36 (21%)[a] |
| 9 | 3.41 | 0.85 (32%) |
| 10 | 35.96 | 2.68 (38%) |
| 11 | 617.56 | 4.11 (41%) |
| 12 | 7039.82 | 5.44 (43%) |

[a] Percentage of quartet
    topologies fixed by $Q^*$

(a)                                    (b)

**Fig. 3.** (a) Speed-up when using $Q^*$ preprocessing. (b) Optimal tree found for a set of 21 *Amanita* species and one outgroup taxon; indicated is the grouping of *Amanita* species into 7 sections and 2 subgenera.

into seven sections supported by Weiß *et al.* [18], who used the distance method neighbor joining, heuristic parsimony methods, and maximum likelihood estimations. Particularily, our grouping is nearly identical to the topology revealed by Weiß *et al.* using maximum likelihood estimation. This topology is well compatible with classification concepts based on morphological characters, e.g., the sister group relationship of sections *Vaginatae* and *Caesareae*, and the monophyly of subgenus *Amanita*.

One might hope that quality of quartet inference techniques will improve in the future. This would lead to instances requiring smaller values of $k$.

## 9   Conclusion

We showed that the Minimum Quartet Inconsistency problem can be solved in worst case time $O(4^k n + n^4)$ when parameter $k$ is the number of faulty quartet topologies. This means that the problem is fixed parameter tractable. Several ideas for tuning the algorithm show that the practical performance of the algorithm is much better that the theoretical bound given above. This is clearly expressed by our experimental results. Note that there is an ongoing discussion about the usefulness of quartet methods: St. John *et al.* [15] give a rather critical exposition of the practical performance of quartet methods (in particular, quartet puzzling) in comparison with the neighbor joining method, which is in opposition to results reported by Strimmer and v. Haeseler [17].

Concerning future work, we want to extend our experiments to weighted quartet topologies and to other data. Also, the fact that we can obtain all optimal and near-optimal solutions and the usefulness of this deserves further investigation. From a parameterized complexity point of view, it remains an open question to find a so-called reduction to problem kernel (see [1,7,8] for details). The further reduction of the tree size concerning theoretical, as well as experimental bounds, is a worthwhile future challenge.

## Acknowledgment

We are grateful to Michael Weiß from the Biology Department (Systematic Botany and Mycology group), Universität Tübingen, for providing us with the *Amanita* data, supporting us in the interpretation of our results, and many constructive remarks improving the presentation significantly.

# References

1. J. Alber, J. Gramm, and R. Niedermeier. Faster exact solutions for hard problems: a parameterized point of view. *Discrete Mathematics*, 229(1-3):3–27, 2001.
2. H.-J. Bandelt and A. Dress. Reconstructing the shape of a tree from observed dissimilarity data. *Advances in Applied Mathematics*, 7:309–343, 1986.
3. A. Ben-Dor, B. Chor, D. Graur, R. Ophir, and D. Pelleg. Constructing phylogenies from quartets: elucidation of eutherian superordinal relationships. *Journal of Computational Biology*, 5:377–390, 1998.
4. V. Berry and O. Gascuel. Inferring evolutionary trees with strong combinatorial evidence. *Theoretical Computer Science*, 240:271–298, 2000. Software available through `http://www.lirmm.fr/~vberry/PHYLOQUART/phyloquart.html`.
5. P. Buneman. The recovery of trees from measures of dissimilarity. In Hodson *et al.*, eds, *Anglo-Romanian Conference on Mathematics in the Archaeological and Historical Sciences,* pages 387–395, 1971. Edinburgh University Press.
6. B. Chor. From quartets to phylogenetic trees. In *Proceedings of the 25th SOFSEM*, number 1521 in LNCS, pages 36–53, 1998. Springer.
7. R. G. Downey and M. R. Fellows. *Parameterized Complexity*. 1999. Springer.
8. M. R. Fellows. Parameterized complexity: new developments and research frontiers. In Downey and Hirschfeldt, eds, *Aspects of Complexity,* to appear, 2001. De Gruyter.
9. J. Felsenstein. PHYLIP (Phylogeny Inference Package) version 3.5c. Distributed by the author. Department of Genetics, University of Washington, Seattle. 1993. Available through `http://evolution.genetics.washington.edu/phylip`.
10. J. Gramm and R. Niedermeier. Minimum Quartet Inconsistency is fixed parameter tractable. Technical Report WSI-2001-3, WSI für Informatik, Universität Tübingen, Fed. Rep. of Germany, January 2001. Report available through `http://www-fs.informatik.uni-tuebingen.de/~gramm/publications`.
11. T. Jiang, P. Kearney, and M. Li. Some open problems in computational molecular biology. *Journal of Algorithms*, 34:194–201, 2000.
12. T. Jiang, P. Kearney, and M. Li. A polynomial time approximation scheme for inferring evolutionary trees from quartet topologies and its application. To appear in *SIAM Journal on Computing*, 2001.
13. R. Niedermeier and P. Rossmanith. An efficient fixed parameter algorithm for 3-Hitting Set. Technical Report WSI-99-18, WSI für Informatik, Universität Tübingen, October 1999. To appear in *Journal of Discrete Algorithms*.

14. R. Niedermeier and P. Rossmanith. On efficient fixed parameter algorithms for Weighted Vertex Cover. In *Proceedings of the 11th International Symposium on Algorithms and Computation*, number 1969 in LNCS, pages 180–191, 2000. Springer.

15. K. St. John, T. Warnow, B.M.E. Moret, and L. Vawter. Performance study of phylogenetic methods: (unweighted) quartet methods and neighbor-joining. In *Proceedings of the 12th ACM-SIAM Symposium on Discrete Algorithms*, pages 196–205, 2001. SIAM Press.

16. M. Steel. The complexity of reconstructing trees from qualitative characters and subtrees. *Journal of Classification*, 9:91–116, 1992.

17. K. Strimmer and A. von Haeseler. Quartet puzzling: a quartet maximum-likelihood method for reconstructing tree topologies. *Molecular Biology and Evolution*, 13(7):964–969, 1996.

18. M. Weiß, Z. Yang, and F. Oberwinkler. Molecular phylogenetic studies in the genus Amanita. *Canadian Journal of Botany*, 76:1170–1179, 1998.

# Optimally Compact Finite Sphere Packings — Hydrophobic Cores in the FCC

Rolf Backofen and Sebastian Will*

Institut für Informatik, LMU München
Oettingenstraße 67, D-80538 München
{backofen,wills}@informatik.uni-muenchen.de

**Abstract.** Lattice protein models are used for hierarchical approaches to protein structure prediction, as well as for investigating principles of protein folding. The problem is that there is so far no known lattice that can model real protein conformations with good quality, *and* for which there is an efficient method to prove whether a conformation found by some heuristic algorithm is optimal. We present such a method for the FCC-HP-Model [3]. For the FCC-HP-Model, we need to find conformations with a maximally compact hydrophobic core. Our method allows us to enumerate maximally compact hydrophobic cores for sufficiently great number of hydrophobic amino-acids. We have used our method to prove the optimality of heuristically predicted structures for HP-sequences in the FCC-HP-model.

## 1 Introduction

The protein structure prediction is one of the most important unsolved problems of computational biology. It can be specified as follows: Given a protein by its sequence of amino acids, what is its native structure? NP-completeness of the problem has been proven for many different models (including lattice and off-lattice models) [8,9]. These results strongly suggest that the protein folding problem is NP-hard in general. Therefore, it is unlikely that a general, efficient algorithm for solving this problem can be given. Actually, the situation is even worse, since the general principles why natural proteins fold into a native structure are unknown. This is cumbersome since rational design is commonly viewed to be of paramount importance e.g. for drug design, where one faces the difficulty to design proteins that have a unique and stable native structure.

To tackle structure prediction and related problems simplified models have been introduced. They are used in hierarchical approaches for protein folding (e.g., [21], see also the meeting review of CASP3 [15], where some groups have used lattice models). Furthermore, they have became a major tool for investigating general properties of protein folding.

Most important are the so-called lattice models. The simplifications commonly used in this class of models are: 1) monomers (or residues) are represented

---

* Supported by the PhD programme "Graduiertenkolleg Logik in der Informatik" (GKLI) of the "Deutsche Forschungsgemeinschaft" (DFG).

A. Amir and G.M. Landau (Eds.): CPM 2001, LNCS 2089, pp. 257–271, 2001.

using a unified size 2) bond length is unified 3) the positions of the monomers are restricted to lattice positions and 4) a simplified energy function.

In the literature, many different lattice models (i.e., lattices and energy functions) have been used. Examples of how such models can be used for predicting the native structure or for investigating principles of protein folding were given [20,1,11,19,12,2,17,21]. Of course, the question arises which lattice and energy functions has to be preferred. There are two (somewhat conflicting) aspects that have to be evaluated when choosing a model: 1) the accuracy of the lattice in approximating real protein conformations, and the ability of the energy function to discriminate native from non-native conformations, and 2) the availability and quality of search algorithm for finding minimal (or nearly minimal) energy conformations.

While the first aspect is well-investigated in the literature (e.g., [18,10]), the second aspect is underrepresented. By and large, there are mainly two different heuristic search approaches used in the literature: 1) Ad hoc restriction of the search space to compact or quasi-compact conformations (a good example is [20], where the search space is restricted to conformations forming an $n \times n \times n$-cube). The main drawback here is that the restriction to compact conformation is not biologically motivated for a complete amino acid sequence (as done in these approaches), but only for the hydrophobic amino acids. In consequence, the restriction either has to be relaxed and then leads to an inefficient algorithm or is chosen to strong and thus may exclude optimal conformations. 2.) Stochastic sampling like Monte Carlo methods with simulated annealing, genetic algorithms etc. Here, the degree of optimality for the best conformations and the quality of the sampling cannot be determined by state of the art methods.[1]

On the other hand, there are only three exact algorithms known [23,4,6] which are able to enumerate minimal (or nearly minimal) energy conformations, all for the cubic lattice. However, the ability of this lattice to approximate real protein conformations is poor. For example, [3] pointed out especially the parity problem in the cubic lattice. This drawback of the cubic lattice is that every two monomers with chain positions of the same parity cannot form a contact.

In this paper, we follow the proposal by [3] to use a lattice model with a simple energy function, namely the HP (hydrophobic-polar) model, but on a better suited lattice (namely the face-centered cubic). There are two reasons for this approach:

1) The FCC can model real protein conformations with good quality (see [18], where it was shown that FCC can model protein conformations with coordinate root mean square deviation below 2 Å)

2) The HP-model models the important aspect of hydrophobicity. Essentially it is a polymer chain representation (on a lattice) with one stabilizing interaction each time two hydrophobic residues have unit distance. This enforces compactification while polar residues and solvent is not explicitly regarded. It follows the

---

[1] Despite there are mathematical treatments of Monte Carlo methods with simulated annealing, the partition function of the ensemble (which is needed for a precise statement) is in general unknown.

assumption that the hydrophobic effect determines the overall configuration of a protein (for a definition of the HP-model, see [16,10]).

Once a search algorithm for minimal energy conformations is established for this *FCC-HP-model*, one can employ it as a filter step in an hierarchical approach. This way, one can improve the energy function to achieve better biological relevance and go on to resemble amino acid positions more accurately.

*Contribution of the Paper.* In this paper, we present the first algorithm for enumerating maximal compact hydrophobic cores in the face-centered cubic lattice. For a given conformation of the FCC-HP-model, the hydrophobic core is the set of all positions occupied by hydrophobic (H) monomers. A hydrophobic core is maximally compact if the number of contacts between neighbored positions is maximized. Thus, a conformation which has a maximally compact hydrophobic core has minimal energy in the HP-model.

There are mainly two applications of the algorithm for finding hydrophobic cores. The first is that it provides a method to check minimality of conformations found by an heuristic algorithm. We have used an heuristic algorithm described earlier [7]. For the first time, we were able to find minimal energy conformations (and to prove their optimality) for HP-sequences in the FCC-HP-model. So far, the only known results for the FCC-HP-models were approximation results with an guaranteed ratio of 60% ([3], [13] provides a general approximation scheme for HP-models on arbitrary lattices; [14] gives an approximation scheme for the HP-model on the cubic lattice).

The second application is that the hydrophobic cores are a promising intermediate step for an algorithm to enumerate *all* minimal energy conformations. This technique has already been used successfully in [23].

## 2    Preliminaries

For a vector $p$, we denote with $p_x$ (resp. $p_y$ or $p_z$) its $x$-coordinate (resp. $y$- or $z$-coordinate). We use a transformed representation of the FCC-lattice (for a detailed description, see [5]. We define the FCC-isomorphic lattice $D_3'$ to be the lattice that consists of the following sets of points: $D_3' = \{ \begin{pmatrix} x \\ y \\ z \end{pmatrix} \mid \begin{pmatrix} x \\ y \\ z \end{pmatrix} \in \mathbb{Z}^3 \text{ and } x \text{ even} \} \uplus \{ \begin{pmatrix} x \\ y+0.5 \\ z+0.5 \end{pmatrix} \mid \begin{pmatrix} x \\ y \\ z \end{pmatrix} \in \mathbb{Z}^3 \text{ and } x \text{ odd} \}$. The first set consist of the points in even x-layers, the second of the points in odd x-layers. The set $N_{D_3'}$ of *minimal vectors* connecting neighbors in $D_3'$ is given by $N_{D_3'} = \{ \begin{pmatrix} 0 \\ \pm 1 \\ 0 \end{pmatrix}, \begin{pmatrix} 0 \\ 0 \\ \pm 1 \end{pmatrix} \} \uplus \{ \begin{pmatrix} \pm 1 \\ \pm 0.5 \\ \pm 0.5 \end{pmatrix} \}$. The vectors in the second set are the vectors connecting neighbors in two successive x-layers. Two points $p$ and $p'$ in $D_3'$ are *neighbors* if $p - p' \in N_{D_3'}$.

A *coloring* is a function $f : D_3' \to \{0,1\}$, where $f^{-1}(1) \neq \emptyset$. We will identify a coloring $f$ with the set of all *points colored by $f$*, i.e. $\{ p \mid f(p) = 1 \}$. Hence, for colorings $f_1, f_2$ we will use standard set notation for size $|f_1|$, union $f_1 \cup f_2$, disjoint union $f_1 \uplus f_2$, and intersection $f_1 \cap f_2$. Given a coloring $f$, we define the *number of contacts of $f$* by $\mathrm{con}(f) := \frac{1}{2} |\{ (p, p') \mid f(p) \wedge f(p') \wedge (p - p') \in N_{D_3'} \}|$.

A coloring $f$ is called a *coloring of the plane* $x = c$ if $f(x, y, z) = 1$ implies $x = c$. We say that $f$ is a *plane coloring* if there is a $c$ such that $f$ is a coloring of plane $x = c$. We define $\mathrm{Surf}_{pl}(f)$ to be the surface of $f$ in the plane $x = c$, i.e. $\mathrm{Surf}_{pl}(f) = |\{(\boldsymbol{p}, \boldsymbol{p}') \mid (\boldsymbol{p} - \boldsymbol{p}') \in N_{D_3'} \wedge f(\boldsymbol{p}) \wedge \neg f(\boldsymbol{p}') \wedge \boldsymbol{p}'_x = c\}$. With $\min_x(f)$ we denote the integer $\min\{\boldsymbol{p}_x \mid \boldsymbol{p} \in f\}$. $\max_x(f)$, $\min_y(f)$, $\max_y(f)$, $\min_z(f)$ and $\max_z(f)$ are defined analogously.

## 3   Description of the Method

Our aim is to determine maximally compact hydrophobic cores. A *hydrophobic core* is just a coloring $f$. A *maximally compact* hydrophobic core for $n$ points is a coloring $f$ of $n$ points that maximizes $\mathrm{con}(f)$. Without loss of generality, we can assume that $\min_x(f) = 1$. Let $k = \max_x(f)$. Then, we partition $f$ into plane colorings $f_1, \ldots, f_k$ of the layers $x = 1, \ldots, x = k$. For searching a maximal coloring $f$, we do a branch-and-bound search on $k$ and $f_1 \ldots f_k$.

Of course, the problem is to give good bounds that allow us to cut off many $k$ and $f_1 \ldots f_k$ that will not maximize $\mathrm{con}(f_1 \uplus \ldots \uplus f_k)$. For this purpose, we distinguish between contacts in a single layer ($= \mathrm{con}(f_i)$ for $1 \leq i \leq k$), and interlayer contacts $\mathrm{IC}_{f_i}^{f_{i+1}}$ for $1 \leq i < k$ between two successive layers (i.e., pairs $(\boldsymbol{p}, \boldsymbol{p}')$ such that $\boldsymbol{p}$ and $\boldsymbol{p}'$ are neighbors, $\boldsymbol{p} \in f_i$ and $\boldsymbol{p}' \in f_{i+1}$). We then give two different bounds on the layer and interlayer contacts, provided some parameters restricting the $f_i$'s.

For every plane coloring $f_i$, these parameters are the size $n_i$ of $f_i$, the number $a_i$ of rows that contain a point of $f_i$, and the number $b_i$ of columns that contain a point of $f_i$. Given these parameters, it is known [23] that the layer contacts of $f_i$ are given by $2n_i - a_i - b_i$. In this paper, we present for *any set of parameters* $n_i, a_i, b_i$ and $n_{i+1}$ an upper bound on the number of interlayer

$$\text{contacts } B_{n_i, a_i, b_i}^{n_{i+1}} \geq \max \left\{ \mathrm{IC}_{f_i}^{f_{i+1}} \,\middle|\, \begin{array}{l} f_i \text{ satisfies } n_i, a_i, b_i \\ \text{and } |f_{i+1}| = n_{i+1}. \end{array} \right\}$$

So far, the only related bound was given in our own work [5]. Although there a bound $B_{n_i, a_i, b_i}^{n_{i+1}}$ was given, this bound does not hold for arbitrary sets of parameters $n_i, a_i, b_i$ and $n_{i+1}$. Instead, the bound is valid for sufficiently filled plane colorings (called normal), which was sufficient for the purpose of [5].

The bound $B_{n_i, a_i, b_i}^{n_{i+1}}$ is used in searching for a maximally compact core for $n$ H-monomers as follows. Instead of directly enumerating $k$ and all possible colorings $f_1 \uplus \ldots \uplus f_k$, we search through all possible sequences of parameters $((n_1, a_1, b_1) \ldots (n_k, a_k, b_k))$ with the property that $n = \sum_i n_i$. By using the $B_{n_i, a_i, b_i}^{n_{i+1}}$, only a few layer sequences have to be considered further. For these optimal layer sequences, we then search for all admissible colorings $f_1 \uplus \ldots \uplus f_k$.

For calculating the bound $B_{n_i, a_i, b_i}^{n_{i+1}}$, we need to introduce additional parameters, namely the number of non-overlapping and unconnected rows in layer $x = i$. These additional parameters allow us to determine the maximal number of interlayer contacts between layer $x = i$ and $x = i + 1$. Further note that only few combinations of $(n_i, a_i, b_i)$ and these additional parameters are admissible. Thus, for every $(n_i, a_i, b_i)$, we search through all admissible numbers of non-overlapping rows in layer $x = i$ to determine $B^{n_{i+1}}_{n_i, a_i, b_i}$.

In Section 4, we define the parameters of a plane coloring and determine which combinations of parameters are admissible. In Section 5, the number of interlayer contacts is given provided the parameters and the number of points with three interlayer contacts, called 3-points, is fixed. In the following section, we determine the number of 3-points that maximizes the interlayer contacts.

# 4  Properties of Overlapping and Non-overlapping Colorings

Let $f$ be a coloring of plane $x = c$. A *horizontal caveat in $f$* is a k-tuple of points $(\boldsymbol{p_1}, \ldots, \boldsymbol{p_k})$ such that $\forall 1 \leq j < k : ((\boldsymbol{p_{j+1}} - \boldsymbol{p_j})_y = 1)$, $\{\boldsymbol{p_1}, \boldsymbol{p_k}\} \in f$ and $\forall 1 < j < k : \boldsymbol{p_j} \notin f$. A *vertical caveat in $f$* is defined analogously satisfying $\forall 1 \leq j < k : ((\boldsymbol{p_{j+1}} - \boldsymbol{p_j})_z = 1)$ instead. We say that $f$ contains a *caveat* if there is at least one horizontal or vertical caveat in $f$. $f$ is called *caveat-free* if it does not contain a caveat. We will handle only caveat-free colorings. The methods can be extended to treat caveats as well, but we suppress them for simplicity.

We now introduce the parameters of a plane coloring $f$ that will allows us to determine layer *and* to bound interlayer contacts. The first set of parameters are the rows and columns occupied by $f$. For an arbitrary plane coloring $f$ of $x = c$ define $\mathrm{occz}(f, z) := \exists y : f(c, y, z)$ and $\mathrm{occy}(f, y) := \exists z : f(c, y, z)$. Furthermore, we define $\mathrm{oylines}(f) := \left| \{ y | \mathrm{occy}(f, y) \} \right|$ and $\mathrm{ozlines}(f) := \left| \{ z | \mathrm{occz}(f, z) \} \right|$. For notational convenience define $\mathrm{olines}(f) := (\mathrm{oylines}(f), \mathrm{ozlines}(f))$. For a coloring $f$, we call rows $z$, where $\mathrm{occz}(f, z)$ holds, and columns $y$, where $\mathrm{occy}(f, y)$, *occupied*, and *unoccupied* otherwise.

For a plane coloring $f$, we define the *layer contacts* $\mathrm{LC}_f$ to be $\mathrm{con}(f)$. We define

$$\mathrm{LC}_{n,a,b} := \max \left\{ \mathrm{LC}_f \middle| \begin{array}{l} f \text{ is a coloring of plane } x = c \\ \wedge f \text{ has lines } (a, b) \wedge |f| = n \end{array} \right\}.$$

**Proposition 1.** *For every caveat-free coloring $f$ with $\mathrm{olines}(f) = (a, b)$, we get $\mathrm{LC}_{n,a,b} = 2n - \frac{1}{2}\mathrm{Surf}_{pl}(f)$ and $\mathrm{Surf}_{pl}(f) = 2(a + b)$.*

**Proof (Sketch).** Each of the $n$ points colored by $f$ has 4 neighbors, which are either occupied by another point, or by a surface point. Hence, $4n = 2\mathrm{LC}_{n,a,b} + \mathrm{Surf}_{pl}(f)$. For the second claim, note that by definition, every occupied row and column must generate 2 surface contacts, and, by caveat-free, there can be no more than 2. $\qquad \square$

The second set of parameters are the number of unconnected and non-overlapping rows. Let $f$ be a coloring of plane $x = c$. We define a row $z$ to be *non-overlapping in $f$* if $z$ is occupied, there is an occupied row $z' > z$, and there is no $y$ such that $f(c, y, z) \wedge f(c, y, z + 1)$. A row $z$ is called *unconnected* if it is non-overlapping and not $\exists y, y' : f(c, y, z) \land f(c, y', z+1) \land |y - y'| \leq 1$. The number of non-overlapping rows is denoted by #non-overlaps($f$) and the number of unconnected rows by #non-connects($f$).

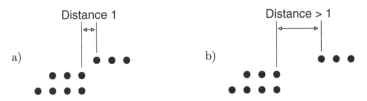

**Fig. 1.** a) Non-overlapping vs. b) unconnected.

To illustrate the terms, Figure 1a) shows a coloring with #non-overlaps($f$) = 1 and #non-connects($f$) = 0, whereas the coloring in Figure 1b) satisfies that #non-overlaps($f$) = 1 and #non-connects($f$) = 1.

We will call a coloring $f$ with #non-overlaps($f$) = 0 *overlapping* (otherwise *non-overlapping*). A coloring with #non-connects($f$) = 0 is called *connected* (otherwise *unconnected*).

In the rest of this section, we give precise bounds on the number of colored points, given the parameters of the plane coloring. We will first state some properties of colorings with respect to olines($f$), #non-overlaps($f$) and #non-connects($f$).

**Proposition 2.** *For every caveat-free coloring f, we have* $|f| \geq \max(\text{olines}(f))$.

Since by definition the maximal occupied row $z$ can not be non-overlapping we immediately get that #non-overlaps($f$) is less than oylines($f$). The next lemma states in addition that #non-overlaps(f) is less than ozlines($f$). Intuitively, this is a consequence of the (non-trivial) fact that every non-overlapping row produces exactly one non-overlapping column.

**Lemma 1.** *For a caveat-free coloring f, we get*

$$\#\text{non-overlaps}(f) < \min(\text{olines}(f)).$$

A caveat-free coloring can be split at non-overlapping rows into sub-colorings with the nice property that the parameters of the coloring can be calculated from the sub-colorings in a simple way. This fact will be employed for inductive arguments. Given a plane coloring $f$ and a row $\min_z(f) \leq z_s < \max_z(f)$, we define $f_{\theta z_s} = \{(c, y, z) \in f \mid z \theta z_s\}$ for $\theta \in \{\leq, >\}$. Note that the restriction on $z_s$ is required, since splitting at row $z_s = \max_z(f)$ would produce an empty sub-coloring $f_{>z_s}$. Further note that this restriction is trivially satisfied by any non-overlapping row.

**Lemma 2 (Split).** *Let f be a caveat-free coloring of the plane* $x = c$ *with* #non-overlaps($f$) $\geq 1$, *and let* $z_s$ *be a non-overlapping row. Then,*

*1.* $f = f_{\leq z_s} \uplus f_{>z_s}$ *and the sub-colorings* $f_{\leq z_s}$ *and* $f_{>z_s}$ *are caveat-free*

2. $\mathrm{olines}(f) = (\mathrm{oylines}(f_{\leq z_s}) + \mathrm{oylines}(f_{>z_s}), \mathrm{ozlines}(f_{\leq z_s}) + \mathrm{ozlines}(f_{>z_s}))$
3. $\#\mathrm{non\text{-}overlaps}(f) = \#\mathrm{non\text{-}overlaps}(f_{\leq z_s}) + \#\mathrm{non\text{-}overlaps}(f_{>z_s}) + 1.$

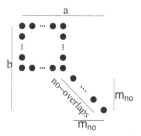

**Fig. 2.** Coloring with maximal number of elements.

There is a dependency of the admissible numbers $\#\mathrm{non\text{-}overlaps}(f)$ and the number of elements in a coloring $f$, given the number of occupied lines in $y$ and $z$ direction. Think of $(a, b)$ (resp. $\mathrm{m_{no}}$) as representing $\mathrm{olines}(f)$ (resp. $\#\mathrm{non\text{-}overlaps}(f)$). We define $n_{\max}(a, b, \mathrm{m_{no}}) := \mathrm{m_{no}} + (a - \mathrm{m_{no}})(b - \mathrm{m_{no}})$ and $n_{\min}(a, b, \mathrm{m_{no}}) := a + b - 1 - \mathrm{m_{no}}$. The idea of the definition of $n_{\max}(a, b, \mathrm{m_{no}})$ is that the number of elements is maximized if we have one big overlapping region and waste as little space as possible for the non-overlapping region. Hence, in this maximal coloring, all of the non-overlapping rows contain exactly one point. Such a coloring is shown in Figure 2.

**Lemma 3.** *All caveat-free colorings $f$ satisfy $|f| \leq n_{\max}(a, b, \mathrm{m_{no}})$, where $\mathrm{m_{no}} = \#\mathrm{non\text{-}overlaps}(f)$ and $(a, b) = \mathrm{olines}(f)$.*

**Lemma 4.** *For all caveat-free colorings $f$ holds $n_{\min}(a, b, \mathrm{m_{no}}) \leq |f|$, where $(a, b) = \mathrm{olines}(f)$ and $\mathrm{m_{no}} = \#\mathrm{non\text{-}overlaps}(f)$.*

**Proof (Sketch).** For the case $\mathrm{m_{no}} = 0$, a coloring $f$ of plane $x = c$ with minimal number of points and $\mathrm{olines}(f) = (a, b)$ is given by the coloring that has $b$ points $(c, 1, 1) \ldots (c, 1, b)$ in the column $y = 1$, and $a$ points $(c, 1, b) \ldots (c, a, b)$ in the row $z = b$. Clearly, $f$ has $a + b - 1$ points since $(c, 1, b)$ is in the first column *and* last row. For $\mathrm{m_{no}} > 0$, the claim follows by induction using the split lemma 2, Claim 3. $\qquad\square$

For convenience, we define the following bounds on the number of non-overlapping rows:

$$\mathrm{no_{min}}(n, a, b) := \min\{\mathrm{m_{no}} \mid 0 \leq \mathrm{m_{no}} \leq \min(a, b) - 1 \wedge n \geq n_{\min}(a, b, \mathrm{m_{no}})\}$$

$$\mathrm{no_{max}}(n, a, b) := \max\{\mathrm{m_{no}} \mid 0 \leq \mathrm{m_{no}} \leq \min(a, b) - 1 \wedge n \leq n_{\max}(a, b, \mathrm{m_{no}})\}$$

**Proposition 3.** *For any caveat-free coloring $f$ with $\mathrm{olines}(f) = (a, b)$ and $|f| = n$ holds $\mathrm{no_{min}}(n, a, b) \leq \#\mathrm{non\text{-}overlaps}(f) \leq \mathrm{no_{max}}(n, a, b)$.*

## 5   Number of $i$-Points for Cavent-Free Colorings

In the next two sections, we will provide a bound on interlayer contacts. For this purpose, we calculate for a coloring $f$ of plane $c$ the numbers of points having 4,3,2, and 1 contacts to $f$ (in the following called $i$-points). Theorem 1 will state that we can achieve the maximal number of interlayer contacts between $x = c$ and $x = c + 1$ if we fill the 4-points first, then (if points are left) the 3-points and so on. Before, we need some definitions and auxiliary lemmata.

In the following, let $f$ be a plane coloring of plane $x = c$ and $f'$ a plane coloring of plane $x = c'$, where $c \neq c'$. We define the *number of interlayer contacts of $f$ and $f'$* by $\mathrm{IC}_f^{f'} = \mathrm{con}(f \uplus f') - \mathrm{LC}_f - \mathrm{LC}_{f'}$. We define contacts$_{\max}(f, n)$ as

$$\max \left\{ \mathrm{IC}_f^{f'} \,\middle|\, f' \text{ is a plane coloring of } x = c + 1 \text{ with } |f'| = n \right\}.$$

A point $\boldsymbol{p}$ is called a *4-point for $f$* if $\boldsymbol{p}$ is in plane $x = c+1$ or $x = c-1$ and $\boldsymbol{p}$ has 4 neighbors $\boldsymbol{p_1}, \dots, \boldsymbol{p_4} \in f$. Analogously, we define 3-points, 2-points and 1-points. Furthermore, we define $\#4_{c-1}(f) = |\{\boldsymbol{p} \mid \boldsymbol{p}$ 4-point for $f$ in $x = c - 1\}|$. Analogously, we define $\#4_{c+1}(f)$ and $\#i_{c\pm1}(f)$ for $i = 1, 2, 3$. We will show that the number of $i$-points for every $i \in \{1, 2, 3, 4\}$ depend only on the number of non-overlaps, the number of non-connects, and the number of x-steps. An *x-step for a plane coloring $f$* is a triple $(\boldsymbol{p_1}, \boldsymbol{p_2}, \boldsymbol{p_3})$ such that $f(\boldsymbol{p_1}) = 0$, $f(\boldsymbol{p_2}) = 1 = f(\boldsymbol{p_3})$, $\boldsymbol{p_1} - \boldsymbol{p_2} = \pm \left(\begin{smallmatrix}0\\1\\0\end{smallmatrix}\right)$ and $\boldsymbol{p_1} - \boldsymbol{p_3} = \pm \left(\begin{smallmatrix}0\\0\\1\end{smallmatrix}\right)$. With xsteps$(f)$ we denote the number of x-steps of $f$. Now we can define the number of $i$-points, depending on $n = |f|$, $s = \mathrm{Surf}_{pl}(f)$, $m_x = \mathrm{xsteps}(f)$, $m_{no} = \#\text{non-overlaps}(f)$ and $m_{nc} = \#\text{non-connects}(f)$:

$$\#4\left(\begin{smallmatrix}n,s\\m_{no},m_{nc},m_x\end{smallmatrix}\right) = n - \frac{1}{2}s + 1 + m_{no}$$

$$\#2\left(\begin{smallmatrix}n,s\\m_{no},m_{nc},m_x\end{smallmatrix}\right) = s - 4 - 2\#3\left(\begin{smallmatrix}n,s\\m_{no},m_{nc},m_x\end{smallmatrix}\right) - 3\,m_{no} - m_{nc}$$

$$\#3\left(\begin{smallmatrix}n,s\\m_{no},m_{nc},m_x\end{smallmatrix}\right) = m_x - 2(m_{no} - m_{nc})$$

$$\#1\left(\begin{smallmatrix}n,s\\m_{no},m_{nc},m_x\end{smallmatrix}\right) = \#3\left(\begin{smallmatrix}n,s\\m_{no},m_{nc},m_x\end{smallmatrix}\right) + 2\,m_{no} + 2\,m_{nc} + 4$$

For preparation, we state two lemmas that investigate how to calculate the $i$-points of $f$ from the two sub-colorings generated by splitting $f$ at a non-overlapping or unconnected row.

**Lemma 5 (Split 3-Points).** *Let $f$ be a caveat-free coloring of plane $x = c$ with #non-overlaps$(f) \geq 1$, and let $z_s$ be a non-overlapping row. Then, $\#3(f) = \#3(f_{\leq z_s}) + \#3(f_{>z_s})$.*

**Proof (Sketch).** We can show that neither $f_{\leq z_s}$, nor $f_{>z_s}$, nor $f$ has a 3-point that lies between rows $z_s$ and $z_s + 1$. This implies that every 3-point for $f$ is either below $z_s$ and is therefore also a 3-point for $f_{\leq z_s}$, or above $z_s + 1$ and is therefore also a 3-point for $f_{>z_s}$. $\qquad\square$

**Lemma 6 (Split at Minimal Unconnected Row).** *Let $f$ be a caveat-free coloring of plane $x = c$ with #non-connects$(f) \geq 1$, and let $z_s$ be the minimal unconnected row. Then, #non-connects$(f_{\leq z_s}) = 0$, #non-connects$(f_{>z_s}) = $ #non-connects$(f) - 1$ and*

$$\text{xsteps}(f_{\leq z_s}) + \text{xsteps}(f_{>z_s}) = \text{xsteps}(f), \tag{1}$$
$$\forall i \in \{1,2,3,4\}: \ \#i(f_{\leq z_s}) + \#i(f_{>z_s}) = \#i(f). \tag{2}$$

**Proof (Sketch).** The first two claims are trivial. For claims (1) and (2), one shows that if $z_s$ is unconnected, the $y$-distance between points in $f_{\leq z_s}$ and $f_{>z_s}$ is always greater than 1. This implies that the sets of $i$-points and $x$-steps of $f_{\leq z_s}$ and $f_{>z_s}$ are disjoint. $\qquad\square$

**Lemma 7.** *Let $f$ be a caveat-free coloring. Then*

$$\forall i \in \{1,2,3,4\}: \#i(f) = \#i\left(\begin{smallmatrix} |f|, \text{Surf}_{pl}(f) \\ \#\text{non-overlaps}(f), \#\text{non-connects}(f), \text{xsteps}(f) \end{smallmatrix}\right).$$

**Proof (Sketch).** The case #non-connects$(f) = 0$ is equivalent to the formula already proven in [5]. For the case #non-connects$(f) = m_{nc} > 0$, we do induction on $m_{nc}$. The claim for $\#4(f)$, $\#3(f)$ and $\#1(f)$ follow from the Split-Lemmata 2, 5 and 6 by simple calculation (recall that by definition, every unconnected line is also non-overlapping). For $\#2(f)$, the claim follows by simple calculation from the equation $4\#4(f) + 3\#3(f) + 2\#2(f) + 1\#1(f) = 4|f|$. This equation holds since the sum of all interlayer contacts between $f$ and the next plane is $4|f|$. $\qquad\square$

## 6  Maximal Number of 3-Points

Due to the last lemma, if we consider colorings with given $n, a, b, m_{no}$, and $m_{nc}$, then $m_x$ does not affect the number of 4-points, but increases the number of 3-points and 1-points, while decreasing the number of 2-points. The increase of 3- and 1-points is 1 per x-step, the decrease of 2-points is 2 per 3-point. This pattern grants that we maximize the possible number of interlayer contacts to a second plane with a given number of elements, if we maximize the number of 3-points in the first plane. For this purpose, we first show that we need not to distinguish between unconnected and non-overlapping rows for the number of 3-points. The reason is that number of 3-points does not change if one transforms a non-overlapping row into into a unconnected row. Consider as an example the two colorings

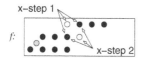

Then both $f$ and $f'$ have one 3-point (indicated in grey). By transforming the non-overlapping row in $f$ into a unconnected row, $f'$ looses two x-steps. Thus, the effects of increasing #non-connects($\cdot$) by 1 are diminished by decreasing xsteps($\cdot$) by 2.

Note that such a bound for the interlayer contacts using a bound for 3-points that does not distinguish between non-overlapping and unconnected rows slightly overestimates, since we assume the best case for the number of 2- and 1-points (note that in contrast to the number of 3-points, the number of 2- and 1-points depend on the exact number of unconnected rows).

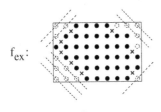

$f_{ex}$:

**Fig. 3.** Detailled Frame.

We start with the extension of the bound for 3-points, as given in [5] in the case of "sufficiently filled" and overlapping colorings, to arbitrary overlapping colorings. We need to recall some definitions from [5]. For an overlapping coloring $f$ with $\overline{lines}(f) = (a, b)$, $a$ and $b$ are the side lengths of the minimal rectangle around the points in $f$ (called frame($f$) in the following). The detailed frame of a coloring $f$ is the tuple $(a, b, i_{lb}, i_{lu}, i_{rb}, i_{ru})$, where $(a, b)$ is the frame of $f$ and $i_{lb}$ is the number of diagonals that can be drawn from the left-bottom corner. $i_{lu}, i_{rb}, i_{ru}$ are defined analogously. For a coloring $f$ with detailed frame $(a, b, i_{lb}, i_{lu}, i_{rb}, i_{ru})$, we call $i = (i_{lb}, i_{lu}, i_{rb}, i_{ru})$ the *indent vector of $f$*. As shown in [5], the indent vector gives a precise bound on the #3($f$), since in this case, xsteps($f$) = #3($f$) and xsteps($f$) = $i_{lb} + i_{lu} + i_{rb} + i_{ru} - $ diagcav($f$). Here, diagcav($f$) counts the number of *diagonal caveats*, which are defined analogous to vertical and horizontal caveats. For example, consider the plane coloring $f_{ex}$ as given in Figure 6. Then the detailed frame of $f_{ex}$ is $(6, 9, 3, 2, 1, 2)$. The number of 3-points (indicated by $\times$) for $f_{ex}$ is $8 = 3 + 2 + 1 + 2$, since $f_{ex}$ does not contain diagonal caveats.

In the overlapping case, we search for a given number of points $n$ and a frame $(a, b)$ the maximal number of x-steps. For this purpose, we define for some indent vector $i = (i_1, i_2, i_3, i_4)$, vol$(a, b, i) := ab - \sum_{1 \leq j \leq 4} \frac{i_j(i_j+1)}{2}$. vol$(a, b, i)$ is the maximal number of points that can be colored by any $f$ that has indent vector $i$ and frame $(a, b)$. $i = (i_1, i_2, i_3, i_4)$ is called *maximal* for $(a, b)$ iff $\sum_{1 \leq j \leq 4} i_j = 2(\min(a, b) - 1)$. For example, if $b \leq a$, then the indent vector $i$ is maximal for $(a, b)$ if every coloring with frame $(a, b)$ and indent vector $i$ has exactly one colored point in the first and last column.

vol$(a, b, i)$ can now be used to calculate the maximal number of x-steps that can be achieved given $n$ colored points and frame $(a, b)$. The maximal number of x-steps is achieved if we make the indents as uniform as possible. For this purpose, define edge$(n, a, b) = \max\{k \in \mathbb{N} \mid $ vol$(a, b, (k, k, k, k))\}$. $k = $ edge$(n, a, b)$ defines the maximal possible uniform indent. Then $r = $ ext$(n, a, b) = \lfloor \frac{ab - 4\frac{k(k+1)}{2} - n}{k+1} \rfloor$ defines the number of times $r$ we can extend the uniform indent by 1. $n$ is called *normal for $(a, b)$* if either $4k + r < 2(a - 1)$, or $4k + r = 2(a - 1)$ and $ab - 4\frac{k(k+1)}{2} - r(k+1) = n$.

Now there are two upper bounds that can be given for the number of x-steps, given $n$ colored points and frame $(a, b)$. The first is given by the indent vector. The second by the fact, that in caveat-free and overlapping colorings, there may be at most between every two successive lines 2 x-steps, which gives at most $2 \min(a, b) - 1$. Thus, the bound given in [5] is as follows:

$$\text{xsteps}_{\text{bnd}}(n, a, b) = \min(4 \, \text{edge}(n, a, b) + \text{ext}(n, a, b), 2(\min(a, b) - 1)).$$

We improve the bound in the case of quadratic frames $(a, a)$ and $n$ is not normal for $(a, a)$. Here, we show that we have an upper bound of $2a - 3$ instead of $2a - 2$ if there is no maximal indent $i$ with $n = \text{vol}(a, a, i)$. We show in this case, that there must be a diagonal caveat.

**Lemma 8.** *For every overlapping caveat-free coloring $f$ we get*

$$\#3(f) \leq \#3_{\text{bound}}(|f|, a, b),$$

*where $(a, b) = \text{frame}(f)$ and*

$$\#3_{\text{bound}}(n, a, b) := \begin{cases} \text{xsteps}_{\text{bnd}}(n, a, b) & n \text{ is normal for frame } (a, b) \\ 2 \min(a, b) - 2 & \text{else if } a \neq b \\ 2a - 2 & \text{else if } \exists i : i \text{ are maximal indents} \\ & \qquad \text{for } (a, a) \wedge n = \text{vol}(a, a, i) \\ 2a - 3 & \text{otherwise} \end{cases}$$

For the general case of possibly non-overlapping colorings, Lemmata 3, 4, and 1 imply that any coloring $f$ with $\text{olines}(f) = (a, b)$ and $\#\text{non-overlaps}(f) = m_{\text{no}}$ satisfies $\text{valid}(n, a, b, m_{\text{no}}) := (m_{\text{no}} < \min(a, b) \wedge n_{\min}(a, b, m_{\text{no}}) \leq n \leq n_{\max}(a, b, m_{\text{no}}))$. Hence, we define $\#3_{\text{bound}}(n, a, b, m_{\text{no}})$ to be $-\infty$ in the case that $\text{valid}(n, a, b, m_{\text{no}})$ does *not* hold. Otherwise, we define $\#3_{\text{bound}}(n, a, b, m_{\text{no}})$ by $\#3_{\text{bound}}(n, a, b)$ if $m_{\text{no}} = 0$ and

$$\max \left\{ \begin{array}{l} \#3_{\text{bound}}(n', a', b', 0) \\ + \#3_{\text{bound}}(n - n', a - a', \\ \quad b - b', m_{\text{no}} - 1) \end{array} \middle| \begin{array}{l} 1 \leq n' \leq n - 1, \\ 1 \leq a' \leq a - 1, \\ 1 \leq b' \leq b - 1, \end{array} \right\},$$

otherwise.

**Lemma 9.** *For every caveat-free coloring $f$, holds $\#3(f) \leq \#3_{\text{bound}}(n, a, b, m_{\text{no}})$, where $n = |f|$, $(a, b) = \text{olines}(f)$, and $m_{\text{no}} = \#\text{non-overlaps}(f)$.*

**Proof (Sketch).** The case $m_{\text{no}} = 0$ is treated in Lemma 8. For $n, a, b$ and $m_{\text{no}} > 0$ with $\text{valid}(n, a, b, m_{\text{no}})$, we can split a coloring $f$ at the minimal non-overlapping line $z_s$ and into $f_{\leq z_s}$ and $f_{>z_s}$ and get $\#3(f) = \#3(f_{\leq z_s}) + \#3(f_{>z_s})$ by Lemma 5. Considering all possible rows for splitting will give the second case of $\#3_{\text{bound}}(n, a, b, m_{\text{no}})$. □

The bound on the number of 3-points can now be used to derive a bound on the number of interlayer contacts for arbitrary colorings. Summarizing, we get the following bound:

$$\text{BNMIC}^{n_2}_{n_1,a_1,b_1}(m_{no1}) := 4\min(n_2, \#4) + 3\min(\#3, \max(n_2 - \#4), 0)$$

$$+ 2\min(\#2, \max(n_2 - \#4 - \#3, 0)) + \min(\#1, \max(n_2 - \#4 - \#3 - \#2, 0))$$

where $\#4 = n - a_1 - b_1 + 1 + m_{no1}$          $\#3 = \#3_{bound}(n_1, a_1, b_1, m_{no1})$
          $\#2 = 2(a_1 + b_1) - 4 - 2\#3 - 3m_{no1}$          $\#1 = \#3 + 2m_{no1} + 4.$

**Theorem 1.** *Let $f_1$ and $f_2$ be coloring of planes $x = c$ and $x = c + 1$, respectively. Let $n_1 = |f_1|$, $\text{olines}(f_1) = (a_1, b_1)$, $|f_2| = n_2$ and $\text{olines}(f_2) = (a_2, b_2)$. Then $\text{IC}^{f_2}_{f_1} \leq \min(\text{BNMIC}^{n_2}_{n_1,a_1,b_1}, \text{BNMIC}^{n_1}_{n_2,a_2,b_2}).$*

## 7   Constructing the Compact Cores

We will now show how to compute the optimally compact cores for a given number of elements, thereby employing the given bound on interlayer contacts, for a branch-and-bound approach. Due to space restrictions, we have to omit many details of the approach.

W.l.o.g, let a coloring $f$ be decomposed into plane colorings $f_1 \uplus \cdots \uplus f_k$. A dynamic programming algorithm allows one to efficiently compute bounds $\text{BMC}(n, n_1, a_1, b_1)$ such that for every coloring $f = f_1 \uplus \cdots \uplus f_k$, it holds that $\text{BMC}(n, n_1, a_1, b_1) \geq \text{con}(f)$, where $|f| = n$, $|f_1| = n_1$, and $\text{olines}(f_1) = (a_1, b_1)$. From this algorithm we get immediately a maximal number of contacts in any coloring with $n$ elements. Further, let a *layer sequence* be a sequence of triples $(n_i, a_i, b_i)$. A coloring $f$ is called *s-compatible*, if every plane restriction $f_i$ of $f$ is compatible to $s_i = (n_i, a_i, b_i)$, i.e. $|f_i| = n_i$ and $\text{olines}(f_i) = (a_i, b_i)$.

By traceback from the above dynamic programming algorithm one efficiently obtains the set of all layer sequences $s$, where there may exist (by our bound) an $s$-compatible coloring $f$ with $b$ contacts. That is, we define this set of sequences by $S(n, b) := \{s \text{ layer sequence} | \text{bound for } s \text{ greater or equal } b\}$.

To find optimally compact colorings it remains to search by constraint based search through the colorings of candidate layer sequences.

Now, we assume that the sets $S(n, b)$ are already precomputed by the dynamic programming algorithm. To find one optimally compact coloring with $n$ elements do the following. Let $b_n$ be the contacts bound for colorings with $n$ elements. For ascending $i \geq 0$, iteratively search for a coloring $f$ with $b_n - i$ contacts in all layer sequences $s \in S(n, b_n - i)$. Clearly, the first coloring $f_b$ found by this procedure has maximal contacts. To find all colorings with a given number $k$ of contacts (e.g. all best colorings) we perform an analogous search in all layer sequences $s \in S(n, b)$.

**Table 1.** Search for one optimally compact core with $n$ elements, given a layer sequence. We give the number of contacts, as well as nodes and time of the constraint search.

| $n$ | # contacts | # search-nodes | time in s |
|-----|-----------|----------------|-----------|
| 23  | 76        | 15             | 0.1       |
| 60  | 243       | 150            | 0.7       |
| 89  | 382       | 255            | 2.1       |
| 100 | 436       | 82             | 1.2       |

**Table 2.** Sequences $L_1$-$L_5$ (taken from [22]) with absolute walks of optimal conformations in FCC-HP-model. The steps of the walk are given by points of the compass. The $+$ and $-$ indices indicate an additional $45°$ walk out of the plane.

$L_1$ HPPPPHHHHPPHPHPHHHHPHPPHHPPH :
$N_e^- E N_e^+ S_e^+ S N_w^- S_w^+ S_w^- S_w^- S_e^+ N_e^- N_w^- N_e^+ S_e^- N_e^+ S_w^+ E S N_w^- S W N_e^+ N_w^+ S_e^+ E N_w^-$

$L_2$ HPPPHHHHPHPHHPPPHPHPHHHPHPPPHP :
$S_e^- S_w^- N S_w^+ N_w^+ N_e^- N_w^- N S_e^+ N_e^- S W N_e^+ N_w^- N_w^- S S_w^+ S_e^- S_e^- W S_e^+ S_w^+ N_e^+ E N N_w^+$

$L_3$ HPHHPPHHHPPHHHHHPPPPPPHHHPPH :
$S_w^+ E E N_e^- N_w^- S_w^+ N E N_w^+ W S_w^- S_e^+ W S_w^+ S_e^+ E N_e^- S_e^+ N_e^- N S_w^- N_w^+ W W S_e^+ S_e^-$

$L_4$ HHPHHPHHPHHHHHHHPPHHHHHHPHHHHHHH :
$S_e^+ N_w^+ N_e^- N_w^- N_w^- E E N_e^- S S S_w^+ S_w^+ N_w^- N_e^- N_e^- N S_w^+ S_e^+ S_e^- N S_e^+ N_e^+ N_w^- S_e^- S_e^- W S_e^+ S_w^+ S_e^- N_e^-$

$L_5$ PHPPHPPHPPHPPPHPHPPHPPPHPHPPPHP :
$S_e^+ S_w^- S_w^+ E S N_w^+ E S_e^- E N_w^+ N_e^- W N_w^+ N_w^- E S_e^+ N_e^- S_e^+ S_w^+ E N W N_e^- W S_w^- N_w^+ S_e^+ S_w^- S_e^+ E N_w^- S S_w^- N S_w^+$

# 8   Results

We have computed all sets of layer sequences $S(n, b)$ for $n \leq 100$ in about 10 days on a standard PC. For a given layer sequence one optimally compact core is usually found within a few seconds by our constraint based search program. Some results are shown in Table 1.

We present some of the optimal cores for $n = 60$ and $n = 100$ elements in Figures 4 and 5. The cores are shown as plane sequence representation. This representation shows a coloring by the sequence of its occupied $x$-layers in the lattice $D_3'$. For each $x$-layer $x = x_0$ the lower left corner of the grid has coordinates $(x_0, 0, 0)$. The grid-lines have distance 1. The core points in each $x$-layer are shown as filled circles. There is a noteworthy difference between layers $x = x_0$, where $x_0$ is even and those where it is odd. In the latter ones the points have non-integer $y$ and $z$ coordinates.

Further, we folded some proteins of the FCC-HP-model using a program from [7] to their now proven optimum. The results are shown in Table 2.

# References

1. V.I. Abkevich, A.M. Gutin, and E.I. Shakhnovich. Impact of local and non-local interactions on thermodynamics and kinetics of protein folding. *Journal of Molecular Biology*, 252:460–471, 1995.

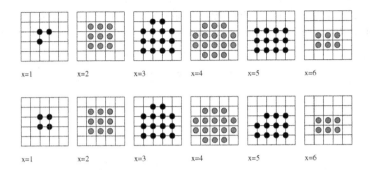

**Fig. 4.** Plane sequence representation of two optimally compact coloring with $n = 60$ elements.

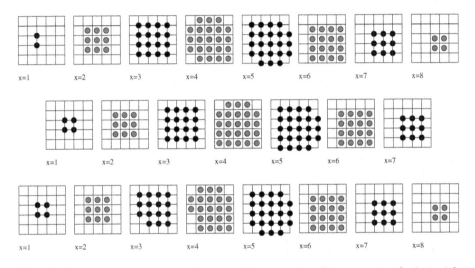

**Fig. 5.** Plane sequence representations of three optimally compact coloring with $n = 100$ elements.

2. V.I. Abkevich, A.M. Gutin, and E.I. Shakhnovich. Computer simulations of prebiotic evolution. In Russ B. Altman, A. Keith Dunker, Lawrence Hunter, and Teri E. Klein, editors, *PSB'97*, pages 27–38, 1997.

3. Richa Agarwala, Serafim Batzoglou, Vlado Dancik, Scott E. Decatur, Martin Farach, Sridhar Hannenhalli, S. Muthukrishnan, and Steven Skiena. Local rules for protein folding on a triangular lattice and generalized hydrophobicity in the HP-model. *Journal of Computational Biology*, 4(2):275–296, 1997.

4. Rolf Backofen. The protein structure prediction problem: A constraint optimisation approach using a new lower bound. *J. Constraints*, 2000. accepted for publication, special issue on 'Constraints in Bioinformatics/Biocomputing'.

5. Rolf Backofen. An upper bound for number of contacts in the HP-model on the face-centered-cubic lattice (FCC). In Raffaele Giancarlo and David Sankoff, ed-

itors, *Proc. of the 11th Annual Symposium on Combinatorial Pattern Matching (CPM2000)*, volume 1848 of *Lecture Notes in Computer Science*, pages 277–292, Berlin, 2000. Springer-Verlag.

6. Rolf Backofen, Sebastian Will, and Erich Bornberg-Bauer. Application of constraint programming techniques for structure prediction of lattice proteins with extended alphabets. *J. Bioinformatics*, 15(3):234–242, 1999.

7. Rolf Backofen, Sebastian Will, and Peter Clote. Algorithmic approach to quantifying the hydrophobic force contribution in protein folding. In Russ B. Altman, A. Keith Dunker, Lawrence Hunter, and Teri E. Klein, editors, *Pacific Symposium on Biocomputing (PSB 2000)*, volume 5, pages 92–103, 2000.

8. B. Berger and T. Leighton. Protein folding in the hydrophobic-hydrophilic (HP) modell is NP-complete. In *Proc. of the Second Annual International Conferences on Compututational Molecular Biology (RECOMB98)*, pages 30–39, New York, 1998.

9. P. Crescenzi, D. Goldman, C. Papadimitriou, A. Piccolboni, and M. Yannakakis. On the complexity of protein folding. In *Proc. of STOC*, pages 597–603, 1998. Short version in *Proc. of RECOMB'98*, pages 61–62.

10. K.A. Dill, S. Bromberg, K. Yue, K.M. Fiebig, D.P. Yee, P.D. Thomas, and H.S. Chan. Principles of protein folding – a perspective of simple exact models. *Protein Science*, 4:561–602, 1995.

11. Aaron R. Dinner, Andreaj Šali, and Martin Karplus. The folding mechanism of larger model proteins: Role of native structure. *Proc. Natl. Acad. Sci. USA*, 93:8356–8361, 1996.

12. S. Govindarajan and R. A. Goldstein. The foldability landscape of model proteins. *Biopolymers*, 42(4):427–438, 1997.

13. William E. Hart and Sorin Istrail. Invariant patterns in crystal lattices: Implications for protein folding algorithms. *J. of Universal Computer Science*, 6(6):560–579, 2000.

14. William E. Hart and Sorin C. Istrail. Fast protein folding in the hydrophobid-hydrophilic model within three-eighths of optimal. *Journal of Computational Biology*, 3(1):53 – 96, 1996.

15. Patrice Koehl and Michael Levitt. A brighter future for protein structure prediction. *Nature Structural Biology*, 6:108–111, 1999.

16. Kit Fun Lau and Ken A. Dill. A lattice statistical mechanics model of the conformational and sequence spaces of proteins. *Macromolecules*, 22:3986 – 3997, 1989.

17. Hao Li, Robert Helling, Chao Tnag, and Ned Wingreen. Emergence of preferred structures in a simple model of protein folding. *Science*, 273:666–669, 1996.

18. Britt H. Park and Michael Levitt. The complexity and accuracy of discrete state models of protein structure. *Journal of Molecular Biology*, 249:493–507, 1995.

19. Ron Unger and John Moult. Local interactions dominate folding in a simple protein model. *Journal of Molecular Biology*, 259:988–994, 1996.

20. A. Šali, E. Shakhnovich, and M. Karplus. Kinetics of protein folding. *Journal of Molecular Biology*, 235:1614–1636, 1994.

21. Yu Xia, Enoch S. Huang, Michael Levitt, and Ram Samudrala. Ab initio construction of protein tertiary structures using a hierarchical approach. *Journal of Molecular Biology*, 300:171 – 185, 2000.

22. Kaizhi Yue and Ken A. Dill. Sequence-structure relationships in proteins and copolymers. *Physical Review E*, 48(3):2267–2278, September 1993.

23. Kaizhi Yue and Ken A. Dill. Forces of tertiary structural organization in globular proteins. *Proc. Natl. Acad. Sci. USA*, 92:146 – 150, 1995.

# Author Index

# Lecture Notes in Computer Science

For information about Vols. 1–1998
please contact your bookseller or Springer-Verlag

Vol. 2040: W. Kou, Y. Yesha, C.J. Tan (Eds.), Electronic Commerce Technologies. Proceedings, 2001. X, 187 pages. 2001.

Vol. 2041: I. Attali, T. Jensen (Eds.), Java on Smart Cards: Programming and Security. Proceedings, 2000. X, 163 pages. 2001.

Vol. 2042: K.-K. Lau (Ed.), Logic Based Program Synthesis and Transformation. Proceedings, 2000. VIII, 183 pages. 2001.

Vol. 2043: D. Craeynest, A. Strohmeier (Eds.), Reliable Software Technologies – Ada-Europe 2001. Proceedings, 2001. XV, 405 pages. 2001.

Vol. 2044: S. Abramsky (Ed.), Typed Lambda Calculi and Applications. Proceedings, 2001. XI, 431 pages. 2001.

Vol. 2045: B. Pfitzmann (Ed.), Advances in Cryptology – EUROCRYPT 2001. Proceedings, 2001. XII, 545 pages. 2001.

Vol. 2047: R. Dumke, C. Rautenstrauch, A. Schmietendorf, A. Scholz (Eds.), Performance Engineering. XIV, 349 pages. 2001.

Vol. 2048: J. Pauli, Learning Based Robot Vision. IX, 288 pages. 2001.

Vol. 2051: A. Middeldorp (Ed.), Rewriting Techniques and Applications. Proceedings, 2001. XII, 363 pages. 2001.

Vol. 2052: V.I. Gorodetski, V.A. Skormin, L.J. Popyack (Eds.), Information Assurance in Computer Networks. Proceedings, 2001. XIII, 313 pages. 2001.

Vol. 2053: O. Danvy, A. Filinski (Eds.), Programs as Data Objects. Proceedings, 2001. VIII, 279 pages. 2001.

Vol. 2054: A. Condon, G. Rozenberg (Eds.), DNA Computing. Proceedings, 2000. X, 271 pages. 2001.

Vol. 2055: M. Margenstern, Y. Rogozhin (Eds.), Machines, Computations, and Universality. Proceedings, 2001. VIII, 321 pages. 2001.

Vol. 2056: E. Stroulia, S. Matwin (Eds.), Advances in Artificial Intelligence. Proceedings, 2001. XII, 366 pages. 2001. (Subseries LNAI).

Vol. 2057: M. Dwyer (Ed.), Model Checking Software. Proceedings, 2001. X, 313 pages. 2001.

Vol. 2059: C. Arcelli, L.P. Cordella, G. Sanniti di Baja (Eds.), Visual Form 2001. Proceedings, 2001. XIV, 799 pages. 2001.

Vol. 2060: T. Böhme, H. Unger (Eds.), Innovative Internet Computing Systems. Proceedings, 2001. VIII, 183 pages. 2001.

Vol. 2062: A. Nareyek, Constraint-Based Agents. XIV, 178 pages. 2001. (Subseries LNAI).

Vol. 2064: J. Blanck, V. Brattka, P. Hertling (Eds.), Computability and Complexity in Analysis. Proceedings, 2000. VIII, 395 pages. 2001.

Vol. 2065: H. Balster, B. de Brock, S. Conrad (Eds.), Database Schema Evolution and Meta-Modeling. Proceedings, 2000. X, 245 pages. 2001.

Vol. 2066: O. Gascuel, M.-F. Sagot (Eds.), Computational Biology. Proceedings, 2000. X, 165 pages. 2001.

Vol. 2068: K.R. Dittrich, A. Geppert, M.C. Norrie (Eds.), Advanced Information Systems Engineering. Proceedings, 2001. XII, 484 pages. 2001.

Vol. 2070: L. Monostori, J. Váncza, M. Ali (Eds.), Engineering of Intelligent Systems. Proceedings, 2001. XVIII, 951 pages. 2001. (Subseries LNAI).

Vol. 2071: R. Harper (Ed.), Types in Compilation. Proceedings, 2000. IX, 207 pages. 2001.

Vol. 2072: J. Lindskov Knudsen (Ed.), ECOOP 2001 – Object-Oriented Programming. Proceedings, 2001. XIII, 429 pages. 2001.

Vol. 2073: V.N. Alexandrov, J.J. Dongarra, B.A. Juliano, R.S. Renner, C.J.K. Tan (Eds.), Computational Science – ICCS 2001. Part I. Proceedings, 2001. XXVIII, 1306 pages. 2001.

Vol. 2074: V.N. Alexandrov, J.J. Dongarra, B.A. Juliano, R.S. Renner, C.J.K. Tan (Eds.), Computational Science – ICCS 2001. Part II. Proceedings, 2001. XXVIII, 1076 pages. 2001.

Vol. 2075: J.-M. Colom, M. Koutny (Eds.), Applications and Theory of Petri Nets 2001. Proceedings, 2001. XII, 403 pages. 2001.

Vol. 2077: V. Ambriola (Ed.), Software Process Technology. Proceedings, 2001. VIII, 247 pages. 2001.

Vol. 2078: R. Reed, J. Reed (Eds.), SDL 2001: Meeting UML. Proceedings, 2001. XI, 439 pages. 2001.

Vol. 2081: K. Aardal, B. Gerards (Eds.), Integer Programming and Combinatorial Optimization. Proceedings, 2001. XI, 423 pages. 2001.

Vol. 2082: M.F. Insana, R.M. Leahy (Eds.), Information Processing in Medical Imaging. Proceedings, 2001. XVI, 537 pages. 2001.

Vol. 2083: R. Goré, A. Leitsch, T. Nipkow (Eds.), Automated Reasoning. Proceedings, 2001. XV, 708 pages. 2001. (Subseries LNAI).

Vol. 2084: J. Mira, A. Prieto (Eds.), Connectionist Models of Neurons, Learning Processes, and Artificial Intelligence. Proceedings, 2001. Part I. XXVII, 836 pages. 2001.

Vol. 2085: J. Mira, A. Prieto (Eds.), Bio-Inspired Applications of Connectionism. Proceedings, 2001. Part II. XXVII, 848 pages. 2001.

Vol. 2089: A. Amir, G.M. Landau (Eds.), Combinatorial Pattern Matching. Proceedings, 2001. VIII, 273 pages. 2001.

Vol. 2091: J. Bigun, F. Smeraldi (Eds.), Audio- and Video-Based Biometric Person Authentication. Proceedings, 2001. XIII, 374 pages. 2001.

Vol. 2092: L. Wolf, D. Hutchison, R. Steinmetz (Eds.), Quality of Service – IWQoS 2001. Proceedings, 2001. XII, 435 pages. 2001.

Vol. 2096: J. Kittler, F. Roli (Eds.), Multiple Classifier Systems. Proceedings, 2001. XII, 456 pages. 2001.

Vol. 2097: B. Read (Ed.), Advances in Databases. Proceedings, 2001. X, 219 pages. 2001.

Vol. 2099: P. de Groote, G. Morrill, C. Retoré (Eds.), Logical Aspects of Computational Linguistics. Proceedings, 2001. VIII, 311 pages. 2001. (Subseries LNAI).

Vol. 2110: B. Hertzberger, A. Hoekstra, R. Williams (Eds.), High-Performance Computing and Networking. Proceedings, 2001. XVII, 733 pages. 2001.